WATER

A COMPREHENSIVE TREATISE

Volume 3

Aqueous Solutions of Simple Electrolytes

WATER

A COMPREHENSIVE TREATISE

Edited by Felix Franks

WATER
A COMPREHENSIVE TREATISE

Edited by Felix Franks
Unilever Research Laboratory
Sharnbrook, Bedford, England

Volume 3
Aqueous Solutions
of Simple Electrolytes

PLENUM PRESS • NEW YORK–LONDON • 1973

Library of Congress Catalog Card Number 78-165694
ISBN 0-306-37183-9

© 1973 Plenum Press, New York
A Division of Plenum Publishing Corporation
227 West 17th Street, New York, N.Y. 10011

United Kingdom edition published by Plenum Press, London
A Division of Plenum Publishing Company, Ltd.
Davis House (4th Floor), 8 Scrubs Lane, Harlesden, London, NW10 6SE, England

Printed in the United States of America

Preface

The chapters making up this volume had originally been planned to form part of a single volume covering solid hydrates and aqueous solutions of simple molecules and ions. However, during the preparation of the manuscripts it became apparent that such a volume would turn out to be very unwieldy and I reluctantly decided to recommend the publication of separate volumes. The most sensible way of dividing the subject matter seemed to lie in the separation of simple ionic solutions.

The emphasis in the present volume is placed on ion–solvent effects, since a number of excellent texts cover the more general aspects of electrolyte solutions, based on the classical theories of Debye, Hückel, Onsager, and Fuoss. It is interesting to speculate as to when a theory becomes "classical." Perhaps this occurs when it has become well known, well liked, and much adapted. The above-mentioned theories of ionic equilibria and transport certainly fulfill these criteria.

There comes a time when the refinements and modifications can no longer be related to physical significance and can no longer hide the fact that certain fundamental assumptions made in the development of the theory are untenable, especially in the light of information obtained from the application of sophisticated molecular and thermodynamic techniques.

There is now general agreement that the description of ionic solutions in terms of only a simple electrostatic potential function modified by a bulk dielectric permittivity can hardly represent the true situation and we are now beginning to see the development and acceptance of the second generation of electrolyte theories. The present state of flux is reflected in most of the following chapters, which might suggest that the timing of the publication of this volume is premature. However, within the framework of this treatise properties of electrolyte solutions play an important part, and I hope that bringing together some of the more recent experimental and theoretical results in one volume will advance the development of useful working models for ion solvation and ionic interactions.

As before I should like to express my gratitude to the contributing authors for their enthusiastic cooperation which is mainly responsible for the short interval between the publication of this and the preceding volume.

F. FRANKS

Biophysics Division
Unilever Research Laboratory Colworth/Welwyn
Colworth House, Sharnbrook, Bedford

Contents

Chapter 1

Thermodynamics of Ion Hydration

H. L. Friedman and C. V. Krishnan

Chapter 2

Thermodynamics of Aqueous Mixed Electrolytes

Henry L. Anderson and Robert H. Wood

Chapter 3

Hydration Effects and Acid–Base Equilibria

Loren G. Hepler and Earl M. Woolley

Chapter 4

Ionic Transport in Water and Mixed Aqueous Solvents

Robert L. Kay

Chapter 5

Infrared Spectroscopy of Aqueous Electrolyte Solutions

Ronald Ernest Verrall

Chapter 6

Raman Spectroscopy of Aqueous Electrolyte Solutions

T. H. Lilley

Chapter 7

Nuclear Magnetic Relaxation Spectroscopy

H. G. Hertz

Chapter 8

Dielectric Properties

Reinhard Pottel

Contents of Volume 1: The Physics and Physical Chemistry of Water

Contents of Volume 2: Water in Crystalline Hydrates; Aqueous Solutions of Simple Nonelectrolytes

Contents of Volume 4: Aqueous Solutions of Macromolecules; Water in Disperse Systems

CHAPTER 1

Thermodynamics of Ionic Hydration

H. L. Friedman and C. V. Krishnan

Department of Chemistry
State University of New York
Stony Brook, New York 11790

1. THE THERMODYNAMIC THEORY OF SOLVATION

1.1. General Remarks

In this section we outline the thermodynamic theory which is employed in extracting coefficients relating to solvation from measured quantities. The points which are emphasized are the role of extrathermodynamic equations for the concentration dependence of thermodynamic functions, the arbitrary nature of standard states, and the additional complication encountered in passing from the study of solvation in solutions of non-electrolytes to solvation in ionic solutions.

Ionic solutions are more difficult to study than solutions of non-electrolytes because of two consequences of the long range of the Coulomb interaction of the ions. One is the appearance of terms proportional to $\sqrt{c}$ and $c \ln c$ in the concentration (c) dependence of thermodynamic coefficients such as the logarithm of the activity coefficient, the molar heat of dilution, or the apparent molar volume. For solutions of nonelectrolytes such coefficients may have a $\ln c$ ideal term but all the rest is expressible as a power series in c. The other consequence of the Coulomb interactions is that ionic solutions are very nearly electrically neutral,[300] so that the concentrations of positive and negative ions cannot be independently varied. Thus while the simplest ionic solution has three components—solvent, cations, and anions—it has only one concentration variable in

1

thermodynamic operations at fixed temperature and pressure. It is necessary to qualify this statement: In some experiments the effects of small deviations from electrical neutrality can be measured either in thermodynamic processes with surface effects or in nonthermodynamic processes. This topic is reserved for discussion in Section 7; otherwise the discussion will be restricted to bulk thermodynamic properties and for these the electrical neutrality requirement is exact.*

The general plan of the outline which follows is first to define and discuss each problem for solutions of nonelectrolytes and then to consider the additional complications required for application to ionic solutions. The many aspects which are well described in textbooks on chemical thermodynamics[300] are assumed to be familiar to the reader.

1.2. Resolution of Data into Solvation and Excess Properties

Consider first the following process involving a nonelectrolyte A and a solvent W:

$$A(\text{gas, concentration } c_g) \rightarrow A(\text{in W at concentration } c_s) \qquad (1)$$

and imagine that we can measure ΔH for this process directly in an isothermal calorimeter. Also, we express ΔH as the enthalpy change per mole of A transferred from the gas phase to the solution. Then

$$\Delta H^{\ominus} \equiv \lim_{c_s \to 0} \lim_{c_g \to 0} \Delta H \qquad (2)$$

is called the solvation enthalpy of A in W. Clearly $\Delta H^{\ominus}/6.023 \times 10^{23}$ is the enthalpy change accompanying the process in which one molecule of A is immersed in a macroscopic amount of W. So $\Delta H^{\ominus}$ is a fundamental quantity from the point of view of one interested in the solvent properties of W.

As a practical matter it requires little thermodynamic theory to evaluate $\Delta H^{\ominus}$ from measurements. Most often it is feasible to make measurements at small enough values of c_s and c_g so that the extrapolations needed to get the limits in eqn. (2) are easy to perform. But we notice that it is not so easy to get $\Delta G^{\ominus}$, the corresponding change in Gibbs free energy.

Of course, the first obstacle is that we never measure ΔG but only quantities related to partial molar free energies. Now we suppose that such measurements are made and textbook thermodynamics applied to convert

* This is explained, for example, by Friedman and Ramanathan.[244]

the data to ΔG for reaction (1), expressed as the free energy change per mole of A in this process when only an infinitesimal amount of A is transferred to the solution from the gas phase. This free energy change is

$$\Delta G = \mu_A(\text{in W}, c_s) - \mu_A(\text{gas}, c_g) \tag{3}$$

where the μ's are chemical potentials.

Next we recall that for a nonelectrolyte solution or for an un-ionized gas we have

$$\mu_A = \mu_A{}^\ominus + RT(\ln c + B_2 c + B_3 c^2 + \cdots + B_n c^{n-1} + \cdots) \tag{4}$$

where $\mu_A{}^\ominus$ does not depend on concentration and where the B_n are "virial coefficients" of the chemical potential of A. For a gas eqn. (4) can be deduced from the virial equation of state for the gas and B_n is given by an explicit formula involving the temperature and the forces among a cluster of n molecules of A.[372] The McMillan–Mayer theory[372,549] shows that the *same* equation applies to a nonionic solution but with a different $\mu_A{}^\ominus$ and with B_n given by an explicit formula involving the forces among n molecules of A distributed in a body of pure solvent W.

In the ideal case the B_n are all zero. In general we separate eqn. (4) into an ideal part

$$\mu_A^{\text{id}} \equiv \mu_A{}^\ominus + RT \ln c \tag{5}$$

and another part, called the excess or configurational part,

$$\mu_A{}^E = B_2 c + \cdots \tag{6}$$

Because we are mainly concerned with ion hydration in this chapter, we shall not discuss the higher terms in the excess part. But the $B_2 c$ term is conveniently included in the discussion of the extrapolation needed to study the thermodynamics of solvation. Furthermore, in some cases it is expected to have a simple relation to solvation itself, as we shall see.

For either the gas or the solution the significance of $\mu_A{}^\ominus$ may be seen in the following way. By comparing eqns. (5) and (6), it is easy to see that to have $\mu_A = \mu_A^{\text{id}}$ to any desired accuracy, one need only take c small enough. Thus at small enough c we have

$$\mu_A{}^\ominus = \mu_A - RT \ln c \tag{7}$$

so that $\mu_A{}^\ominus$ is the chemical potential of A in the *hypothetical* "standard" state with $c = 1$ and with $\mu_A{}^E$ "turned off," i.e., with no forces acting among the A molecules. It is conventional to express this by saying that $\mu_A{}^\ominus$ is the chemical potential of A in the "hypothetical $c = 1$ state" and that in this state the A molecules do not see each other but do see the solvent.

The preceding analysis is applicable even though the concentration scales, c, c_s, and c_g have been left unspecified. One does not have a completely free choice of concentration scales; the restriction is that at low c the concentration must become proportional to the number of A molecules per unit volume. This is quite true for any of the scales in common use.

Because the choice of concentration scales is arbitrary, so are μ_A^{id}, $\mu_A{}^E$, and $\mu_A{}^\ominus$. For example, if we write eqn. (5) for a different concentration scale,

$$\mu_A^{id} = \mu_A^{**} + RT \ln c', \qquad c' \to 0$$

then we find that

$$\mu_A^{**} = \mu_A{}^\ominus + RT \ln(c/c')_{c=0} \tag{8}$$

Thus the hypothetical $c = 1$ standard state is not the same as the hypothetical $c' = 1$ standard state. So it is clear that the choice of standard state is equivalent to a choice of concentration scales and both are equally arbitrary.

Now, using eqn. (4) for both the gas and the solution in W, we may find the explicit concentration dependence of ΔG for the process in Eqn. (1) as given in eqn. (3):

$$-\Delta G = \mu_A^{\ominus g} - \mu_A^{\ominus s} + RT[\ln(c_g/c_s) + B_{2g}c_g - B_{2s}c_s + \cdots] \tag{9}$$

where indices s and g specify the solution in W or the gas, respectively. But upon comparing this with eqn. (2) we see that if we extrapolate c_g to zero and c_s to zero independently, as we did for ΔH, then we get $\Delta G^\ominus = \pm\infty$. This limit is apparently not useful for investigating solvation.

Instead, it is conventional to define

$$\Delta G^\ominus \equiv \mu_A^{\ominus s} - \mu_A^{\ominus g} \tag{10}$$

This leads to the problem which seems unavoidable: $\Delta G^\ominus$ depends not only on intermolecular forces associated with solvation but also on the concentration scales c_s and c_g. This problem does not appear with ΔH because in general

$$H = \partial(G/T)/\partial(1/T)$$

and applying this equation to eqn. (9) to get ΔH shows that the logarithm term disappears. Of course, the logarithm term is associated with the so-called "ideal entropy of mixing." Thus, applying $S = -\partial G/\partial T$ to calculate ΔS from eqn. (9), we find that the logarithm term persists and therefore $\Delta S^\ominus$ depends on the choice of concentration scales or standard states.

In the same way, by applying $V = \partial G/\partial P$ to calculate ΔV from eqn. (9), we find that $\Delta V^\ominus$ for the process in reaction (1) does not depend on the choice of concentration scales through the logarithmic term, but of course

$$\partial \mu_A^{\ominus g}/\partial P = V_A^{\ominus g} = RT/P$$

where P is the pressure of the hypothetical $c_g = 1$ standard state of the gas, which does depend on c_g.

More generally, for solutions one can obtain the partial molar derivatives

$$\bar{V}_{As} \equiv \partial \mu_{As}/\partial P$$

from experimental data and compare with the derivative of eqn. (9),

$$\bar{V}_{As} = \bar{V}_{As}^\ominus + c(\partial B_{2s}/\partial P) + \cdots$$

to determine

$$\bar{V}_{As}^\ominus \equiv \partial \mu_{As}^\ominus/\partial P$$

This may be compared with the volume associated with A in various other situations which may seem relevant for elucidating the solvation effects in $\bar{V}_{As}^\ominus$.

In summary, it is noted that there is a hierarchy of difficulty associated with thermodynamic solvation functions:

For G one must compare two states and must choose concentrations scales, i.e., standard states.

For H one must compare two states but need not be concerned with concentration scales.

For V one has neither complication: $\bar{V}_{As}^\ominus$ can be independently measured for each solution of interest.

As previously noted, S resembles G in this regard. The higher thermodynamic derivatives of G, namely the heat capacity C_P, the isothermal compressibility K_T, the adiabatic compressibility K_S, the coefficient of thermal expansion α, and others, resemble V: Solvation functions can be determined which are both independent of concentration scales and which can be measured on the solution *per se*.

Finally we can consider the additional complications encountered in the study of ion solvation.

First we note that the simplest process which may be considered to get free energy and enthalpy functions is, in place of the process in eqn. (1),

$$\nu_+ A^{z+}(\text{gas},\, c_{Ag}) + \nu_- B^{z-}(\text{gas},\, c_{Bg}) \rightarrow \nu_+ A^{z+}(\text{in W},\, c_{As}) + \nu_- B^{z-}(\text{in W},\, c_{Bs})$$
(11)

with $\nu_+ z_+ + \nu_- z_- = 0$ and $c_A z_+ + c_B z_- = 0$, because we cannot study the bulk thermodynamic properties of solutions of A^{z+} or B^{z-} alone owing to the electrical neutrality condition. The process in eqn. (11) cannot be studied directly; instead ΔG, ΔS, and ΔH are obtained from a body of experimental data using thermodynamic cycles.[74,496] It is still necessary to extrapolate in an appropriate way to low concentrations to evaluate $\Delta G°$, $\Delta S°$, and $\Delta H°$.

For an ionic solution or a plasma (an ionized gas) the relevant equation of state has the following form, in place of eqn. (4):

$$\mu_\pm \equiv (\nu_+ \mu_A + \nu_- \mu_B)/(\nu_+ + \nu_-)$$
$$= \mu_\pm^* + RT \ln c + S\sqrt{c} + Ec \ln c + B_2(\varkappa)c + \cdots \quad (12)$$

where $c \equiv c_A/\nu_+ = c_B/\nu_-$ and $\varkappa$ is the Debye kappa function. This may be compared with eqn. (4) as follows:

1. Since μ_A and μ_B cannot be measured independently, we now require only the mean ionic chemical potential $\mu_\pm$.

2. The excess part now has terms in $\sqrt{c}$ and $c \ln c$ arising from the fact that even at very low concentrations the Coulomb interactions have an appreciable effect. The coefficients S and E are given by the Debye–Hückel theory or equivalent theories in terms of only the ionic charges, the solvent dielectric constant, and the temperature.[242] Other characteristics of the interionic force do not enter. The coefficient E vanishes if $z_+ + z_- = 0$.

3. $B_2(\varkappa)$ still depends on the entire force function between each pair of ions in the solution, but it now depends on the concentration through the Debye $\varkappa$, i.e., it depends on the ionic strength.[242]

The experimental methods involved in establishing the energetics of the ions in the gas phase mostly involve exceedingly low concentrations, at which the excess terms in eqn. (9) are negligible. For solutions the situation depends very much on the system and the thermodynamic property being measured. If the Debye–Hückel terms are important, the recommended procedure[673] is to calculate them from the theory, subtract them from the measurement, and extrapolate the remaining thermodynamic function as one would for a nonelectrolyte. In this way one can obtain the standard free energy of ionic solvation

$$\Delta G_\pm{}^\circ = \mu_{\pm,s}^* - \mu_{\pm,g}^* \tag{13}$$

and its various thermodynamic derivatives.

1.3. Standard States

In Section 1.2 the interdependence of standard states and concentration scales was emphasized. For gases, in almost all cases, the choice has been hypothetical 1 atm standard states, corresponding to expressing c_g in units of atmospheres. Of course, the atmosphere is not exactly a unit of concentration, but the pressure is proportional to concentration if the gas is sufficiently dilute, so it is acceptable as a measure of concentration according to the criterion stated in Section 1.2.

If one chooses $c_s = x_A$, the mole fraction of A in solution, and $c_g = P_A$, the pressure (atmospheres) of A in the gas, then $\Delta G^\ominus$ for the solvation of a nonelectrolyte is the free energy change in the process

$$A(g, \text{hyp. 1 atm}) \rightarrow A(\text{in W, hyp. } x_A = 1) \tag{14}$$

which has a formal similarity to the process for the liquefaction of gaseous A, which may be written, for *any* solvent W,

$$A(g, \text{hyp. 1 atm}) \rightarrow A(\text{in W, } x_A = 1) \tag{15}$$

because the final state is the pure liquid A (assuming the solution is liquid).

The hypothetical 1 atm standard state for the gas is no more fundamental than any other, but its use here may be justified on the basis that it is universally used in tables of thermodynamic properties of gases. The hypothetical $x_A = 1$ standard state for the solution is the one most frequently chosen, and it has been used in some key papers on the thermodynamics of solvation.[224,230,342] This seems to be sufficient reason to continue its use, but it must be noted that, from a theoretical point of view, it does not have an advantage over other standard states for solutions except in special cases: Isotopic mixtures and substitutional solid solutions are examples. A contrary view is implied in many thermodynamics texts which reserve the name "rational activity coefficients" for activity coefficients related to the hypothetical $x_A = 1$ standard states, as though other standard states had less intellectual merit.

However, it is easy to see that in general there is nothing that makes mole-fraction concentration scales more "rational" than any other. For example, if W is water, a highly associated liquid, it is difficult to justify the nominal mole fraction

$$x_A = N_A/(N_A + N_W)$$

as preferable to the effective mole fraction, which may be defined as

$$x_A{}' = N_A/[N_A + (N_W/\bar{n})]$$

where $\bar{n}$ is the mean association number for the water molecules. One might estimate $\bar{n}$ from the entropy of vaporization of water, for example. For another example, even for models for substitutional solid solutions, if A or W occupies more than one lattice site, then the nominal x_A is not the fundamental one.[219,385–387] A similar "correction" must apply in fluid solutions when A and W are molecules of different sizes, so that it does not seem possible to show that x_A is a more rational concentration unit than the volume fraction, defined as

$$f_A = N_A V_A/(N_A V_A + N_W V_W)$$

where V_i is the molal volume of pure species i, which is also widely used.[368]

These considerations are relevant to the important exercise in which one deduces molecular-level information from thermodynamic coefficients for solvation just by analyzing a set of data, without the benefit of a statistical mechanical theory. If the conclusions *depend* on the choice of hypothetical $x_A = 1$ standard states for the solutions, then they are not sound. If the conclusions do not depend on the choice of standard states, then they are as sound as the weakest of the other elements of the analysis! Fortunately, the latter is usually the case.

One other important choice of standard states must be described. If one chooses $c_s = C_s$, the molarity of A in the solution, and $c_g = C_g$, the concentration of the gas in moles per liter, then $\Delta G^{\ominus}$ is the free energy change for the process

$$A(\text{gas, hyp. } C_g = 1) \rightarrow A(\text{in W, hyp. } C_s = 1) \tag{16}$$

and then $\Delta G^{\ominus}/6.023 \times 10^{23}$ is the free energy change for the process in which one molecule of A is taken from an empty box of some definite macroscopic volume V and placed in volume V of the solvent W. In this case we also have

$$\Delta G^{\ominus} = -RT \ln S_0$$

where S_0 is the so-called Ostwald absorption coefficient, which is often tabulated for gases. At an elementary level, the choice of standard states in eqn. (16) eliminates the translational entropy contribution to $\Delta G^{\ominus}$, but a deeper analysis shows that this is not really so. Thus it may be noted that in the initial state the entire volume V is accessible to the molecule of A, while in the final state much of the space is taken up by molecules of W. Efforts to correct for this are aimed at estimating the "free volume" V_f in the liquid W which is actually available for occupancy by a solute particle. Then one can identify a contribution[224,377] $RT \ln(V/V_f)$ in $\Delta G^{\ominus}$. This is adequate to show that there is nothing really fundamental about this choice of standard states either. Unfortunately, the investigation of the free volume and a related concept, the communal entropy,[377] does not seem to provide a basis for further advances in the quantitative understanding of ionic hydration.

Of course, if one had a proper theory of a solvation process and wished to compare it with experiment, the standard states chosen for tabulation of the experimental data would be unimportant, as long as they were exactly specified. This is illustrated by an example below.

Until now the discussion of the choice of standard states has been focused on solutions of nonelectrolytes. When electrolytes are considered everything is the same except that, in addition, one can measure thermodynamic coefficients only for a process as in eqn. (11) which involves no net transport of charge from one phase to another. Having an experimental value for $\Delta G^{\ominus}$ for the solvation of, say $H^+ + Cl^-$, it is intriguing to consider the possibility of separating this into individual ionic contributions. This is considered in Section 1.7.

To elucidate the role of standard states, it seems useful to consider an example of the comparison of a theory of solvation with experimental data. The very simplest such theory is that due to Born for ionic solutes.[68]

Given a sphere of radius r^* bearing a charge e and immersed in a medium which has dielectric constant ε everywhere, the free energy of the electric field *outside* of the sphere is

$$F^{\text{field}} = e^2/2r^*\varepsilon \tag{17}$$

This result from electrostatic theory is valid whether the charge e is at the center of the sphere, uniformly distributed at the surface of the sphere, or distributed in any other centrally symmetric way inside the sphere. Also, the sphere itself may be a dielectric medium with arbitrary dielectric constant, with the charges imbedded in it. Thus the following considerations are valid for any model for an ion with a centrally symmetric charge distribution entirely inside a sphere of radius $r*$.

For such a model, assuming also that the change in electric field free energy is the only contribution to the solvation free energy of an electrolyte $A_{\nu+}B_{\nu-}$, we have

$$\Delta G^{\ominus} = \tfrac{1}{2}e_0^2(\varepsilon^{-1} - 1)[(z_+^2\nu_+/r_+*) + (z_-^2\nu_-/r_-*)]6.023 \times 10^{23} \qquad (18)$$

where e_0 is the electronic charge and r_i* is the radius of ion i. This is Born's model calculation of $\Delta G^{\ominus}$. Now it is necessary to find out whether a comparison with experiment should make use of $\Delta G^{\ominus}$ for the standard state in eqn. (14) or eqn. (16) or ... ? The answer clearly is that comparison with eqn. (16) is appropriate, since the model calculation corresponds to each ion being in a box with definite macroscopic volume V in the initial state and then letting the solvent in to provide the final state. Otherwise a term $RT \ln(V_i/V_f)$ would have to be added to the change in field free energy.

Finally we consider the relation between $\Delta G^{\ominus}$ for eqns. (14) and (16) for an electrolyte.

It is helpful to begin by noting that if, in the process of eqn. (16), we choose

$$C_s = C_g = 1 \text{ atm}/RT$$

in the standard states, we do not change this $\Delta G^{\ominus}$ although now the process of eqn. (16) has the same initial state as that of eqn. (14). Then it may be observed that for the process

$$\nu_+A^{z+}(\text{in W, hyp. } C_s = 1 \text{ atm}/RT) \rightarrow \nu_+A^{z+}(\text{in W, hyp. } x_A = 1)$$

we have

$$\Delta G = \nu_+RT \ln[C_s(x_A = 1)/(1 \text{ atm}/RT)] = \nu_+RT \ln(RT/V_W \cdot l \text{ atm})$$

because, at low concentrations, we have $C_s = x_A/V_W$, where V_W is the molar volume of the pure solvent. After considering the corresponding process for species B we deduce that

$$\Delta G^{\ominus}[\text{eqn. (14)}] = \Delta G^{\ominus}[\text{eqn. (16)}] + (\nu_+ + \nu_-)RT \ln[RT/(V_W \cdot l\,\text{atm})] \quad (19)$$

At 25°C the fraction in the logarithm is $24.5l/V_W$. At 25°C if the solvent W is water, we have, from eqn. (19),

$$\Delta G^{\ominus}[\text{eqn. (14)}] = \Delta G^{\ominus}[\text{eqn. (16)}] + (\nu_+ + \nu_-)4.25 \quad \text{kcal mol}^{-1}$$

$$\Delta S^{\ominus}[\text{eqn. (14)}] = \Delta S^{\ominus}[\text{eqn. (16)}] - (\nu_+ + \nu_-)14.25 \quad \text{cal mol}^{-1}\,\text{deg}^{-1}$$

$$\Delta H^{\ominus}[\text{eqn. (14)}] = \Delta H^{\ominus}[\text{eqn. (16)}]$$

1.4. Reference Solvents

The processes in eqns. (1) and (11) are the simplest processes related to solvation of nonelectrolytes and electrolytes, respectively, but sometimes the thermodynamic coefficients for the processes cannot be measured experimentally, even indirectly, because the solute of interest cannot be studied in the gas phase. Important examples are the nonelectrolyte glucose and the electrolyte tetrabutylammonium bromide. While one may hope that eventually these experimental problems will be overcome, in the meantime the study of solvation of these solutes may be carried forward by replacing the gaseous initial states by solution states in a *reference solvent*.

Of course the ideal reference solvent is one which does not interact with the solutes of interest; the ideal reference solvent is a vacuum. But a reference solvent for the study of hydration needs to satisfy only a very much weaker condition; compared to water, it should exhibit only relatively regular interaction with the solutes of interest. The interaction should be, compared to water, less sensitive to the chemical nature of the solute, although it may depend in a regular way on the size and charge distribution of the solute species.

Propylene carbonate,

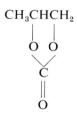

proves to be a useful reference solvent for the study of ionic hydration.

Its dielectric constant is high enough so that 1–1 electrolytes in it are completely ionized at concentrations about 0.001 M, at which thermodynamic coefficients related to ionic solvation are readily measurable. Evidence that its solvation properties are markedly more regular than those of water is shown in Figs. 1 and 2, which also illustrate two general techniques for investigating this point.

Other nonhydrogen-bonded liquids with high dielectric constant, low acidity, low basicity, and good chemical stability may also be expected to be useful as reference solvents. The properties of a number of promising reference solvents are listed in Table I. It is regrettable that at this time the thermodynamic data on electrolytic solutions in these solvents are so sparse that one can seldom use them to elucidate ionic hydration.

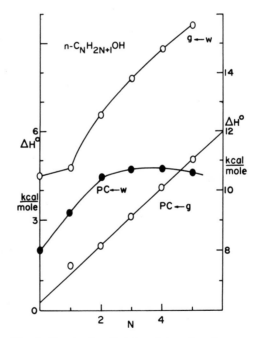

Fig. 1. Standard enthalpies of transfer of normal alcohols for the processes shown. (●) Left scale. (○) Right scale. These data show that the use of propylene carbonate is suitable as a reference solvent for the study of hydration of solutes having alkyl chains. Data from Krishnan and Friedman.[471]

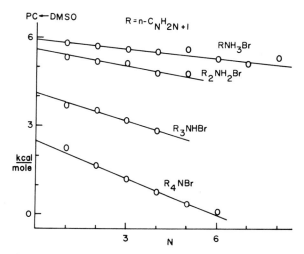

Fig. 2. Enthalpies of transfer to propylene carbonate from DMSO. The slopes are in proportion to the number of alkyl chains and the slope per alkyl chain is the same as found for the normal alcohols, showing that either of these solvents is suitable as a reference solvent in the study of the hydration of solutes with alkyl chains. Data from Krishnan and Friedman.[475]

A simple notation which will be employed here is that for the process

$$X(\text{standard state in } a) \rightarrow X(\text{standard state in } b)$$

we have

$$\Delta G^{\ominus} = (X)^{G}_{b \leftarrow a}, \qquad \Delta H^{\ominus} = (X)^{H}_{b \leftarrow a}, \qquad \Delta S^{\ominus} = (X)^{S}_{b \leftarrow a}$$

and so forth.[471]

1.5. Pair Interaction Coefficients

The coefficient B_2 in eqn. (4) for the concentration dependence of the chemical potential can be shown to have a simple interpretation in terms of solvation. To facilitate the discussion, a model is invoked, so one might expect the topic to appear in Section 2. However, we believe that the general conclusions reached here are independent of the model.

TABLE I. Properties of Solvents for Electrolytes[a]

Solvent	Dielectric constant	Kirkwood g-factor	Dipole moment, D	Molecular polarizability, $Å^3$	Z Values, kcal mol^{-1}	Desolvation enthalpy of the hydrogen bond,[477] kcal mol^{-1}
Acetonitrile	36.0	—	3.97	4.45	71.3	6.05
Dimethylacetamide	37.8	1.1[36]	3.79	9.65	66.9	8.17
Dimethylformamide	36.7	1.0[36]	3.86	7.91	68.5	7.83
Dimethylsulfoxide	45.0	0.95[12]	3.96	7.97	71.1	8.29
Hexamethylphosphoramide	29.6	—	5.37	18.8	62.8	—
Nitromethane	36.7	—	3.46	4.95	71.5[477]	5.59
Propylene carbonate	65.1	1.18[734]	4.94	8.56	71.9[477]	6.33
3-Methylsulfolane	43.0[b]	—	4.3	11.38[477]	77.5[b]	6.35

[a] The data (unless otherwise specified) are taken from the compilation given by Parker.[624]
[b] Value for sulfolane.

The coefficient B_2 appearing in eqn. (4) has the form*

$$B_2 = -\tfrac{1}{2} \int_0^\infty [\exp(-u_{ij}(r)/kT) - 1]4\pi r^2 \, dr \qquad (20)$$

where $u_{ij}(r)$ is the potential of the force between particles i and j (cf. Section 1.2). It depends on the species of these particles and on the distance r between their centers. If they are not spherical, it also depends on their orientations, but the orientation dependence is not shown in eqn. (20). For a one-component gas, particles i and j are any two of the gas molecules. For a one-solute solution particles i and j are any two of the solute molecules. This case, a solution of a single nonionic component, will be discussed first.

The potential function u_{ij} may be related to solvation phenomena in the following way. Consider first a model system in which the solute particles interact like two hard spheres of radius r_i^* and r_j^* in the gas phase, thus with the potential function shown in Fig. 3(a). It may be noted that this simple model does not correspond exactly to the physical situation in which one has two hard spheres immersed in a real liquid, in which case the potential function is more complicated than that in Fig. 3(a) due to effects of solvent–solvent interactions.[548,736,885,886]

Now suppose that the solute molecules each tend to bind one molecular layer of solvent quite strongly but the model is otherwise like the one which gives the potential function in Fig. 3(a). In this case the potential function might be like that in Fig. 3(b) or 3(c), where d^* is the diameter of a solvent molecule and A_{ij} is the energy (really free energy) required to disrupt the solvation layers as solute particles i and j are brought together. Whether the potential function in Fig. 3(b) is more realistic than that in Fig. 3(c) depends on further details of the solute–solvent interaction, but is not important in a qualitative discussion because the dependence of B_2 upon A_{ij} is qualitatively the same in either case. Also by examining eqn. (20),

---

* This is strictly true only if c is the concentration in units of particles per unit volume and if μ_A is the chemical potential of A in the solution at osmotic equilibrium with the solvent, the pressure on the solvent being the same as that in the state for which $\mu_A^{\ominus}$ is defined. There are essentially two parts to the conversion of a thermodynamic function to this state. The major part of the correction[241,242] to $\mu_A - \mu_A^{\ominus}$ in one of the usual concentration scales to conform to the McMillan–Mayer concentration scale is the change in concentration units in the ideal term, which is easily made. The remaining part, associated with the compressibility of the solutions, is usually negligible compared to uncertainties in the experimental data.

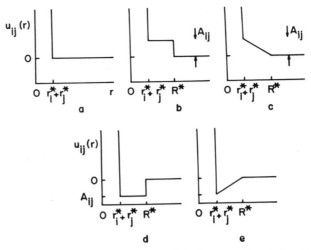

Fig. 3. Solute–solute pair potentials for several models. We define $R^* = r_i^* + r_j^* + 2d^*$, where d^* is the diameter of a solvent molecule.

one can see that if $A_{ij} \gg kT$, then B_2 for the model in Fig. 3(b) becomes the same as B_2 for the model in Fig. 3(a) but with $r_i^* + r_j^*$ replaced by R^*. It follows that the thermodynamic effect of increasing A in the model in either Fig. 3(b) or Fig. 3(c) is qualitatively the same as the thermodynamic effect of increasing the size of the solute particles, i.e., increasing the volume excluded to the solvent.

In Figs. 3(d) and 3(e) the modification of the hard-sphere potential is opposite to that described in the preceding paragraph; now the modification is to introduce an extra attraction between the solute particles. This may be due to direct forces, such as the London dispersion force, between i and j, but the point of interest here is that it also may be due to solvation. This would be the case if the solute particles affect the structure of the solvent nearby so as to make the solvent–solvent attractive forces less effective. In this case when two solute particles come close together the disruption is minimized, causing a net i–j attraction as shown in Figs. 3(d) and 3(e).

A useful way to picture these solvation effects is shown in Fig. 4. The cospheres, shown in Fig. 4 as shaded areas, were introduced by Gurney[302] and Frank and Evans[230] for the discussion of solvation effects (Section 2.5). The cospheres cover the region occupied by solvent that is markedly affected by the presence of solute molecules. Of course, for the most delicate considerations one would need to allow all of the solvent to be in the cospheres, but it will be assumed that the significant effects come from only the few solvent molecules that are directly next to the solute particles.

When two solute particles come close enough together so that their cospheres overlap, as shown in Fig. 4(a), some of the cosphere material is displaced. The A_{ij} parameter in Fig. 3 is directly related to the free energy change in the process in which the cosphere material relaxes to the state of the normal bulk solvent. In this type of analysis solvation [Fig. 3(b,c)] is represented by having the cosphere material in a state of lower free energy per molecule than the bulk solvent, while the converse effect [Fig. 3(d,e)]

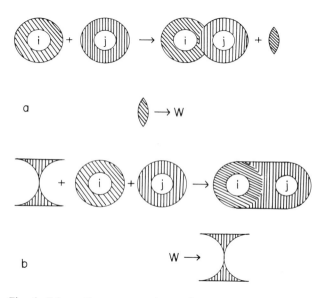

Fig. 4. Schematic representations of process in which cospheres overlap, assuming either (a) mutually destructive interaction or (b) mutually constructive interaction. The cosphere material is represented by shading in such a way as to allow for the fact that the cosphere on solute particle i may be different from that on solute particle j. The representations here assume the cosphere material of j "dominates" the cosphere material of i, which is only one possibility within the scope of the model. In both (a) and (b) the process by which the cosphere material relaxes to the state of the solvent W or is made from the solvent is written as distinct from the overlap process only to emphasize the concept of resolving the process into two steps and not to imply that the model requires any particular sequence of events as two ions come together.

is represented by having the cosphere material in a state of higher free energy per molecule than the bulk solvent.

While the molecular origins of the cospheres may be mysterious, it is now easy to see that the connection between solvation thermodynamic properties and excess thermodynamic properties is fixed by what happens when the cospheres overlap.

The overlap effect represented in Fig. 4(a) may be described as mutual destruction of the cospheres, although if i and j are different species, one cosphere may be disrupted before the other. While the destructive behavior may seem reasonable, the converse behavior, represented in Fig. 4(b), is conceivable as well. This shows a mutual constructive interaction as the cospheres come together. It is easy to imagine, for example, that this might happen as the first stage in forming a clathrate-hydrate crystal from an aqueous solution. However, in cases in which the data are currently available, when only two cospheres overlap the effect is mutually destructive,[226,228] although other things may happen in the larger clusters needed for the analysis of B_3, B_4, etc.

Now turning to electrolyte solutions, we find that $B_2(\varkappa)$ in eqn. (12) is very similar to B_2 in eqn. (20), so all of the preceding discussion is applicable, although some complications need to be noted.

First one always has more than one solute species in an electrolyte solution and then it is necessary to use the relations

$$c^2 B_2(\varkappa) = \sum_i \sum_j c_i c_j B_{ij}(\varkappa), \qquad c \equiv \sum_i c_i \qquad (21)$$

where the sums are over solute species. Since the concentrations c_i cannot all be independently varied in an ionic solution, one can only measure certain combinations of the $B_{ij}(\varkappa)$.

Second, $B_{ij}(\varkappa)$ is given* by an equation very close to eqn. (20), but there are some differences in detail due to the long range of the Coulomb interaction. As the notation shows, B_{ij} for ionic solutions depends on $\varkappa$.

Third, the concentration dependence of the various terms is such that one cannot easily extract the in-principle measurable combinations of $B_{ij}(\varkappa)$ from thermodynamic data for comparison with calculation and it becomes necessary to take into account some effects from the higher terms of eqn. (12).

Still, beginning with ionic solution models in which the pair potentials have any of the forms given in Fig. 3 *plus* a Coulomb term $e_i e_j / \varepsilon r$, one

---

* See the footnote describing the conditions under which eqn. (20) applies.

can calculate the thermodynamic excess properties for comparison with experiment to find which models and which sets of A_{ij} are consistent with the data.[670]

By taking appropriate thermodynamic derivatives of eqn. (4), one can obtain the corresponding equations for the other partial molal thermodynamic functions of the solute. As an example, consider the partial molal enthalpy

$$\bar{H}_A = \partial(\mu_A/T)/\partial(1/T) = \bar{H}_A{}^{\ominus} + Rc[\partial B_2/\partial(1/T)] + \cdots$$
$$= \bar{H}_A{}^{\ominus} + \bar{H}_A{}^E \tag{22}$$

where $\bar{H}_A{}^E$ is the same as L_A, the partial molal relative enthalpy, in the familiar Harned and Owen notation.[317] From eqn. (20) it is easy to see that for the models in Figs. 3 and 4 the derivative of B_2 appearing in eqn. (22) depends on A_{ij} as well as on the corresponding energy

$$E_{ij} \equiv \partial(A_{ij}/T)/\partial(1/T) \tag{23}$$

which is directly related to the energy change in the process in which the cosphere material relaxes to the state of the pure bulk solvent.

Thus in terms of the model in Figs. 3 and 4(a) the leading term in $\bar{H}_A{}^E$, the term proportional to the concentration, depends on the free energy *and* the energy change in the process in which cosphere material relaxes to the state of the bulk solvent. A similar analysis shows that the leading term in $\bar{V}_A{}^E$ depends on the free energy *and* the volume change of the same process.

Now there are many studies of the enthalpies and volumes of solutions in which investigators have compared $\bar{H}_A{}^E$ or $\bar{V}_A{}^E$ for different solutes A and then drawn conclusions which are equivalent to conclusions about the solute dependence of E_{ij} or of $V_{ij} = \partial A_{ij}/\partial P$ for the model in Fig. 4(a). The preceding analysis shows that even with the simple models discussed here this is possible without making model calculations only if the solute particle radius r^* and the free energy parameter A_{ij} are independent of the solute species. Of course, this is not usually known to be the case.

An additional cautionary note comes from examining second derivatives of the chemical potential. For example, by differentiating eqn. (22) with respect to temperature to obtain an expression for the partial molal heat capacity $(\bar{C}_P)_A$ of the solute, one finds that the leading term of $(\bar{C}_P)_A{}^E$ depends on the energy change of the process in which the cosphere material relaxes as well as on its free energy and heat capacity changes. Thus if $E_{ij} \neq 0$, there is a contribution of cosphere overlap to $(\bar{C}_P)_A{}^E$ even if

$dE_{ij}/dT = 0$. In practical terms this makes it quite hopeless to interpret $(\bar{C}_P)_A{}^E$ data without making use of $\bar{H}_A{}^E$ data. This is easily understood in terms of a simple chemical analogy as follows.

The initial and final states of the process in which two solute particles come from infinite separation to some *definite* separation r at which there is cosphere overlap may be taken as the two states of a reversible process

$$A + A \rightleftharpoons A_2 \qquad (24)$$

to which the principles of chemical thermodynamics may be applied. Now consider a system in which the equilibrium in eqn. (24) is established. There is a contribution to the heat capacity of the system from this process if ΔH for the process is nonzero, even if the ΔC_P for the process is zero. As is well known in some cases this even leads to *negative* heat capacities for the overall system, although not negative values of C_P.[300]

In summary the following claims may be made for the analysis of the leading terms of thermodynamic excess functions in terms of the Gurney cosphere model for solvation:

1. It shows clearly how excess properties may possibly be related to solvation phenomena.

2. While it is not a very realistic picture of solute–solute interactions in solutions, its arbitrary features are merely simplifications of reality. For example, in real solutions the McMillan–Mayer theory shows that $u_{ij}(r)$ in eqn. (20) is in general a function of temperature; in applying the Gurney model, the temperature dependence is represented in terms of a parameter E_{ij} which is associated with the energy change in the process in which solute particles i and j approach each other.

3. It enables one to combine the power of chemical thermodynamics with the insight provided by a "molecular" model.*

The statistical mechanical background for the analysis of the leading terms in $\mu_A{}^E$, $\bar{H}_A{}^E$, $\bar{V}_A{}^E$, $(\bar{C}_P)_A{}^E$, etc., is well known[242,450,549] and the reader may wonder that applications, even to the simple Gurney model, are nonexistent (with one exception[459]) for nonelectrolyte solutions. One problem is that in the systems which have been most studied the solute particles

---

* Of course the model is only molecular in terms of the coordinates of the solute particles, while the coordinates of the solvent particles are suppressed so that the solvent is treated as a structureless fluid, although it may be in various different "states." The pair potentials in the model may be formulated in terms of a more detailed model in which the coordinates of the solvent particles *do* appear explicitly.[242,549]

have parts of different chemical structures so that the associated cospheres cannot be expected to be spherical, which makes the treatment of the models rather complicated. Examples of such solutes are C_2H_5OH and glucose. The internal motion of such molecules also may be taken as a reason to suspect that simple cosphere models may not be accurate enough to be interesting. The other problem is the scarcity of data for excess thermodynamic functions, especially the excess free energy, for solutions of relatively spherical solutes, such as aqueous solutions of xenon, methane, and even pentaerythritol.

The above problems are minimized for ionic solutions, at least in water, since the excess thermodynamic functions are available and many well-studied electrolytes have ions which are spherical, or nearly so. However, for ionic solutions one has eqn. (12) instead of eqn. (4) and eqn. (21) instead of eqn. (20) and this leads to new complications: the appearance of temperature or pressure derivatives of $\varkappa$ and of the coefficients of the $\sqrt{c}$ and $c \ln c$ terms in eqn. (12) when one obtains $\bar{H}_A^E$, $\bar{V}_A^E$, etc. from eqn. (12). Thus for ionic solutions the simple Gurney model described above may be used for comparison with experiment only by carefully calculating the thermodynamic properties of the model corresponding to those determined by experiment.

1.6. Temperature and Pressure

Solution standard states at 25°C and 1 atm are employed here. Data at conditions so extreme that they cannot be represented accurately in terms of the free energy and its T and P derivatives through several orders at 25° and 1 atm are beyond the scope of this review.

1.7. Single-Ion Properties

The fact that thermodynamic coefficients for ionic solvation are made up of additive ionic contributions is very useful in assessing the quality of experimental data, as well as in overcoming the difficulties of working with a particular electrolyte which might be of interest but has too low a stability or solubility for convenient study. Both of these purposes are served by *conventions* for the thermodynamic coefficients for single-ion solvation or other transfer processes.

The most widely used such convention is that the standard potential $\mathscr{E}^0$ of the hydrogen electrode is zero. Its familiar applications illustrate both of the purposes mentioned above.

The fact that thermodynamic coefficients for ionic solvation are made up of additive ionic contributions has also led to many efforts to determine the single-ion properties in a less arbitrary way using experiment or theory. Depending on the thermodynamic coefficient and on the system, there now seem to be several ways in which single-ion properties may be determined, but all require the investigator to leave the royal road provided by the theory of the thermodynamics of bulk systems. Fortunately, in some cases different methods of determining the same real single-ion property lead to the same result.

To measure *real* single-ion properties one needs to circumvent the electroneutrality of bulk systems. This leads to measurement of volta potential differences[160,524,626,664] and of the so-called ionic vibration potentials which accompany the passage of sound through ionic solutions.[877] It has also been proposed that one could measure the electric quadrupole radiation from a rotating electrode of appropriate shape.[614] Many aspects of the experimental and theoretical methods for the investigation of electrified interfaces are discussed in a clear and detailed way in Chapter 7 of the text by Bockris and Reddy.[61]

Of course, the objective of determining absolute single-ion properties is to elucidate the role of ionic charge in ionic solvation, since operations which do not correspond to changing the net charge in each phase can be adequately handled by conventional single-ion properties. However it may be argued that the additional significance of absolute as compared to conventional single-ion properties can only be appreciated in the framework of the discussion of some particular molecular model for the solvation structures or processes. For example, given that the hydration free energy of F^- is more negative than that of Cs^+, one would next ask about the relative sizes of the ions, which implies a model in which the ion is a particle of definite size immersed in the water. These considerations have led to the use of model concepts or calculations to obtain *absolute* single-ion thermodynamic coefficients for solvation.

Some calculations of this kind which derive from the Born solvation theory (Section 1.2) are discussed in Section 2. For large polyatomic ions the Born contribution to solvation is negligible compared to other effects and then one may assume, for example, that the solvation of Ph_4As^+ is the same as that of Ph_4B^-,[297] where $Ph- \equiv C_6H_5-$, that the solvation of Cp_2Fe^+ is the same as that of Cp_2Co^-,[760] where $Cp- \equiv c-C_5H_5-$, or that the solvation of $i-Am_4N^+$ is the same as that of $i-Am_4B^-$,[116,117] where $i-Am- \equiv i-C_5H_{11}-$. It seems that these assumptions are all quite accurate in aprotic solvents but recent work[411,414,473,575] shows that the solvation of Ph_4B^-

is different from that of Ph_4As^+ both in methanol and in water, as though the phenyl groups do not insulate the ionic charge from the solvent effectively. One must suspect that the same difficulty will arise with the cyclopentadiene and other aromatic derivatives. The assumption cannot be checked for the tetraalkyl ions in water because i-Am_4B^- reacts too fast with water.

Closely related to the preceding is a method introduced by Conway et al.[132] in which one assumes that the solvation of $(C_nH_{2n+1})_4N^+$ has only a negligible contribution from the Born electrostatic effects and that the significant contribution is proportional to n. Then a plot of a solvation thermodynamic coefficient, $\bar{V}_+^{\ominus} + \bar{V}_-^{\ominus}$ in their case, for $(C_nH_{2n+1})_4NX$ as a function of n (or molecular weight of the cation) must give a straight line with the absolute $\bar{V}_-^{\ominus}$ as the $n = 0$ intercept. In their application to aqueous solutions at 25°C this gives $\bar{V}^{\ominus}$ of Br^- in good agreement with the value obtained from ionic vibrations potentials by Zana and Yeager.[877] While the method seems to work well for ionic enthalpies of transfer to a reference solvent (propylene carbonate) from nonaqueous solvents, the enthalpies of transfer of tetraalkylammonium salts to propylene carbonate from water are markedly nonlinear in n, so the method cannot be applied.[472,473]

It is well known that, in the terminology used here, *real* single-ion properties include contributions from the surface potential of the solvent which are absent in *absolute* single-ion properties. Thus it is necessary in making comparisons to distinguish sharply among real, absolute, and all of the conventional single-ion properties as well as the various choices of standard states.

The whole subject of the determination of single-ion thermodynamic coefficients for hydration has recently been covered in an excellent review by Desnoyers and Jolicoeur,[174] while in a somewhat earlier review Rosseinsky provided a very complete account of studies of single-ion hydration free energies.[697] Both of these reviews are drawn on heavily here, but the emphasis on the determination of absolute single-ion properties is not followed.

Instead, only conventional single-ion properties are tabulated here. But whenever possible the convention, rather than being arbitrary, such as some property of H^+ set equal to zero, is chosen to agree with the "best value" of the absolute single-ion property being tabulated. By this strategy we tabulate "conventional" rather than "absolute" single ion properties so the user who has not studied the determination of single-ion properties will not be misled as to their significance, while the chosen conventions which are probably close to the absolute values seem to have many advantages over arbitrary conventions.

The thermodynamic excess functions can also be formulated as thermodynamic functions for the transfer of the solute from pure solvent to a solution. In the case of ionic solutes the determination of single-ion free energies of transfer for this process is equivalent to the determination of single-ion activity coefficients. While such concepts seem useful in elucidating the properties of solutions of polyelectrolytes,[552] they have not led to improvement in the understanding of solutions of simple electrolytes and will not be used here.

2. MOLECULAR INTERPRETATION

2.1. Introduction

The emphasis here is on theories of ionic hydration which begin with models and deduce the thermodynamic coefficients of hydration of hypothetical systems which correspond to the models. Studies of three kinds of models are discussed: In Sections 2.2 and 2.3 the models neglect the molecular structure of the solvent to the extent that the theory of the electrostatic field may be used to calculate the thermodynamic properties of the models. In Section 2.4 the concern is with "Hamiltonian models," mostly with those in which there is a particle representing the ion and many particles representing the solvent molecules, all interacting according to specified laws of force (i.e., specified force as a function of distance or specified potential energy function.) Such models must be treated by the methods of statistical mechanics in order to yield the thermodynamic properties. In Section 2.5 the concern is with chemical models; in these some part of the water, of course next to an ion, is specified as being in a different chemical state than the state of the liquid water in the absence of ions. With such models the methods of chemical thermodynamics are adequate for calculating the properties of the overall system.

It must be admitted that there are some arbitrary features in this classification of models. For example, in Hamiltonian models one may incorporate features of the other two classes, as illustrated in the discussion in Section 1.5.

As remarked in Section 1.7, to compare a theory of ionic solvation with experiment does not require data for single-ion properties. Rather, it is the other way around: If one has a theory, by comparing theory and experiment, the "best" single-ion results can be deduced from the data, in the framework of the theory. This has led to much theoretical work with the objective of deducing single-ion thermodynamic coefficients from data as a first step in the molecular interpretation of the data. As shown in Fig. 5, which is adapted from a figure in a recent study by Halliwell and Nyburg,[307] the values for the hydration free energy of H^+ arrived at in numerous studies over a period of almost 50 years show a variation of about 4% with no great improvement in the consistency of different results as time goes by.

Those entries in Fig. 5 represented as hexagons depend on data from experiments in which volta potentials are measured or controlled. As is

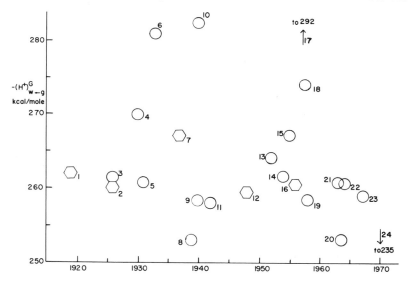

Fig. 5. Chronological survey of determinations of the absolute or real free energy of hydration of H^+ at 25°C for hypothetical 1 M standard states in both the gas phase and in solution. This figure is adapted from Halliwell and Nyburg.[307] The sources of the data follow. In some cases investigators determined $\Delta H^\ominus$ rather than $\Delta G^\ominus$ but their results have been converted to $\Delta G^\ominus$ by correction for the absolute entropy of hydration of H^+, assumed to be -25.2 gibbs mol^{-1}, corresponding to -5.5 gibbs mol^{-1} for the absolute entropy of H^+(aq). (1) Fajans.[208] (2) Latimer.[495] (3) Webb.[824] (4) Garrick.[259,260] (5) Van Arkel and de Boer.[786] (6) Bernal and Fowler.[47] (7) Klein and Lange.[453] (8) Latimer et al.[498] (9) Verwey.[798] (10) Baughan.[40] (11) Verwey.[800] (12) Hush.[391] (13) Miscenko.[581] (14) Azzam[31] (15) Oshida and Horiguchi.[616] (16) Randles.[664] (17) Laidler and Pegis.[484] (18) Van Eck.[788] (19) Brady.[73] (20) Halliwell and Nyburg.[307] (21) Blandamer and Symons.[58] (22) Noyes.[603] (23) Millen and Watts.[569] (24) Salomon.[708]

well known,[160,524,626,664,877] the single-ion free energies obtained in this way, called the real free energies, have an additive contribution equal to the product of the ionic charge and the surface potential of the phase. This contribution is not present in the single-ion free energies, called the absolute (or chemical) free energies. Unfortunately, the surface potential, the electrical potential difference between a point within the phase and a point in the vacuum outside, is not even well-defined, let alone measurable. But only when one can measure the real single-ion free energies and has a reliable theory for obtaining the absolute single-ion free energies from thermodynamic data for neutral systems, can one deduce the surface potentials. In this way Case et al.[95] found the surface potential of acetonitrile (0.1 V more negative inside the liquid) using the "theory" that large organic ions of similar structure but opposite charge should have the same solvation free energy.

The best guess seems to be that the surface potential of water at 25°C is about 0.5 V, with the inside of the liquid phase more negative than the empty space outside of it. It must be noted that this makes a difference of about 10 kcal mol^{-1} between the real and the absolute free energies of hydration of singly charged ions, with cations shifted one way and anions the other.[697,877]

An excellent introductory account of many of the topics in Section 2 may be found in Chapter 2 of the text by Bockris and Reddy.[61]

2.2. The Born Model

In the simplest model for ionic hydration, due to Born,[68] the ions are represented as charged hard spheres and the solvent as a fluid with a dielectric constant ε which is uniform even in the presence of ionic fields. The solvation free energy for this model may be derived as described in Section 1.3 with the result for a neutral electrolyte given in eqn. (18) and the result for an ion i having charge z_i and radius r_i^* given by

$$\Delta G_i^{\ominus} = -(6.023 \times 10^{23} z_i^2 e_0^2 / 2r_i^*)[1 - (1/\varepsilon)] \tag{25}$$

$$\Delta G_i^{\ominus} = -z_i^2 (163.9 \text{ kcal mol}^{-1})(1 \text{ Å}/r_i^*) \quad \text{for water at } 25°C \tag{26}$$

To compare Born's model with experiment, one may use eqn. (18) and compare with data for solvation of a neutral salt. Alternatively, one may calculate

$$\Delta G_i^{\ominus} - \Delta G_j^{\ominus}$$

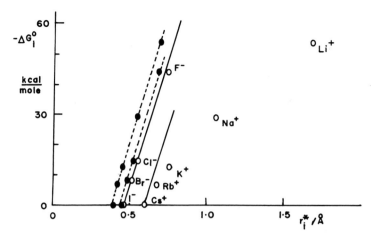

Fig. 6. Free energies of hydration of gaseous ions; cations relative to Cs^+, anions relative to I^-. ($\bigcirc$) Plotted for $r_i^* =$ Pauling crystal radius. ($\bullet$) Plotted for $r_i^* = r_i^P + \Delta$ as described in Section 2.3.1. The straight lines are all drawn with the Born-model slope, 164 kcal mol^{-1} Å. Adapted from Latimer et al.[498]

for pairs of ions with $z_i = z_j$ and compare with thermodynamic data, as shown in Fig. 6. Another approach is to choose a value of $\Delta G^\ominus$ for the process

$$H^+(g, 1\ M) \rightarrow H^+(aq, 1\ M)$$

such that eqn. (1) will represent the resulting free energies of hydration for other single ions as accurately as possible. Figure 7 illustrates a comparison of this sort which has the advantage that ions of various charges can be included in a single comparison. It will be noted that in these two examples, which are chosen from the literature, the ionic radii are approximated differently. In fact, the choice of the r_i^* turns out to be a major problem in applications of this model. Perhaps this is inevitable; real ions are not hard spheres, so there is bound to be some uncertainty as to which value for the hard-sphere radius best represents a given ion in a given situation. This aspect is discussed further in Sections 2.3 and 2.4.

From the study of Figs. 6 and 7 it may be concluded that Born's model is good enough to represent the major part of the hydration free energy and its variation with ionic charge and size. However, the errors are not insignificant; it turns out that the model is very limited in its ability to represent differences in solvation free energy among ions, differences accompanying changes in temperature and pressure (i.e., solvation entropies and volumes), and differences in solvation free energy of a given ion in two solvents. An example is given in Fig. 8. The enthalpy data here bear little relation to the Born-model curve obtained from eqn. (25) by taking the difference in solvation free energies for the two solvents and calculating the corresponding enthalpy change $\Delta H = \partial(\Delta G/T)/\partial(1/T)$ using data for ε and $d\varepsilon/dT$ for the two solvents.

Quite recently a new set of ionic radii based on electron density maps in crystalline NaCl has been proposed by Gourary and Adrian.[288] Blandamer and Symons[58] pointed out that if these radii are used for r_i^*, then in graphs like that shown in Fig. 6 the larger alkali metal and halide ions

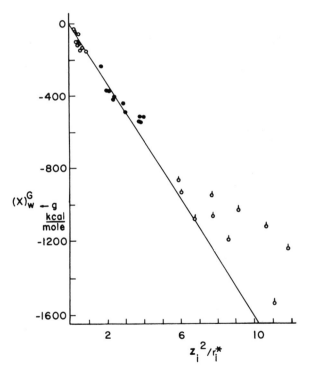

Fig. 7. Free energies of hydration of ions based on convention that the free energy of hydration of H⁺ is -292 kcal mol⁻¹. Here $r_i^* = 1.25 \times$ Goldschmidt crystal radius. (——) Line representing eqn. (26). (○) Monovalent ions, (●) divalent ions, (◌) trivalent ions. Adapted from Laidler and Pegis.[484]

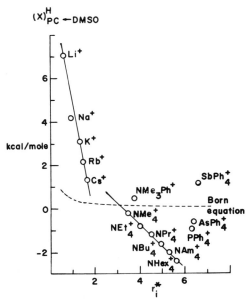

Fig. 8. Ionic enthalpies of transfer to PC (propylene carbonate) from DMSO (dimethylsulfoxide)
at 25°C. These data depend on the convention that
the enthalpy of transfer is the same for Ph_4As^+ as
for Ph_4B^-. From Krishnan and Friedman.[472]
Here r_i* is the Pauling crystal radius for the
monatomic ions and an estimate from models for
the others.

fall on a single line. This suggests that the Gourary and Adrian radii are
the appropriate ones for the study of ionic solvation if one may neglect
the effect of the sign of the ionic charge (but see Section 2.4.3). On this
scale Rb^+ and Cl^- are of the same size. Assuming that their hydration free
energies are equal, we are led to a scale of single-ion hydration free energies
on which the value for H^+ is -260.7 kcal mol^{-1} (cf. Fig. 5). The behavior
described here is not consistent with the Born equation, however, because
as $1/r*$ gets smaller the curve on which the data points lie approaches a
straight line with the equation

$$\Delta G_i^{\ominus} = 15 - (101/r_i*)$$

rather than eqn. (26). But then there is no basis for us to neglect the effect
of the sign of the ionic charge. Moreover, the Born model with the Gourary

and Adrian radii is no better than the same model with the Pauling radii in comparison with data of the kind shown in Fig. 8 as well as with some of the data shown in Section 2.3.

An intriguing fact is that the directly measured $\Delta G^\ominus$ data for the gas reaction

$$\text{ion} \cdot 3H_2O + H_2O \rightarrow \text{ion} \cdot 4H_2O, \qquad \text{ion} = Na^+, K^+, F^-, Cl^-, Br^-, I^-$$

plotted as a function of single-ion hydration free energies $\Delta G^\ominus_{\text{ion}}$ fall on a single curve when the $\Delta G^\ominus_{\text{ion}}$ are fixed by the Gourary and Adrian radii but not when they are fixed by the Pauling radii or the $r_i^P + \Delta$ radii discussed in Section 2.3.[438] It seems that the full significance of this can only be understood in terms of models in which the water next to an ion is represented as discrete molecules. However, calculating the thermodynamic coefficients of solvation on the basis of such models has proved to be exceedingly difficult, as discussed in Section 2.4.

The Born model is so simple and the results obtained with it so promising that there have been a great many efforts to refine it. Some of these studies are described in the remainder of Section 2. The general conclusion is that it is hard to improve on the Born model for estimating the part of the solvation energetics that is peculiar to ionic solutes but that the remainder is very important if one seeks a molecular interpretation of the phenomena!

2.3. The Debye–Pauling Model

2.3.1. Empirical Aspects

Voet[803] has shown that the hydration enthalpies of alkali metal ions, alkaline earth ions, and closed-shell 3+ ions were all consistent with the Born model when r_i^* was given by*

$$r_i^* = r_i^G + \Delta \tag{27}$$

where r_i^G is the Goldschmidt crystal radius of the ion and $\Delta = 0.7$ Å for all the ions mentioned. He interpreted $r_i^G + \Delta$ as the distance from the center of a cation to the center of the electric dipole in a neighboring water molecule.

---

* Goldschmidt and Pauling radii differ little except for Li$^+$. For tables of these and other commonly used radii see, for example, the review by Desnoyers and Jolicoeur.[174]

Subsequently Latimer *et al.*[498] made a corresponding study of the free energies and entropies of hydration of the alkali halides. They used

$$r_i^* = r_i^P + \varDelta \tag{28}$$

with r_i^P the Pauling crystal radius and $\varDelta = \varDelta_+ = 0.85\ \text{Å}$ for cations and $\varDelta = \varDelta_- = 0.10\ \text{Å}$ for anions. With eqns. (26) and (28) and these parameters the data in Fig. 6 fall on the broken lines. These lines both have the 164 kcal mol^{-1} slope specified by eqn. (26). By inspecting the figure, one sees that if the single-ion hydration free energy of I$^-$ is taken to be 9 kcal mol^{-1} more negative than that of Cs$^+$, then the single-ion hydration free energies of all the ions will be given by eqn. (26). On this scale the hydration free energy of H$^+$ is -253 kcal mol^{-1} (cf. Fig. 5).

2.3.2. *The Model*

It seems important to note that the behavior found by Voet[803] and Latimer *et al.*[498] corresponds exactly to what may be expected for a model for ionic solutions based on an earlier investigation of the range of validity of the Debye–Hückel limiting law.[161] In this model each ion is supposed to be a rigid sphere of radius r_i° and charge $z_i e_0$ distributed spherically within it. The dielectric constant of the solution in the neighborhood of the ion is supposed to vary as shown in Fig. 9. This behavior is supposed to represent the effect of saturation of the dielectric in the high field next to the ion.

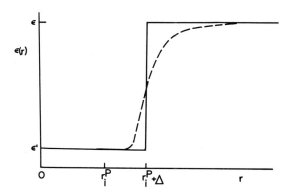

Fig. 9. Local dielectric constant $\varepsilon(r)$ in the neighborhood of an ion i in the Debye–Pauling model. The solid line represents the Debye–Pauling model, the broken line is a refinement discussed in Section 2.3.7.

To calculate the solvation free energy of the ion in the Debye–Pauling model we calculate the electric field free energy about the ion in solution, as did Frank[223] in his study of the higher terms in the Debye–Hückel theory for this model. Following the calculation in Section 1.3, it is easily seen that the solvation free energy of a single ion is

$$\frac{1}{2}(z_i e_0)^2 \left[\frac{1}{\varepsilon} \frac{1}{r_i^\circ + \varDelta} + \frac{1}{\varepsilon'} \left(\frac{1}{r_i^\circ} - \frac{1}{r_i^\circ + \varDelta} \right) - \frac{1}{r_i^\circ} \right]$$

where the first term in brackets comes from the field free energy in solution outside of the $r_i^\circ + \varDelta$ sphere, the second term (in $1/\varepsilon'$) comes from the field free energy in solution in the annular space between r_i° and $r_i^\circ + \varDelta$, and the last term comes from the field free energy about the ion in the vacuum. If we take $\varepsilon' = 1$, this equation may be simplified and it then reduces to eqn. (26) with $r_i^* = r_i^\circ + \varDelta$. The derivation has been given here because it is sometimes asserted that the use of the radius of eqn. (28) in the Born eqn. (26) corresponds to a physical model in which the charged sphere representing the ion has a different size in the vacuum and in solution.[483,484]

The Debye–Pauling model with the parameters introduced by Latimer et al.[498] is very successful in representing the hydration free energies and entropies of ions, even multiply charged ions.[438,497] Of course it also is quite satisfactory in representing hydration-enthalpy data. However, it now seems very clear that this approach is not good enough for the molecular interpretation of the ionic hydration data summarized in this chapter, aside from the difficulty in arriving at a molecular interpretation of the numerical values found for $\varDelta_+$ and $\varDelta_-$. To develop this point, it seems useful to begin by summarizing some studies of solvation in nonhydrogen-bonded solvents. If the Debye–Pauling model is not a satisfactory basis for molecular interpretation of solvation in the absence of hydrogen bonds, it can scarcely be expected to be satisfactory for hydration.

2.3.3. Example of Failure of the Model

Referring to Fig. 8, the equation for the single-ion enthalpy of transfer for the Debye–Pauling model for this system is

$$\varDelta H_i = [0.81/(r_i^P + \varDelta_{i(\text{DMSO})})] - [0.26/(r_i^P + \varDelta_{i(\text{PC})})]$$
$$\leq 0.81/r_i^P \quad \text{if} \quad \varDelta_{i(\text{DMSO})} \geq 0 \quad \text{and} \quad \varDelta_{i(\text{PC})} \geq 0 \qquad (29)$$

so the model cannot give a difference in enthalpy of transfer for Li$^+$ and

Cs^+ greater than 0.85 kcal mol^{-1}, while the observed difference is 6 kcal mol^{-1}. The discrepancy is small compared to the solvation enthalpy of either ion in one of the solvents, but it is so large that the model cannot be of use in interpreting the difference in solvent properties of DMSO and PC.

2.3.4. *Refinement of the Model*

It has often been suggested that either the Born model or the Debye–Pauling model can at best form a basis for calculating the "electrostatic" part of the ionic solvation energy and that this must be added to another contribution such as one would have for the *discharged* ions.[164,165,243,324,600] That is, the ionic solvation free energy is given by an equation such as

$$\Delta G_i^{\ominus} = \Delta G_{i,\text{el}}^{\ominus} + \Delta G_{i,\text{nel}}^{\ominus} \tag{30}$$

where the first term on the right is the electrostatic contribution and the second is the nonelectrostatic contribution, which may be estimated, for example, on the basis of the solvation energy of an uncharged species that is isoelectronic with ion i (and isostructural as well if ion i is polyatomic).

This elaboration on the Debye–Pauling model sometimes works very well. For example, it fits the free energies of transfer of the alkali metal perrhenates to nitromethane from water to within 0.2 kcal mol^{-1} when the following parameters are employed[324]:

$$\Delta_+ = 0.85 \text{ Å}, \quad \Delta_- = 0.10 \text{ Å} \quad \text{in} \quad H_2O$$
$$\Delta_+ = 1.09 \text{ Å}, \quad \Delta_- = 0.00 \text{ Å} \quad \text{in} \quad CH_3NO_2$$

The Δ_- parameter is not well established because only a single anion was studied, but the difference in Δ_+ is significant as a single number which, within the framework of the Debye–Pauling model, expresses the difference in ability of the solvents to solvate cations. Note that this difference appears after allowance for the measured difference in $\Delta G_{i,\text{nel}}^{\ominus}$ in the two solvents.

The application of the Debye–Pauling model which was just described illustrates the *most* that one may learn from a theory of ion solvation based on electrostatic models in which the molecular properties of the solvent do not appear: One can relate the thermodynamic coefficients of ion solvation to those of the corresponding uncharged species. In the process one or more new parameters such as Δ_+ must appear to account for the local dielectric properties of the solvent in the neighborhood of the ion. Even this limited advance is dependent upon the assumption that, in the process in which a nonionic solute is charged to make an ion, the term $\Delta G_{i,\text{nel}}^{\ominus}$ in eqn. (30) does not change.

2.3.5. *Failure of the Model in Water*

Now returning to the calculation of solvation energetics from the Debye–Pauling model without allowance for the $\Delta G^{\ominus}_{i,\text{nel}}$ term, it seems clear that such calculations cannot be very accurate in water since they are not accurate in the nonhydrogen-bonding solvents. An example of a case in which such calculations would obviously fail is given by the enthalpies of transfer of ions to D_2O from H_2O in Table XVI (Section 5.3). Both ε and $d\varepsilon/dT$ are practically the same for these two solvents at 25°C, so this solvent-isotope effect could not be explained in terms of the Debye–Pauling model even if one assumed that Δ_+ and Δ_- had different values in D_2O than in H_2O.

2.3.6. *A Related Study*

It seems useful to consider here an investigation of ionic hydration by Noyes.[600,601] Although he did not specifically consider the Debye–Pauling model, his work may be discussed in terms of it.

We rewrite eqn. (26) as follows:

$$\Delta G_i^{\ominus} = -B_1/r_1^* = -B_1/(r_i^P + \Delta)$$
$$= -(B_1/r_i^P)[1 - (\Delta/r_i^P) + (\Delta/r_i^P)^2 - \cdots]$$

or

$$\Delta G_i^{\ominus} = -(B_1/r_i^P) + B_2(r_i^P)^{-2} - \cdots + B_n(-r_i^P)^{-n} + \cdots \qquad (31)$$

with

$$B_n \equiv B_1 \Delta^{n-1}$$

Noyes's basic equation is equivalent to neglecting terms with $n \geq 3$ in eqn. (31). He uses this expression for the $\Delta G^{\ominus}_{i,\text{el}}$ term in eqn. (30) and then, with an estimate for $\Delta G^{\ominus}_{i,\text{nel}}$, compares the resulting expression for $\Delta G_i^{\ominus}$ with experimental data for the sum $G_i^{\ominus} + G^{\ominus}_{\text{H}+}$ in the case of anions and with experimental data for the difference $G_i^{\ominus} - G^{\ominus}_{\text{H}+}$ in the case of cations. Using the data for the alkali and halide ions (except Li^+) and the theoretical value for B_1, he fixes the remaining parameters $B_{2,+}$, $B_{2,-}$, and $\Delta G^{\ominus}_{\text{H}+}$ to make the "calculated" function agree with the data as well as possible. In the present terms $B_{2,+} = B_1 \Delta_+$ and $B_{2,-} = B_1 \Delta_-$ and Noyes's results for the parameters are

$$\Delta_+ = 0.43 \text{ Å}, \qquad \Delta_- = 0.12 \text{ Å}, \qquad \Delta G^{\ominus}_{\text{H}+} = -260.6 \text{ kcal mol}^{-1}$$

The hydration free energy may be compared with the other values in Fig. 5.

All these parameters differ somewhat from those obtained by Latimer *et al.*[498] (cf. Section 2.3.1), partly because of small differences in the data employed and partly because the earlier workers did not include an allowance for $\Delta G_{i,\text{nel}}^{\ominus}$.

The same method has been extended by Noyes to the study of other thermodynamic coefficients of hydration. It may be seen from his results that the parameters selected to fit the data for the other coefficients require that Δ_+ and Δ_- be functions of temperature and pressure. This seems odd because Latimer *et al.* were able to fit the hydration entropies with the values of Δ_+ and Δ_- selected on the basis of free energy data.

2.3.7. Further Refinements of the Debye–Pauling Model

A refinement of the Debye–Pauling model may be obtained by replacing the discontinuous local dielectric constant $\varepsilon(r)$ by a continuous function, as shown by an example in Fig. 9. This corresponds to a physical model in which the solvent is represented by a dielectric medium in which the dielectric constant is a function of the local thermodynamic variables, the electric field strength and the pressure in particular. Of course, this can equally well be regarded as a refinement of Born's model. In any case most of the recent work with refinements of this type depends on Frank's analysis of rigorous thermodynamics applicable to a fluid medium in an electric field.[225]

Booth[67] has extended Kirkwood's theory[449] of the dielectric constant of an assembly of polar and polarizable molecules to the high-field regime and compared the theoretical results for water with measurements of Malsch.[550,551] The result can be expressed in terms of a simplified functional form due to Grahame[289,290]

$$\varepsilon(E) = \{[\varepsilon(0) - n^2]/(1 + bE^2)\} + n^2 \tag{32}$$

for the local dielectric constant as a function of electric field strength E. Here n is the refractive index of the medium. One cannot actually see the functional form of $\varepsilon(E)$ from Malsch's experiments but the effect he measured corresponds to a parameter b that is within about 30% of the value calculated from the properties of water using Booth's theory.

Before discussing some of the studies of ionic hydration based on eqn. (32) or more directly on Booth's theory, it seems important to note that Schellman[722,723] has shown that the theoretical basis is quite shaky, even to the point that one cannot be sure that the theory gives $\varepsilon(E) < \varepsilon(0)$, and also that Malsch apparently measured the adiabatic response of the fluid to an electric field pulse, rather than the isothermal response given by the theory and which is needed in the calculations of hydration thermodynamics. These matters are very difficult and it seems dangerous to rely on intuition even to conclude that one necessarily has dielectric "saturation," i.e., that $\varepsilon(E) < \varepsilon(0)$ if $E^2 > 0$. For example, in a detailed calculation by Levine and Rozenthal it is found that, at least for one definition of the local dielectric constant in the space between two ions in solution, this may be larger than the bulk value.[508]

An interesting feature of the refined Debye–Pauling model which appears in the work of Noyes[600] may be described as follows. We employ eqn. (25) together with the single-ion-hydration free energies based on $\Delta G_{\text{H}^+}^{\ominus} = -260$ kcal mol^{-1}, which ought to be close to the absolute values, judging from Fig. 5. Then in eqn. (25) we take $r_i^* = r_i^{\text{P}}$, the Pauling crystal radius of ion i, and fit the equation to the data by adjusting ε, which we now call ε_i since the value needed to fit the data may be different for each ionic species i. Of course, from a logical point of view this makes as much sense as fitting the Born equation to the data by adjusting r_i^*. Noyes finds, for cations of various charge,

$$\varepsilon_i = 0.946 + 1.376 r_i^{\text{P}} \pm 0.054 \qquad (33)$$

where the uncertainty given is the standard deviation for 18 cations. This result is not consistent with the interpretation of the Debye–Pauling model in terms of dielectric saturation, for two reasons. (a) According to this interpretation, ε_i must be an appropriate average over an $\varepsilon(r)$ function like the broken line in Fig. 9 and it does not make sense that it should be independent of the charge on the ion. (b) For the smaller ions ε_i is less than ~ 1.8, the electronic contribution to the dielectric constant of water. This seems possible in the fields next to ions, which may reach 10^8 V cm^{-1}, but there is no reason to believe that the Booth theory[67] is valid in this range.

These observations reveal serious difficulties in the physical interpretation of ionic hydration in terms of the Debye–Pauling model *if* one further assumes that Δ in the model merely reflects the dependence of the solvent dielectric constant upon the local electric field and pressure. However, it must be noted that the same exercise, but with the Gourary and Adrian radii[288] in place of the Pauling radii used by Noyes, gives values of the effective dielectric constant ε_i which seem more realistic.

Part of what is missing in this interpretation of the Debye–Pauling Δ is allowance for the interplay of repulsive forces between the ion and the solvent and between solvent molecules. These forces tend to result in "void" spaces like those in a close-packed assembly of hard spheres. As shown by Glueckauf,[277] simple geometric considerations show that these void spaces are roughly accounted for by adding a term to the crystal radius of the ion; thus the void spaces make a contribution to the Δ parameter in the Debye–Pauling model. In fact, Glueckauf has investigated a number of variations on this model in which he incorporated various approximations to the local dielectric constant $\varepsilon(r)$ for $r > r_i{}^*$ and in which the results are compared with hydration entropies and volumes as well as free energies.[277,279] These models fit the free energy and entropy data for the alkali metal ions quite well if the model parameters are appropriately chosen, but of course that is not an improvement on the Latimer–Pitzer–Slansky results, except in the sense that introducing a more detailed model has not spoiled the agreement they obtained with the Debye–Pauling model. However, in comparison of his models with hydration entropies and volumes for larger ions, Glueckauf finds discrepencies which point to the need for a more "chemical" interpretation of the interaction of ions with the neighboring water molecules. Perhaps this is sufficient justification to omit further mention of many studies of models centered on the locally varying dielectric constant, beginning with that of Webb[824] and represented very recently by Laidler and Muirhead-Gould,[483] but it may also be noted that all these studies of electrostrictive effects on ionic solution properties have been recently reviewed by Conway.[125]

2.4. Hamiltonian Models

2.4.1. *Introduction*

The solvation models considered in the preceding section comprised one particle, representing the ion, immersed in a "medium" representing the solvent. In a more detailed model we may represent the medium itself by an assembly of a large number of identical particles representing the solvent molecules. As soon as this is done the methods of classical electrostatics employed in Sections 2.2 and 2.3 become inadequate for calculating the properties of the model system; one must use the methods of statistical mechanics.

To proceed, one needs to know, for a classical system, the Hamiltonian function for the assembly of one ion with N solvent particles:

$$H_{N+1}(\mathbf{p}_1, \ldots, \mathbf{p}_{N+1}, \mathbf{r}_1, \ldots, \mathbf{r}_{N+1}) = \sum_{i=1}^{N+1} (\mathbf{p}_i{}^2/m_i) + U_{N+1}(\mathbf{r}_1, \ldots, \mathbf{r}_{N+1})$$

$$(34)$$

for each set of momenta $\mathbf{p}_1, \mathbf{p}_2, \ldots$ and locations $\mathbf{r}_1, \mathbf{r}_2, \ldots$ of the $(N+1)$-particle system. In eqn. (34) U_{N+1} is the potential of the interaction of the $N+1$ particles in the specified configuration. If the particles representing the ion or the solvent molecules are not spherical, then each r_i must specify the orientation as well as the location of the particle. If angular momenta are important, then each p_i must specify the angular momentum as well as the translational momentum, and the kinetic energy term in eqn. (34) must be written to include the contributions from angular momenta. Finally, if quantum mechanical effects are to be accounted for, one employs the standard transcription to convert the Hamiltonian *function* H_{N+1} to the appropriate Hamiltonian operator.[377,488]

A basic theorem in statistical mechanics is that the Helmholtz free energy $A(N+1, V, T)$ of the system of $N+1$ particles in volume V at temperature T is given by

$$\exp[-\beta A(N + 1, V, T)] = \int d\mathbf{p}_1 \int d\mathbf{p}_2 \cdots \int d\mathbf{r}_1 \cdots \int d\mathbf{r}_{N+1}$$
$$\times \exp[-\beta H_{N+1}(\mathbf{p}_1, \ldots, \mathbf{r}_{N+1})] \qquad (35)$$

where $\beta = 1/kT$ and, of course, ΔA for a solvation process like that in eqn. (1) may be obtained by forming the ratio of eqn. (35) applied to the initial and final states of the process. Again there is a well-known transcription of eqn. (35) for use in cases in which quantum mechanical effects are included.[489]

In such a model it is easy enough to use the mass m_i (and, when appropriate, the moment of inertia I_i) corresponding exactly to that of the real ion or solvent molecule, but the potential energy function U_{N+1} is generally not well known and, in order to use eqn. (35), one needs to make some assumption about its form. This way of proceeding, in which one assumes a definite functional form for the Hamiltonian and then proceeds to a more or less exact evaluation of the integral in eqn. (35), may be called the study of Hamiltonian models.

Enormous progress has been made in recent years in the development of approximate but accurate methods of evaluating the integral in eqn. (35), whether directly or indirectly through the use of other exact relations. However, these methods are applicable only to models in which the potential function is pairwise-additive

$$U_{N+1} = \tfrac{1}{2} \sum_{i=1}^{N+1} \sum_{j=1}^{N+1} u_{ij}(\mathbf{r}_i, \mathbf{r}_j) \qquad (36)$$

and are practical especially in the case that the pair potential u_{ij} depends only on the distance between particles i and j. Of course, a model subject to both of these conditions cannot represent a hydration system very realistically. In the real system there are not only orientation-dependent forces involving charge–dipole, dipole–dipole, and higher-order multipole interactions, but the molecular polarizabilities make U_{N+1} not pairwise-additive.[397,744]

Of course, for a still more detailed model in which the particles are the ion nucleus, the nuclei of the N solvent molecules, and all of the electrons, the Hamiltonian will indeed have a potential function that is pairwise-additive, but the possibility of accurate evaluation of the thermodynamic properties of such a model is remote.

We proceed to review what has been done with models in which the system is represented as an assembly of one ion and N solvent particles, as specified in eqn. (35), and then review what has been learned from the study of certain models which are hybrids of Hamiltonian models and "electrostatic" models of the kind discussed in Sections 2.2 and 2.3.

2.4.2. Studies with Hamiltonian Models

Golden and Guttman[286] assume that

$$U_{N+1} = E_0 + \lambda E_1 + \lambda^2 E_2 \tag{37}$$

where λ is a parameter, ranging from zero to one, which is proportional to the *formal* charge on the ion in the system; each factor z_i appearing in U_{N+1} is replaced by $\lambda |z_i|$ and the terms are collected as in eqn. (36). Then, taking $\lambda = 1$ for a cation or $\lambda = -1$ for an anion gives the potential function in the ionic solvation model, while taking $\lambda = 0$ gives the potential in a model for the solvation of an uncharged species for which the "ion"–solvent van der Waals, repulsive, dipole–dipole, ..., interactions are the same as for the ion, that is, the potential for the system in which the ion is formally discharged. By *formally* it is meant to emphasize that the discharging is not accomplished by removing or adding electrons, which would change the electronic structure and so the "ion"– solvent van der Waals and other interactions.

Were the solvent molecules not polarizable, there would be no terms beyond the λE_1 term in eqn. (37). The $\lambda^2 E_2$ term comes, for example, from the mutual interaction of dipoles induced on two solvent molecules by the ion. Terms of order λ^3 and higher of this kind, due to the solvent molecule polarizability, are neglected in the model, but it is asserted by the authors that this does not affect the conclusion.

Golden and Guttman treat the quantum mechanical model corresponding to eqn. (37), so that kinetic energy effects are included. However, for such a general model one cannot expect to get very detailed results. What they find can be expressed in the following way. Define $A(\lambda)$ as the value of $A(N+1, V, T)$ calculated for some particular value of λ, followed by taking the thermodynamic limit, i.e., the limit as N increases to infinity while N/V remains fixed. This is the usual prescription to get bulk thermodynamic properties from calculations for a model with specified N and V. Then $A(\lambda) - A(0)$ is the free energy of charging the ion in solution. In this way of formulating the model the field free energy, emphasized in Sections 2.2 and 2.3, does not appear explicitly, its effects being accounted for in another way. Therefore $A(\lambda) - A(0)$ is also the difference in solvation free energy of the ion with the fraction λ of its full charge and the discharged ion.

For example, for a suitable model $A(1) - A(0)$ is the solvation free energy of a K^+ ion minus that of an Ar atom, since a formally discharged potassium ion must be pretty close to an argon atom, while $A(-1) - A(0)$ is the free energy of solvation of a Cl^- ion minus that of an Ar atom.

Golden and Guttman find that in general $A(\lambda)$ has the form shown in Fig. 10. That is, the curve is always concave downward, which is consistent with the Born model, eqn. (25), but the maximum value of $A(\lambda)$ is greater than $A(0)$ and occurs on one side or another of $\lambda = 0$, depending on the functional form of E_0 [see eqn. (37)]. This remarkable result, applied to the example in the preceding paragraph, means that if a small positive formal charge added to an Ar atom increases its solvation free energy,

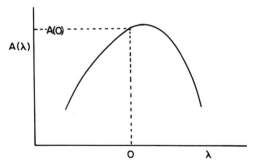

Fig. 10. Solvation free energy as a function of formal charge according to the calculation of Golden and Guttman.

(weaker solvation) then a small negative charge added to an Ar atom reduces its solvation free energy (stronger solvation). This is the case shown in Fig. 10.

As Golden and Guttman point out, a simple interpretation of this effect is that, in this example, the Ar atom in the solvent is at a positive electrical potential and putting positive charge on it is destabilizing until the added charge is large enough to dominate the potential produced by solvating the neutral Ar atom. By comparison with single-ion hydration free energies, they conclude that the case shown in Fig. 10 qualitatively accords with ion solvation in water, with value of λ at the maximum in the range 0.03–0.1 and with the zero-charge potential in the range from 0.3 to 1 V, depending on the scale of ionic radii employed.

Their result may be interpreted in the framework of the Born and Debye–Pauling models if one identifies the nonelectrostatic contribution to the solvation free energy, as in eqn. (30), *and* assumes that $\Delta G_{i,\mathrm{nel}}^{\ominus}$ changes when the state of formal charge of the ion changes. While this is reasonable in terms of the structural interpretation of $\Delta G_{i,\mathrm{nel}}^{\ominus}$ (see Section 2.5), this effect makes it quite impossible to use eqn. (30) for the molecular interpretation of ionic hydration!

Somewhat earlier Marcus[553–556] obtained somewhat more detailed results for the interpretation of experiments involving nonequilibrium polarization by a method similar to that described here. The implications for thermodynamics of ionic hydration are not clear, so Marcus's results are not included here, but it seems that these methods can be used for further insight into the thermodynamics of ionic solvation.

While there do not seem to be any studies of ionic hydration using more detailed Hamiltonian models, recent developments indicate that such studies are feasible for models that are realistic enough to be of help in the molecular interpretation of ionic hydration thermodynamics. One such development, the study by Rahman and Stillinger,[659] treats a model for water in which the potential function is assumed to be pairwise-additive, with the pair potential selected to be appropriate to the liquid rather than to the gas. In this way, as Stillinger has shown,[753] one can include at least part of the effects of nonpairwise interaction due to molecular polarizability and due to quantum mechanical effects in the interactions.[163,311] Rahman and Stillinger use the molecular dynamics method of treating the model, which gives dynamic as well as thermodynamic properties, and find that their results are in excellent agreement with many of the measurable properties of real water. Thus the model is accurate enough to be interesting and it only remains to apply it and the method to systems for ionic hydration, although the technical difficulties and the computing expense are considerable. Judging from the water study,[659] such a calculation would provide many valuable insights into the molecular configurations and processes giving rise to observable hydration phenomena.

The other promising development is Wertheim's analytical (exact) solution for the "mean spherical model" for an assembly of hard spheres with electric multipoles (charges, dipoles, quadrupoles) of various orders.[832] Such a model may not be realistic enough for many purposes and the "mean spherical model" is, in the language employed here, an approximation technique rather than a Hamiltonian model, but it seems well established that results obtained by the mean spherical model approximation for systems at liquid densities are quite accurate. Having accurate results for the correlation functions and thermodynamic coefficients of the Hamiltonian model corresponding to an assembly of hard spheres with multipoles, one can use well-known perturbation methods of statistical mechanics[489,794,880,825,826] to get the thermodynamic coefficients of other Hamiltonian models in which the potentials are more realistic. This general approach has already been worked out for mixtures of nonpolar molecules, in which case the mean-spherical-model (same as Percus–Yevick equation) approximation for an assembly of hard spheres of various sizes is the appropriate basis for the perturbation calculation.[504,557]

2.4.3. Hybrid Models

These are models in which the interaction of an ion and its first-nearest-neighbor solvent molecules (primary solvation sheath) is calculated in terms of intermolecular forces, just as for a Hamiltonian model, while the interaction of this complex with the remaining solvent is calculated as in the theory of the Born model, perhaps with allowance for the non-electrostatic part of the solvation of the complex, as in eqn. (30). There have been very many studies of models of this sort[46,61,83,192,586,791,798–800] which illustrate what effects they may include. Unfortunately, it seems that there are important problems with calculations of this type which need to be overcome before they can provide a reliable interpretation of hydration phenomena more detailed than that from the Debye–Pauling model. These problems may be illustrated by reference to the recent contribution of Muirhead-Gould and Laidler.[586]

The process of hydration of an ion X^z, where z is the positive or negative charge, may be analyzed as shown in eqn. (38). Here the surface potential of water is neglected, so ΔG calculated for the overall process corresponds to the absolute rather than the real free energy of hydration (Sections 1.7 and 2.1). The steps are all assumed to take place at 25°C:

$$nH_2O(liq) \rightarrow nH_2O(g, \text{ hyp. 1 atm})$$
$$\Delta G_a \qquad (38a)$$

$$X^z(g, \text{ hyp. 1 atm}) + nH_2O(g, \text{ hyp. 1 atm}) \rightarrow (X \cdot nH_2O)^z(g, \text{ hyp. 1 atm})$$
$$\Delta G_b \qquad (38b)$$

$$(X \cdot nH_2O)^z(g, \text{ hyp. 1 atm}) \rightarrow X^z(aq, \text{ hyp. 1 } M) + nH_2O(liq)$$
$$\Delta G_c \qquad (38c)$$

In the process in eqn. (38a), n molecules of water are vaporized. Of course, ΔG_a is the standard free energy of vaporization of n moles of water.

In the process in eqn. (38b) the complex of specified composition is formed. Muirhead-Gould and Laidler[586] represent the water molecule by a particle having charges at the locations of the atoms in an actual water molecule and chosen so that the model particle will have the same dipole moment as an isolated real water molecule. They assume that the complex $(X \cdot nH_2O)^z$ is symmetrically constructed, with the distance from the center of X to the center of each O given by the sum of the crystal radius of X and that of the water molecule (1.38 Å). They express ΔG_b as the sum of three terms:

1. $\Delta G_{b,1}$ due to the interaction of the charge on X^z with the charges on the n water molecules in the complex, all calculated from Coulomb's law.

2. $\Delta G_{b,2}$ due to the London dispersion force acting between the ion and each of the n water molecules and acting between nearest-neighbor water molecules, all calculated from London's equation for the $1/r^6$ term in the dispersion potential in terms of the polarizabilities and ionization potentials of the particles.[377]

3. $\Delta G_{b,3}$ due to the repulsive force, assumed to be derived from a potential of the form A/r^{12}, acting between the ion and each of the n water molecules in the complex.

In the process in eqn. (38c) the complex is dissolved in water. It is assumed that at infinite dilution in water the hydration layer about X^z forms a complex of the same structure as that specified for $(X \cdot nH_2O)^z$ in the gas phase so the final state in eqn. (38c) is the same as the state of the complex in the hypothetical 1 M aqueous solution. Muirhead-Gould and Laidler express ΔG_c as the sum of three terms:

1. $\Delta G_{c,1}$ given by eqn. (25) with $r_i{}^* = r_i{}^G + 2.76$ Å. This is the Born model hydration energy of a spherical particle with the same radius as the complex ($r_i{}^G$ is the Goldschmidt crystal radius of the ion).

2. $\Delta G_{c,2}$ estimated as the contribution of $2n$ hydrogen bonds for cations, or n hydrogen bonds for anions, with each hydrogen bond assumed to contribute -3.57 kcal mol^{-1}.

3. $\Delta G_{c,3}$, the standard free energy of hydration of the rare gas isoelectronic with X^z.

In addition, there is a term ΔG_{ss} to correct $\Delta G_{c,1}$ to apply to hypothetical 1 atm standard states in the gas phase and hypothetical 1 M standard states in solution.

For example, the values obtained for $X^z = Rb^+$ and $n = 6$ given in Table II are typical of the results of Muirhead-Gould and Laidler.[586] The most striking aspect is the large number of contributions, according to this analysis, having magnitudes greater than 5 kcal mol^{-1}. Except for ΔG_a and ΔG_{ss}, each of these terms has a substantial uncertainty in terms of the way the parameters are chosen, or a substantial error in terms of the treatment of the model. A few examples are discussed here to illustrate the difficulty inherent in hybrid models.

Consider first the size and distribution of the charges in the model particle that represents a water molecule. If one wishes to represent a gaseous water molecule, the charge distribution in the model should be consistent with what is known about the quadrupole and octupole tensors[190,273,377,594] as well as the electric dipole moment. But it is well known that there are many-body effects which make the effective dipole moment and other parameters of the intermolecular potential different in dense fluids, such as liquid water, than in the isolated molecules.[246,613,753] Thus for calculations of the properties of liquid water from a model with a pairwise-additive Hamiltonian, Rahman and Stillinger used parameters corresponding to a dipole moment of 2.11 D for the water molecule.[45,659] The Ben-Naim–Stillinger[45] model of the water molecule they used is probably much more suitable than a model giving gas-phase properties

TABLE II. Example of Calculation of Free Energy of Hydration from a Hybrid Model

Term of ΔG^a	a	b,1	b,2	b,3	c,1	c,2	c,3	ss	Total	Experimental
kcal mol^{-1}	63.1	-91.2	-9.6	8.8	-38.5	-19.2	3.6	1.9	-81.1	-75.7

^{a} Data for Rb^+ taken from Muirhead-Gould and Laidler.[586]

for the estimation of ion–water interactions as well as water–water interactions. In view of the difference in dipole moments this change would change $\Delta G_{b,1}$ to about -105 kcal mol^{-1}.

In similar fashion the parameters in the calculation of $\Delta G_{b,2}$ and $\Delta G_{b,3}$ could be chosen to be consistent with the second virial coefficient of neon, as assumed by Rowlinson[699] and Rahman and Stillinger.[659]

It is certainly an error to suppose that ΔG_b can be calculated in the way described by Muirhead-Gould and Laidler. For example, since the number of particles changes from $n + 1$ to n in this process, the free energy change for the process must depend upon the concentrations in the initial and final states. What is missing is an account of the translational as well as rotational and vibrational contributions to the free energy. Thus accepting the model as specified by the authors, the calculation of ΔG_b really pertains to the energy change at absolute zero, $\Delta E_0^{\ominus}$. In fact, the way to obtain ΔG_b for a Hamiltonian model is to calculate $\Delta E_0^{\ominus}$ and the partition functions of the reactants and products and then calculate ΔG_b from these functions using textbook statistical mechanics[221,372,563] (cf. Eley and Evans[192]).

Finally, as Vaslow[791] has shown, the potential functions used to represent the ion–water and water–water interactions do not tend to "fix" the geometry in the complex very well; the analogy with chemical binding in which the forces are more directional is not very good. This observation, in addition to making the calculation of ΔG_b more delicate, suggests that the complex may be considerably distorted in the process in (38c), and that this will make a contribution to ΔG_c.

Stokes has pointed out that the gaseous monatomic ions are considerably larger than one might estimate from their crystal radii and that it is preferable to derive their sizes from those of the isoelectronic rare gases by scaling.[756,757] In the same way, the complex $(X \cdot nH_2O)^z$ may be expected to be larger in the gas phase than in water; this is another way to see the need for accounting for the distortion of the complex in the process in eqn. (38c).*

---

* It may be possible to avoid the distortion contributions to ΔG_c by assuming an $(n + 1)$-body potential for the gas phase complex which gives it the same equilibrium geometry as the ion–water complex in solution, assuming that the geometry of the latter is known. However, it remains to be demonstrated that it is possible to do this in a consistent way. For example, when the complex is immersed in water one may expect the induced electric moments of the water molecules of the complex to change. If the geometry of the complex does not adjust to this change, it would seem that the complex is no longer in internal equilibrium. When the contributions of the internal motions (vibrations) of the complex to ΔG_c are not neglected the problem of the distortions is magnified.

Probably the only part of ΔG_c which has been calculated accurately is $\Delta G_{c,1}$. In obtaining $\Delta G_{c,2}$ in Table II, the work required to make a cavity to receive the ion has been neglected, as has an allowance for the water–water hydrogen bonds broken when the water-complex hydrogen bonds are made. Of course $\Delta G_{c,3}$ assumed by Muirhead-Gould and Laidler does not account for these effects.

Very recently there have been several calculations of the binding energy and force constants of complexes $X^z \cdot nH_2O$ like that formed in eqn. (38b), with $X^z = H^+$ or $OH^{-(153,167,595)}$ or other ions.[90,91,169,170] In these calculations the particles are taken to be the nuclei and the electrons of the ions and water molecules, so the Hamiltonian is well known, but very extensive studies are required to control the approximations used. From the results one can calculate ΔG_b for comparison with experimental measurements from the laboratories of Kebarle[438] and Friedman.[169,170] The most complete results have been obtained for complexes with H^+ or OH^- but the consequences for our understanding of ionic hydration in the sense of this chapter, that is, for the overall process in eqn. (38), are not yet clear. Still, it seems likely that quantum mechanical calculations, especially the more refined ones,[595] which stand up well when compared with the experiments[169,170,438] will provide a good basis for constructing effective Hamiltonians for models in which the particles are the ions and the water molecules and which are much easier to handle in the context of model calculations in which much larger numbers of water molecules are allowed to interact with the ions. An example of this may be found in the results of Burton and Daly.[90,91] Using the CNDO/2 approximation method they have calculated $\Delta E_0^{\ominus}$ for the process in eqn. (38b) for $n = 4$ or $n = 6$ and for $X^z = Li^+$, Na^+, Be^{2+}, and Mg^{2+}. In each calculation a configuration of the nuclei in the complex is assumed and the electronic binding energy is then derived. This is done for a series of configurations of fixed symmetry, and fixed relative configuration of the nuclei for each water molecule, in which the water–ion distance is varied. The result is that the most stable complex is that for which the ion–water separation corresponds to an ionic radius about 50% larger than the Pauling crystal radius. The CNDO methods are usually reliable for the investigation of configurational aspects, so this result may be used to fix one of the parameters of the Hamiltonian of any model in which the ion and the water molecules are taken as the particles. On the other hand, the result also suggests that other configurations are more important in the solvation complex in solution, for Bol, using a differential X-ray scattering technique, finds that the $M^{2+}–H_2O$ distance in solutions of a number of divalent ions is very nearly the same as in the crystalline hydrates.[63]

It is not clear whether models in which the water molecules and the ion are the particles need to be treated quantum mechanically. First, it may be recalled that the Rahman–Stillinger calculation,[659] which is for a classical model, is very successful in representing the equilibrium and dynamic properties of water. Second, it may be pointed out that Feder and Taube[214] found that the $H_2{}^{16}O$–$H_2{}^{18}O$ fractionation effects associated with hydration are very small for singly charged ions; since isotope fractionation has a quantum mechanical origin, this suggests that the quantum mechanical effects in hydration are small. Also, the solvent isotope effects $(X)_{D_2O \leftarrow H_2O}^{G}$ and $(X)_{D_2O \leftarrow H_2O}^{H}$, while easily measurable (see Section 5), are very small compared to the overall hydration free energies and enthalpies. A theoretical study of this problem by Sposito and Babcock does not seem very helpful because of the limitations of the model they treat; in effect they take infinite values for two of the moments of inertia of the particle representing the water molecule in the field of an ion.[743]

2.5. A Chemical Model

As indicated in Section 2.4, the contribution of the formation of a solvation complex $X^z \cdot nH_2O$ to the thermodynamics of solvation [the overall process in eqn. (38)] has yet to be deduced for any Hamiltonian model for the complex, however simple. On the other hand, one may assume that the solvation complex forms in solution and then attempt to deduce its properties from experimental data. The procedure is analogous to the well-known one whereby the thermodynamic properties of molecules, especially the energies, were interpreted in terms of chemical bonds, assuming that the bonds make additive contributions to the properties, long before accurate calculations of molecular properties such as binding energies could be made. The empirical approach even allowed the actual non-additivity of chemical bond contributions to be identified in some cases, benzene, for example. Chemical models do not, however, always lead to advances in understanding. For example, the interpretation (before 1923) of the properties of aqueous sodium chloride solution in terms of association of Na^+ and Cl^- to form $NaCl$ is now known to be wrong in its implications for the structure of the solutions, at least under ordinary conditions.

Other chemical models are often proposed for the interpretation of solution properties, including the following. Robinson and Stokes[691,692] and Glueckauf[276] interpret the excess free energies of aqueous solutions of sugars or strong electrolytes in terms of complexes of the solute species with fixed numbers of solvent molecules; except when there are Debye–Hückel effects, these complexes are assumed to mix ideally with each other and with the solvent, the choice of scale for ideal mixing being controversial. Coll et al.,[119] Yagil,[872] Marshall,[558] and others interpret the mass-action constants for solute-association equilibria in terms of the participation of fixed numbers of water molecules in the association reaction, with auxiliary assumptions concerning ideality of mixing of the resulting species and the dependence of the chemical potential of the solvent upon the composition and other thermodynamic variables. Wyatt[870] has very successfully interpreted the thermodynamic properties of H_2O–SO_3 mixtures in the composition region close to pure H_2SO_4 in terms of association reactions among a few species which mix ideally.

In a remarkable contribution in 1945 Frank and Evans[230] proposed a chemical model for the interpretation of hydration phenomena which has proved very successful in providing a consistent molecular interpretation of a great variety of data obtained in the following years. We now describe a model that is very closely related to theirs, although more arbitrary and definite in some features. It also corresponds very closely to the solvation models discussed by Gurney[302] and Samoilov.[713] The model has already been used above (Section 1.5) to explain the relation of thermodynamic excess functions to solvation functions. When employed in that way the model may be applied to the interpretation of thermodynamic excess functions without the need for assuming any sort of ideal mixing.

In the chemical model discussed here we assume that around each solute particle X^z there is a region, the cosphere, having the thickness of one solvent molecule in which the solvent properties are affected by the presence of the solute, and we characterize these effects by the thermodynamics of the process in eqn. (39), where n is the number of solvent molecules in the cosphere, and

$$n[\text{solvent(pure bulk liquid)}] \rightarrow n[\text{solvent(in cosphere state next to } X^z)] \qquad (39)$$

where $z = 0$ for a nonionic solute. Following Frank and Evans,[230] one may estimate the thermodynamic coefficients for eqn. (39) by comparison of the solvation of X^z with the solvation thermodynamics in a "normal" system, solutions of permanent gases in nonpolar solvents, for example. The cosphere effects are apparently very small in the normal systems, as shown by the success of a particular theory of solvation which does not take them into account.[230]

Thus for a solution of argon in water Frank and Evans estimate that for the process in eqn. (39) the entropy change is $\Delta S = -12 \pm 3$ gibbs for the n moles of water required to form the cospheres for a mole of Ar. We estimate $n \simeq 12$, the ratio of the cosphere volume

$$(4\pi/3)[(1.7 + 2.8)^3 - 1.7^3] \text{ Å}^3 = 350 \text{ Å}^3$$

to the mean volume of a water molecule in the liquid,

$$18 \text{ cm}^3 \text{ mol}^{-1} = 30 \text{ Å}^3 \text{ molecule}^{-1}$$

Then for the process in eqn. (39), with $X^z = Ar$, we have $\Delta S = -1.0 \pm 0.3$ gibbs mol^{-1}. While this figure for the *molar* entropy change of the cosphere water depends on the arbitrary way in which n is calculated, it may be noted that the molar entropy change is small even if one takes $n = 6$, as implied by another arbitrary choice for the structure of the cosphere.

Now, considering ionic solutes, and again following Frank and Evans, we compare the standard entropy of solvation of two argon atoms

$$2Ar(g, \text{ hyp. 1 atm}) \rightarrow 2Ar(\text{solution, hyp. 1 } M) \tag{40}$$

with the standard entropy of hydration of a potassium ion and a chloride ion, each of which is isoelectronic with an argon atom,

$$K^+(g, \text{ hyp. 1 atm}) + Cl^-(g, \text{ hyp. 1 atm})$$
$$\rightarrow K^+(\text{solution, hyp. 1 } M) + Cl^-(\text{solution, hyp. 1 } M) \tag{41}$$

The comparisons for solutions in methanol and in water are given in Table III. The result in methanol is qualitatively what is expected if the dominant effect on charging two argon atoms in solution to make a potas-

TABLE III. Some Solvation Entropies in Methanol and in Water[a]

	In methanol	In water
$\Delta S^\ominus$, eqn. (40)	-32	-60.4
$\Delta S^\ominus$, eqn. (41)	-80	-51.9

[a] Entropies in gibbs mol^{-1} at 25°C. Standard states: hyp. 1 atm in gas, hyp. unit mole fraction in solution. From Krishnan and Friedman.[476]

sium ion and a chloride ion is to orient the solvent about the solute particles by means of the electric fields. On the other hand, this is apparently not the dominant effect in aqueous solutions! The anomaly in aqueous solution is considerably larger when one allows for the Born-model contribution to the ionic entropies due to polarization of the solvent outside of the cospheres.[230,476] As Frank and Evans show,[230] a simple model calculation leads to the expectation that there is something like a *double* cosphere around each ion, as represented in Fig. 11.

In the introduction of the above model the entropy changes in the process in eqn. (39) were emphasized. The motivation for this is not only historical. Since in statistical mechanical theories the Helmholtz free energy is most often the thermodynamic function which can be most directly calculated for a model, it might be preferable to introduce the subject by emphasizing ΔA, or what will be numerically nearly the same, ΔG, for the process in eqn. (39). However, this turns out to be quite unsatisfactory in more empirical approaches to the properties of aqueous solutions. The reason is that there seems very often to be a compensation of the ΔH and ΔS for processes such as eqn. (39), that is, $\Delta H \simeq T \Delta S$, so the corresponding effects in the free energy are harder to identify. This phenomenon will now be briefly discussed.

The Barclay–Butler rule

$$\Delta H(x) = \Delta H_0 + T^* \Delta S(x) \tag{42}$$

expresses the fact that for a series of similar processes which differ in some

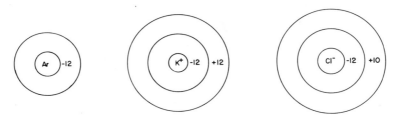

Fig. 11. Cosphere representation of hydration of Ar, K^+, and Cl^- drawn to scale for the chemical model discussed here. The numbers in the cospheres are the total entropy changes [gibbs, eqn. (37)] for the water in a mole of cospheres, deduced by Frank and Evans. According to the present model each of the smaller cospheres holds about 12 water molecules and each of the larger ones about 37 water molecules. The water outside the cospheres is assumed to be in the same state as in the absence of all solutes except for Born-model electric field effects.

minor respect, specified by x, the enthalpy change is linearly related to the entropy change, the constants of the linear equation being ΔH_0 and T^*. Thus x may be a number specifying the length of an alkyl chain in a homologous series, the period of an atom in a given group of the periodic table, or the size of a molecule in a series in which the size is varied. The applicability of the Barclay–Butler rule to solvation phenomena and the basis for the rule in terms of molecular interactions have been extensively studied by Frank[224] and Frank and Evans.[230] Very recently Lumry and Rajender[536] have summarized the extensive evidence, which has been accumulating for many years, that in processes relating to hydration at 25°C one very often finds the Barclay–Butler rule obeyed with T^* slightly less than 300°K; in such cases we have compensation of the kind described in the preceding paragraph. Typically $T^* = 280°K$, and then we have

$$\Delta G(x) = \Delta H(x) - T\,\Delta S(x) = \Delta H_0 - (T - T^*)\,\Delta S(x) \qquad (43)$$

and one finds that at $T = 298°K$ the position of equilibrium in the process is controlled by the sign of the entropy change (a so-called "entropy-controlled process") but that only about 10% of the x dependence of $\Delta H(x)$ or $T\,\Delta S(x)$ shows up in $\Delta G(x)$. The very same processes for which this compensation (sometimes called Lumry's law) is found are ones for which ΔC_P is apt to be large and x-dependent, so T^* depends on the temperature appropriate to the data to which eqn. (42) is applied. That is, in eqn. (42) we should explicitly indicate the T dependence, for example, in the first term as $\Delta H(x, T)$, and note that the coefficients of the linear relation are T-dependent, and strongly so in the case of phenomena related to hydration.

That the enthalpy–entropy compensation holds for the process in eqn. (39) is most clearly demonstrated by examining the data for solvent isotope effects on hydration. This is discussed in Section 5.4.

Beginning with Frank and Evans,[230] it has often been noticed that heat capacity effects associated with cosphere formation may be very large, and that there may be two contributions to the experimental data, one from ΔC_P for the process in eqn. (39) and another for the temperature dependence of the equilibrium constant if ΔH is not zero. This has already been mentioned in Section 1.5. A satisfactory way of separating these contributions has yet to be found. Of course, the effects of cosphere formation upon the volumes and all the other thermodynamic coefficients in solution are fair game for study; some results are described in Section 4.

A substantial problem in nomenclature has developed regarding the interpretation of solution properties in terms of cosphere effects, because of the many different locally defined states of water that must be allowed for to account for the phenomena, even for the three solute species in Fig. 11. The problem is serious enough that it seems to be the main source of some criticisms of the model.[380] For this reason, and to facilitate the application of the model to the discussion of the data, the nomenclature which is currently in fairly wide use is formalized somewhat here (Table IV). Of course, this is a long list of states of cosphere water, and it may be argued that with an equation with enough parameters one can write one's name, but the point of view is taken here that in using a cosphere model one must expect to mimic the essential character of chemistry. For example, one knows not only that the reactivity of a C—Cl bond is different from that of a C—H bond, but also that the reactivity depends on the other bonds in the molecule, through nearest neighbors, next-nearest neighbors, etc. Progress in chemistry has depended very much on the accumulation of data for all of these effects. In the present case there does not seem to be any evidence that II_{al} cosphere water is different from II_{rg} cosphere water, while evidence is just appearing that II_{al} cosphere water is rather different from II_{ar} cosphere water. The nomenclature in Table IV also facilitates the discussion of effects of overlapping cospheres, whether on different

TABLE IV. Nomenclature for the States of Cosphere Water

Hydration of the first kind: States in which the water is oriented by ionic fields or other directional solute–solvent forces.

State I_c: characteristic of the inner cospheres of small cations.

State I_a: characteristic of the inner cospheres of small anions.

State I_{hb}: characteristic of the inner cospheres of hydroxyl groups or R_3NH^+ ions.

Hydration of the second kind: states in which the water is perturbed by the proximity of a solute particle but the effect cannot be ascribed to directional solute–solvent forces.

State II_{rg}: characteristic of the cosphere of a rare gas atom.

State II_{al}: characteristic of the cosphere of an alkyl group.

State II_{ar}: characteristic of the cosphere of an aromatic group.

State II_{sb}: characteristic of the outer cospheres of small ions, the seat of the so-called structure-breaking phenomenon.

parts of the same molecule (some cases in Section 5) or on different molecules as they collide (Section 7).

Ideally, the data on ionic hydration summarized in the remainder of this chapter would be used to fix the thermodynamic coefficients for the formation of various cosphere states. Unfortunately, this sort of quantitative interpretation of the data goes beyond what is usually found in the papers reviewed here, so it has only been possible to point out the qualitative interpretation of the data in terms of the chemical model. In the case of thermodynamic coefficients corresponding to second and higher derivatives of the free energy even the qualitative interpretation is not straightforward, as pointed out above.

An important complication occurs when different types of cospheres form on the different parts of a solute molecule, for example, for the species $(CH_3)_3NH^+$ one may expect to have a type II_{al} cosphere on the methyl groups and a type I_{hb} cosphere on the $—NH^+$ group. It seems quite clear from the data, some of which are discussed in Section 6, that adjacent cospheres interfere with each other so that the overall solvation is less energetic than one might guess from the properties of the separate cospheres. This is completely analogous to the effect of overlap of cospheres on different solute particles when they collide; the effect seems always to be mutually destructive (Sections 1.5, 7).

The molecular interpretation of the effects corresponding to the formation of cosphere states is, of course, a very important objective. More or less clear-cut proposals have been made by many authors, Frank and Evans,[230] Frank and Wen,[232] and Ben-Naim and Stillinger,[754] for example. In every case the proposal is based upon the configurational requirements of well-hydrogen-bonded structures of assemblies of water molecules and upon the forces acting on the water molecules at the interface between water and the surface of a solute particle or between water and a type I cosphere. An especially important development is the determination of the structures of many clathrate hydrates, for these may often be regarded as models for cospheres of types II_{rg} and II_{al} (see Ref. 408 and this treatise, Volume 2, Chapter 3).

While these efforts at molecular interpretation are rather qualitative and lacking in statistical mechanical structure, they are usually supported by consideration of the dynamic properties of the solutions or the way they scatter X rays or neutrons.[706] Since these aspects are outside the scope of this review, the molecular interpretation will be bypassed here in favor of emphasis on the contributions of the various cosphere states to the thermodynamic properties of the solutions.

3. HYDRATION OF GASEOUS IONS

3.1. Free Energies, Enthalpies, and Entropies of Hydration

A selection of "best" single-ion data is given in Table V. The free energies are all based on measured ionization potentials or electron affinities rather than on calculated lattice energies because the latter are usually of unknown accuracy. The construction of this table involves branching chains of calculations based on large amounts of data for various physical measurements, so the references given are to earlier compilations rather than to original experimental papers. The procedure for constructing such a table is detailed in some of the earlier papers.[74,682]

The notation for thermodynamic coefficients for transfer processes in the headings of Table V is explained in Section 1.4. The standard states in Table V are hypothetical 1 atm in the gas phase (g) and hypothetical 1 M in the aqueous phase (w). See Section 1.3 for further information about standard states.

As discussed in Section 2, the principal contribution to all the free energy data in Table V is given by a simple calculation based on the Debye–Pauling model. The remaining contributions, often totaling less than a few kcal mol^{-1}, have their origins in various effects which are more easily investigated in terms of other thermodynamic coefficients in which the Born-charging contribution is relatively smaller and in which the entropy–enthalpy compensation (Section 2.5) is not relevant.

The enthalpies of hydration are numerically very close to the free energies and it seems preferable to discuss the differences, essentially the entropies of hydration. Since the entropies of the gaseous ions can usually be estimated easily and quite accurately by statistical mechanics, the Sackur–Tetrode equation, they become "known" quantities in the discussion of entropies of hydration and it becomes a matter of choice whether one treats entropies of hydration or the single-ion aqueous entropies, together with a more or less accurate allowance for the terms of the Sackur–Tetrode equation which differ from one ion to another.

3.2. Equations for Entropies of Aqueous Ions

Many simple empirical equations have been proposed which account quite accurately for the ionic entropies in aqueous solution, and therefore for the ionic hydration entropies. These equations are often useful for estimating the entropies of ions for which data are lacking. For example,

TABLE V. Free Energies, Enthalpies, and Entropies of Hydration of Monatomic Ions at 25°C[a]

Ion, X	$-(X)^G_{w \leftarrow g}$	$-(X)^H_{w \leftarrow g}$	$-(X)^S_{w \leftarrow g}$	$\bar{S}^{\ominus}_{X(g)}$	$\bar{S}^{\ominus}_{X(aq)}$	Ion radius (Pauling), Å
H+	260.5	269.8	31.3	26.0	−5.3	—
Li+	123.5	133.5[b]	33.7	31.8	−1.9	0.60
Na+	98.3	106.1	26.2	35.3	9.1	0.95
K+	80.8	86.1	17.7	36.9	19.2	1.33
Rb+	76.6	81.0	14.8	39.2	24.4	1.48
Cs+	71.0	75.2	14.1	40.6	26.5	1.69
Cu+	(136.2)	151.1	42.9	38.4	4.5[262]	0.96
Ag+	114.5	122.7	27.6	40.0	12.4	1.26
Tl+	82.0	87.0	16.7	41.8	25.1	1.40
Be2+	582.3[600]	612.6	101.6	—	(−65.6)	0.31
Mg2+	455.5	477.6	74.3	35.5	−38.8	0.65
Ca2+	380.8	398.8	60.8	37.0	−23.8	0.99
Sr2+	345.9	363.5	59.2	39.2	−20.0	1.13
Ba2+	315.1	329.5	48.5	40.9	−7.6	1.35
Ra2+	306.0	319.6	45.6	—	2.4	1.40
Cr2+	444.8	460.3	53.0	—	—	0.84
Mn2+	437.8	459.2	72.1	41.5	−30.6	0.80
Fe2+	456.4	480.2	79.8	43.0	−36.8	0.76
Co2+	479.5	503.3	80	42.8	−37.2	0.74
Ni2+	494.2	518.8	82.4	42.3	−40.1	0.72
Cu2+	498.7	519.7	73.9	42.9	−34.2[262]	0.72
Zn2+	484.6	506.8	74.5	38.4	−36.1	0.74
Cd2+	430.5	449.8	65.2	40.0	−29.5[867]	0.97
Hg2+	(436.3)	453.7[600]	58.4	—	−16.0	1.10
Sn2+	371.4	389.5[600]	60.7	—	−16.5	1.12
Pb2+	357.8	371.9	47.4	41.9	−5.5	1.20
Cr3+	1037	1099.9[600]	143.9	—	−89.4	0.69
Fe3+	1035.5	1073.4	127.5	41.5	−86.0	0.64
Al3+	1103.3	1141.0	126.6	35.8	−90.8	0.50
Ga3+	1106.0	1147.0	137.9	39.10	−98.9	0.62
In3+	973.2	1009.3	117.9	40.0	−77.9	0.81
Tl3+	975.9	1028.0	—	—	(−57.9)	0.95
La3+	—	820.2[d]	97.5	40.7	−56.8[c]	1.15

TABLE V. (*Continued*)

Ion, X	$-(X)^G_{w \leftarrow g}$	$-(X)^H_{w \leftarrow g}$	$-(X)^S_{w \leftarrow g}$	$\bar{S}^{\ominus}_{X(g)}$	$\bar{S}^{\ominus}_{X(aq)}$	Ion radius (Pauling), Å
Ce^{3+}	—	832.4	96.7	44.3	−52.4	1.11
Pr^{3+}	—	842.6	97.8	45.1	−52.7	1.09
Nd^{3+}	—	849.5	100.7	45.4	−55.3	1.08
Pm^{3+}	—	858.1	(102.3)	45.2[469]	(−57.1)	1.06
Sm^{3+}	—	866.9	104.0	44.6	−59.4	1.04
Eu^{3+}	—	874.7	106.8	42.7	−64.1	1.03
Gd^{3+}	—	880.4	109.1	45.2	−63.9	1.02
Tb^{3+}	—	888.6	109.6	46.2	−63.4	1.00
Dy^{3+}	—	896.1	110.9	46.7	−64.2	0.99
Ho^{3+}	—	903.4	111.4	46.9	−64.5	0.97
Er^{3+}	—	909.1	111.3	46.8	−64.5	0.96
Tm^{3+}	—	915.2	112.3	46.4	−65.9	0.95
Yb^{3+}	—	920.5	113.0	45.5	−67.5	0.94
Lu^{3+}	—	925.6	113.3	41.4	−71.9	0.93
Sc^{3+}	929.3	962.7	(112.5)	—	(−70.8)	0.81
Y^{3+}	—	896.6	111.3	39.4	−71.9[c]	0.93
U^{3+}	—	—	—	—	−51.9[648]	1.03
Pu^{3+}	—	849.6[d]	—	—	−54.9	1.00
U^{4+}	—	—	—	—	−99.2	0.93
Pu^{4+}	—	—	—	—	−108.2	0.90
F^-	103.8	113.3	31.8	34.8	3.0	1.36
Cl^-	75.8	81.3	18.2	36.7	18.5	1.81
Br^-	72.5	77.9	14.5	39.1	24.6	1.95
I^-	61.4	64.1	9.0	40.4	31.4	2.16
S^{2-}	303.6	309.8	20.5	24.7	4.2	1.84
Se^{2-}	—	—	—	—	(10.6)	—
OH^-	90.6	101.2	35.6	38.4	2.8	(1.8)

[a] The conventions used here are that $(H^+)^G_{w \leftarrow g} = -260.5$ kcal mol^{-1} and $(H^+)^S_{w \leftarrow g} = -31.3$ gibbs mol^{-1}. Values in parentheses are uncertain. The values for $(H^+)^S_{w \leftarrow g}$ and $(H^+)^G_{w \leftarrow g}$ depend on the initial and final standard states as follows: For H^+(w, hyp. 1 M) ← H^+(g, hyp. 1 atm) we have $(H^+)^S_{w \leftarrow g} = -31.3$ gibbs mol^{-1} and $(H^+)^G_{w \leftarrow g} = -260.5$ kcal mol^{-1}. For H^+(w, hyp. 1 M) ← H^+(g, hyp. 1 M) we have $(H^+)^S_{w \leftarrow g} = -25.0$ gibbs mol^{-1} and $(H^+)^G_{w \leftarrow g} = -262.4$ kcal mol^{-1}. For H^+(w, hyp. unit mole fraction) ← H^+(g, hyp. 1 atm) we have $(H^+)^S_{w \leftarrow g} = -39.3$ gibbs mol^{-1} and $(H^+)^G_{w \leftarrow g} = 258.15$ kcal mol^{-1}. The data (unless otherwise specified) are taken from the compilation of Rosseinsky[697] and Latimer.[496]

[b] The data for all alkali metal ions are taken from Halliwell and Nyburg.[307]

[c] Data for all the rare earths are taken from Bertha and Choppin.[50]

[d] Data for all the rare earths are taken from Morss.[585]

they can be used to estimate $\Delta S^{\ominus}$ of some process in solution for which one has only determined $\Delta G^{\ominus}$ or $\Delta H^{\ominus}$ by experiment. Important applications include the case in which the process of interest is the formation of a transition state from the reactants in a kinetic process, since the absolute rate theory relates the thermodynamics of formation of the transition state to the rate of the overall process. However, as suggested by the work of Gurney[302] and others, equations for the size and charge dependence of the ionic entropy, such as eqns. (44)–(52), are less informative about the molecular origin of the effects than comparison of entropies with various other thermodynamic coefficients and with the viscosity B coefficient.

In this section the following nomenclature will be used.

M_i The gram atomic weight of an ion of species i.

z_i The signed charge of ion i in protonic units.

r_i^{P} The Pauling crystal radius of i in angstrom units.

Δ A length parameter (Å) whose significance is explained for each use.

S_i $= \bar{S}_i^{\ominus}(\mathrm{aq}) - z_i \bar{S}_{\mathrm{H}^+}^{\ominus}$, the conventional single-ion entropy of species i in hypothetical 1 M aqueous solution at 25°C in the units gibbs mol^{-1} of i. Note that here "conventional" has the meaning that $\bar{S}_{\mathrm{H}^+}^{\ominus} = 0$ is the basis of the single-ion values.

A rather long list of empirical equations for S_i or $(i)_{\mathrm{w} \leftarrow \mathrm{g}}^{S}$ will now be discussed. Deviation graphs showing how well some of the equations fit the data in Tables V, VIA, and VIB are given in Fig. 12.

Latimer et al.[498] proposed

$$(i)_{\mathrm{g} \leftarrow \mathrm{w}}^{S} = -9.720 z_i^2 / (r_i^{\mathrm{P}} + \Delta) \tag{44}$$

which is obtained by differentiating eqn. (25) with respect to temperature, using $d(\ln \varepsilon)/dT = -45.88 \times 10^{-4} \deg^{-1}$ for water at 25°C. Here Δ is 0.85 Å for cations and 0.10 Å for anions, as discussed in Section 2.3. Thus this equation follows from the Debye–Pauling model if one assumes that Δ_+ and Δ_- do not depend upon temperature.[497]

Powell and Latimer[648] (see also Scott[730]) proposed

$$S_i = 1.50R(\ln M_i) + 37 - [270 \, | \, z_i \, | / (r_i^{\mathrm{P}} + \Delta)^2] \tag{45}$$

where Δ is 2 Å for cations and 1 Å for anions. To the extent that this equation fits the data well, it suggests that the entropy of the Born-model process is dominated by another effect due to the interaction of the ion

TABLE VIA. Conventional Single-Ion Entropies of Some Diatomic and Polyatomic Ions[a]

Ion, X	$\bar{S}^{\ominus}_{X(g)}$	$\bar{S}^{\ominus}_{X(aq)}$	Interatomic distance r, Å
NO_2^-	56.5	35.2	1.24
ClO_2^-	60.9	29.4	1.59
NO_3^-	58.8	40.3	1.24
ClO_3^-	63.8	44.3	1.48
BrO_3^-	67.5	43.8	1.68
CO_3^{2-}	58.8	−2.1	1.26
SO_3^{2-}	63.4	3.6	1.39
ClO_4^-	63.1	48.5	1.52
ReO_4^-	70.9	55.3	1.90
CrO_4^{2-}	64.7	19.8	1.60
MoO_4^{2-}	70.4	(24.6)[496]	1.83
SO_4^{2-}	62.9	14.7	1.50
SeO_4^{2-}	67.3	16.3	1.65
PO_4^{3-}	63.9	−36.1	1.55
AsO_4^{3-}	67.5	−18.7	1.75
OH^-	38.4	2.8	0.97
CN^-	45.2	33.5	1.10
OCN^-	53.0	36.4	$r_{OC} = 1.13, r_{CN} = 1.21$
SCN^-	55.5	(41.3)[496]	$r_{CN} = 1.17, r_{SC} = 1.61$
HF_2^-	50.8	5.8	2.26
N_3^-	50.7	(37.3)	2.30
BF_4^-	64.3	45.3	1.43
BH_4^-	45.2	30.8	1.26
$Ag(CN)_2^-$	73.6	54.3 63.4[b]	3.29
$Au(CN)_2^-$	73.5	34.8	3.3
NH_4^+	44.5	21.7	1.03
UO_2^+	61.5	6.7[496]	1.6

[a] The convention used here is that $\bar{S}^{\ominus}_{H^+(aq)} = -5.3$ gibbs mol⁻¹. Entropies are given in gibbs mol⁻¹. The data (unless otherwise specified) are taken from Altschuller.[9,10] Values in parentheses are uncertain.
[b] Calculated from the data given by Nancollas.[588]

Ion	$\bar{S}^{\ominus}_{X(aq)}$	Ion	$\bar{S}^{\ominus}_{X(aq)}$	Ion	$\bar{S}^{\ominus}_{X(aq)}$	Ion	$\bar{S}^{\ominus}_{X(aq)}$
H^+	-5.3	$PtCl_6^{2-}$	63.2	AlF_2^+	-27.2	S_2^{2-}	(10.6)
HS^-	19.9	$Fe(CN)_6^{4-}$	89.2	AlF_3°	-6.0	S_3^{2-}	(15.6)
HSe^-	27.3	$Ni(CN)_4^{2-}$	43.6	AlF_4^-	10.0	S_4^{2-}	(20.6)
HSO_3^-	31.3	$Zn(CN)_2^{\circ}$	45.3^b	AlF_5^{2-}	18.0	$S_2O_3^{2-}$	(18.6)
$HSeO_3^-$	35.7	$Zn(CN)_3^-$	73.4^b	AlF_6^{3-}	18.0	$S_2O_5^{2-}$	(35.6)
HCO_3^-	28.0	$Zn(CN)_4^{2-}$	94.5^b	TiF_6^{2-}	(30.6)	$S_2O_6^{2-}$	(40.6)
$HGeO_3^-$	(26.3)	$Cd(CN)_4^{2-}$	107.8^b	SnF_6^{2-}	10.6	$S_2O_8^{2-}$	(45.6)
HSO_4^-	35.6	$Hg(CN)_4^{2-}$	74.6	SiF_6^{2-}	-1.4	$S_3O_6^{2-}$	(43.6)
$HSeO_4^-$	27.3	$Cu(CN)_2^-$	54.3$^{(262)}$	$CdCl^+$	3.0$^{(651)}$	$S_4O_6^{2-}$	(46.6)
$HCrO_4^-$	21.8	$Cu(CN)_3^{2-}$	68.6$^{(262)}$	$CdCl_2^{\circ}$	23.0$^{(651)}$	$S_5O_6^{2-}$	(50.6)
HCO_2^-	27.2	$Cu(CN)_4^{3-}$	65.9$^{(262)}$	$CdCl_3^-$	53.3$^{(651)}$	$S_2O_4^{2-}$	38.6
$HN_2O_4^-$	39.3	$Ag(CN)_2^-$	101.9^b	$FeCl_2^+$	-32.6	$N_2O_2^{2-}$	17.2
HPO_4^{2-}	2.0	$Cu(SCN)^+$	6.2	$FeBr_2^+$	-38.6	UO_2^{2+}	-27.6
$HAsO_4^{2-}$	11.5	$C_2O_4^{2-}$	21.2	FeF_3°	8.0	$U(OH)^{3+}$	-45.9
$H_2BO_3^-$	12.6	WO_4^{2-}	25.6	$Fe(OH)^{2+}$	-33.8	$UO_2(NO_3)_2^{\circ}$	53
$H_2PO_4^-$	26.6	MnO_4^-	50.7	$Fe(NO)^{2+}$	-21.2	$UO_2SO_4^{\circ}$	-13
$H_2AsO_4^-$	33.3	$Cr_2O_7^{2-}$	61.7	$Cr(OH)^{2+}$	-21.7	PuO_2^+	13.7
$H_2PO_3^-$	(24.3)	$AgCl^{\circ}$	37.0	$CrCl_2^+$	24.7	PuO_2^{2+}	23.6
$H_3BO_3^{\circ}$	38.2	$AgCl_2^-$	61.3	$Ag(NH_3)_2^+$	52.5	Br_3^-	(45.3)
$AuCl_4^-$	66.3	$TlCl^{\circ}$	48.0, 45.3^b	$Cu(NH_3)_2^+$	57.7	I_3^-	62.4
$AuBr_4^-$	80.3	$TlBr^{\circ}$	50, 45.5^b	$CuCl_2^-$	54.7	VO_4^-	(53.3)
$HgBr_4^{2-}$	94.6	$SnCl^+$	16.3	ClO^-	15.3	$IrCl_6^{3-}$	58.9
HgI_4^{2-}	100.6	$SnCl_2^{\circ}$	39.0	IO_3^-	33.3	$IrCl_6^{2-}$	63.6
$PdCl_4^{2-}$	46.6	$SnCl_3^-$	84.5	TeO_3^{2-}	(8.6)	$RhCl_6^{3-}$	(65.9)
$PdBr_4^{2-}$	(54.6)	$SnBr^+$	16.5	BeO_2^{2-}	16.4	$[Co(NH_3)_5Cl]^{+2}$	85.5
$PtCl_4^{2-}$	52.6	$SnBr_2^{\circ}$	47.0	BO_2^-	25.3	$[Co(NH_3)_5H_2O]^{3+}$	57.5
$PtBr_4^{2-}$	(59.6)	$SnBr_3^-$	86.1	AlO_2^-	30.3	—	—
$PdCl_6^{2-}$	(62.6)	AlF^{2+}	-56.2	SeO_3^{2-}	14.5	—	—

^{a} The convention used here is that $\bar{S}^{\ominus}_{H^+(aq)} = -5.3$ gibbs mol^{-1}. Entropies are given in gibbs mol^{-1} at 25°C. The data (unless otherwise specified) are taken from Latimer[496] and from references given by Cobble.[114]
^{b} Calculated from the data given by Nancollas.[588]

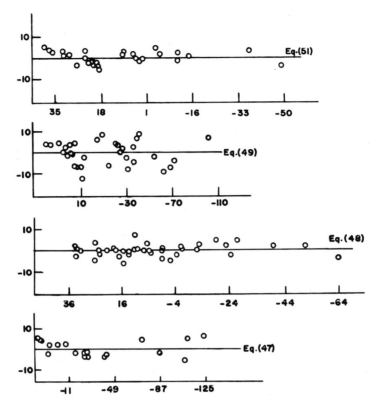

Fig. 12. Deviation plots for entropies of aqueous ions. The deviations in gibbs mol^{-1} are on the y axis. The x axes are $\bar{S}_i^{\ominus} - 1.5R \ln M_i$ for eqn. (47), $S_i - 1.5R \ln M_i$ for eqn. (48), $S_i - n\bar{S}_w^{\ominus}$ for eqn. (49), and $\bar{S}_i^{\ominus} -1.5R \ln M_i$ for eqn. (51).

with the neighboring water dipoles. The empirical performance of this equation can be improved somewhat by adding a term on the right to account for the entropy associated with the degeneracy of the ground state of the gaseous ion for ions such as Fe^{3+}.

Powell proposed[647] the simpler equation

$$S_i = 47 - [154 \mid z_i \mid /(r_i^P + \Delta)^2] \tag{46}$$

where Δ is 1.3 Å for cations and 0.4 Å for anions.

Laidler proposed the equation[481,482]

$$\bar{S}_i{}^{\circ} = 10.2 + 1.50 M(\ln M_i) - 11.6(z_i^2/r_i^u) \tag{47}$$

for the single-ion entropy based on $\bar{S}_{H+}^{\ominus} = -6$ gibbs mol^{-1}, where r_i^u is Pauling's univalent (monatomic) radius for species i.

Of course the comparison of eqns. (44)–(47), which can best be applied to monatomic ions, shows once again that the choice of ionic size parameters is very critical (Section 2.2). A corollary is that the "best" choice of ionic radii cannot be determined empirically from solution data, since all these equations fit the data quite well.

For oxyanions Cobble proposed[113]

$$S_i = 1.50R(\ln M_i) + 66 + 81(z_i f_i/r_{io}) \tag{48}$$

where f_i is a "structural" factor in the range from 0.68 to 1.55 depending mostly on the geometry of the ion, and r_{io} is the distance from the center of the ion to the oxygen nucleus.

For inorganic complex ions Cobble proposed[113]

$$S_i = 49 - 99(z_i f_i/r_{iL}) + n\bar{S}_w^{\ominus} \tag{49}$$

where r_{iL} is the distance from the center of the ion to the center of a ligand, n is the number of water molecules displaced when complex ion i is formed from the aquated metal ion, and $\bar{S}_w^{\ominus}$ is the partial molar entropy of water. The structural factor f_i is 1.00 if the ligands are monatomic or hydroxyl; 0.65 if the ligands are NO, NH_3, CN$^-$, or SO_3^{2-}. Unfortunately, eqn. (49) does not apply if $z_i = 0$; a different set of coefficients is then required.

Connick and Powell[123] (see also George[264,265]) proposed the following for oxyanions:

$$S_i = 43.5 + 46.5 \mid z_i \mid + 13.0n \tag{50}$$

where n is the number of oxygen atoms in the oxyanion, not counting the oxygen in hydroxyl groups, if any. This equation can also be used to estimate the entropies of some binuclear ions if one allows an additional -18.0 gibbs mol^{-1} for dimerization. For example, the value for dichromate ion $Cr_2O_7^-$ is obtained as $2(CrO_3^-) - 18 = 2 \times 36 - 18 = 54$ gibbs mol^{-1}.

Couture and Laidler[137] proposed the following for oxyanions:

$$\bar{S}_i^{\circ} = 40.2 + 1.50R(\ln M_i) - 109[z_i^2/n(r_{io} + 1.40)] \tag{51}$$

where $\bar{S}_i^{\ominus}$ is the ionic entropy relative to -5.5 gibbs mol^{-1} for the absolute ionic entropy of H$^+$(aq), r_{io} is the distance from the center of the ion to the oxygen nucleus, and n is the number of charge-bearing ligands.

Also for the oxyanions Altshuller proposed the equation[9,10]

$$(i)_{w \leftarrow g}^{S} = -5 - (29 - 4.5n)z_i^2 \tag{52}$$

where n is the number of oxygen atoms in the anion. This work is based on Altshuller's estimates of $\bar{S}^{\ominus}_{i(g)}$ given in Table VIA, together with the experimental data for S_i. Equation (52) is based on $\bar{S}^{\ominus}_{H^+} = 0$.

3.3. The Chemical Model Interpretation of Ionic Entropies

Following Frank and Evans,[230] we employ the equation

$$(X)^S_{w \leftarrow g} = R \ln(V_f/V_g) + S_{X,I} + S_{X,II} + S_{X,Born} \tag{53}$$

where on the right the first term allows for the fact that the free volume V_f accessible to the solute in solution is not the same as the volume accessible to the particle in the gas (Section 1.3). For hypothetical unit mole fraction standard states in solution and hypothetical 1 atm standard states for the gases this term is estimated by Frank and Evans as -20 gibbs mol^{-1}. For hypothetical 1 M standard states in both phases it becomes -6 gibbs mol^{-1}. The estimate is not very certain, as they point out, but this term is independent of solute species and does not affect the relative values of the cosphere II terms we seek to evaluate. The term $S_{X,I}$ is a contribution for water forming type I cosphere states (see Table IV) and is taken as zero for nonionic species and -12 gibbs mol^{-1} of ions for monatomic ions. The term $S_{X,II}$ is the contribution from water in type II cosphere states; it is determined from the data by the use of this equation. The last term is the Born-model contribution for the solvent outside of the first cosphere, given by [cf. eqn. (44)]

$$S_{X,Born} = -9.7z_X^2/(r_X^P + 2.8) \quad \text{gibbs mol}^{-1}$$

where r_X^P is the Pauling radius of X in angstrom units and 2.8 Å is the diameter of a water molecule.

Thus eqn. (53) is a straightforward chemical model calculation except for the inclusion of the Born effect for the polarization of the solvent beyond the first cosphere. Inclusion of the Born-model effects for the outer cosphere seems physically reasonable because the cosphere II structural effects are so small when calculated on a per mole of water basis (see Fig. 11) that one may suppose that the dielectric properties of this cosphere material are the same as those of pure water. At any rate, $S_{X,Born}$ varies little from one ion to another when z_X is fixed.

The calculation of $S_{X,II}$ for a large number of solute species by applying eqn. (53) to the data in Tables V, VIA, and VIB is summarized in Table VII. The figures differ slightly from those of Frank and Evans[230] mainly due to

TABLE VII. Cosphere II Entropies Deduced from Eqn. (53) for Aqueous Solutes at 25°C^a

Solute, X	$S_{X,Born}$	$S_{X,II}$	Solute, X	$S_{X,Born}$	$S_{X,II}$
He	0	−7	IO_3^-	−1.6	−8
Ne	0	−9	CN^-	−2.0	14
Ar	0	−10	SO_4^{2-}	−7.2	−17
Kr	0	−12	CrO_4^{2-}	−6.9	−14
Xe	0	−14	Mg^{2+}	−11.6	−39
Rn	0	−14	Ca^{2+}	−10.4	−26
H^+	−3.6	−4	Sr^{2+}	−10.0	−25
Li^+	−2.9	−7	Ba^{2+}	−9.6	−15
Na^+	−2.7	0	La^{3+}	−22.8	−51
K^+	−2.4	9	Ce^{3+}	−23.0	−50
Rb^+	−2.3	12	Pr^{3+}	−23.1	−51
Cs^+	−2.2	12	Nd^{3+}	−23.2	−54
NH_4^+	−2.3	14	Sm^{3+}	−23.4	−57
F^-	−2.4	−5	Eu^{3+}	−23.5	−59
Cl^-	−2.2	8	Gd^{3+}	−23.6	−62
Br^-	−2.1	12	Tb^{3+}	−23.7	−62
I^-	−2.0	17	Dy^{3+}	−23.7	−63
ClO_4^-	−1.7	11	Ho^{3+}	−23.9	−64
ReO_4^-	−1.6	8	Er^{3+}	−23.9	−63
NO_3^-	−1.8	8	Tm^{3+}	−24.0	−64
ClO_3^-	−1.8	8	Yb^{3+}	−24.1	−65
BrO_3^-	−1.7	4	Lu^{3+}	−24.1	−65

^{a} Entropies in gibbs mol^{-1} of solute.

the use here of a corrected value for the Born coefficient and, where possible, more recent data.

The quantity $S_{X,II}$ is the entropy change for the process

$$n_{X,II}H_2O(\text{pure liq.}) \rightarrow n_{X,II}H_2O(\text{in cosphere II about solute X}) \quad (54)$$

per mole of X. Here $n_{X,II}$ is the number of water molecules in cosphere II about solute X. As described in Section 2.5, in the cosphere II state one has hydration of the second kind, so this applies to the inner cosphere of rare gas atoms and is assumed to apply to the outer cospheres of ions like those in Table VII. However, for highly charged ions the $S_{X,I}$ estimate used here, which Frank and Evans[230] obtained for singly charged ions, may not be nearly negative enough owing to the larger number of water molecules polarized as the ionic charge increases. At this time it seems best to let this defect appear as a contribution to $S_{X,II}$ for these ions, as we have done in Table VII.

Frank and Evans point out that the $(X)^S_{w \leftarrow g}$ values for the ions are strongly correlated with the effects of these ions upon the viscosity of the solution. In terms of the Jones–Dole B coefficients, which appear in the equation for the viscosity of a solution as a function of molarity m,

$$\eta(m)/\eta(0) = 1 + A\sqrt{m} + (v_+ B_+ + v_- B_-)m + \cdots$$

one finds that B_X becomes more positive as the hydration entropy of X becomes more negative. The correlation of $S_{X,II}$ with B_X is equally strong. In many cases plots of $S_{X,II}$ as a function of B_X are accurately linear for ions X of a given charge type. Especially striking is the fact that those few ions for which $S_{X,II}$ is positive have negative values of B. Therefore they are called structure-breaking ions. Their cosphere II water is said to be in a structure-broken state; it is denoted here as state II_{sb} water (see Table IV). It seems reasonable to suppose that whatever effect produces II_{sb} water on the second cosphere of the alkali metal ions will also produce it in the *third* cosphere of the more highly charged cations, but in such cases there is even less basis for estimating the contribution of $S_{X,I}$ in eqn. (53).

For tetraalkylammonium ions and other large ions with no polar groups on the surface there is strong evidence that the type I cosphere is absent and the water in the type II cosphere is in a state more like II_{rg}, the state in the cospheres of the rare gases, rather than in state II_{sb} (see Sections 5.3, 6.3). However, the implications of this for the hydration entropies of these ions cannot be tested for lack of data.

3.4. Transition and Rare Earth Metal Ions at Infinite Dilution

3.4.1. *Enthalpies of Hydration*

The two humps in the plot (Fig. 13) of enthalpies of hydration of divalent and trivalent transition metal ions versus atomic number are ascribed to the peculiar effects arising from the repulsive field of the lone pairs of the oxygen atoms of neighboring water molecules acting more strongly on the d orbitals directed along the x, y, z, axes than on those directed between the axes. When corrected for this type of interaction of the water molecules with the d orbitals the hydration enthalpies exhibit a

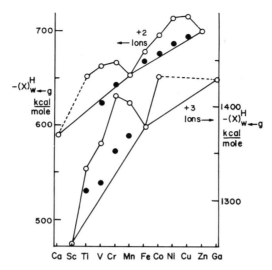

Fig. 13. Enthalpies of hydration of transition metal ions. (○) Experimental values based on the assumption that the heat of formation of H^+ in aqueous solution is zero; (●) values after subtracting water field stabilization energy. Data from George and McClure.[266]

normal radius dependence. This effect, called ligand field stabilization, is also important for many complex ions and has been extensively reviewed.[28,103,266,588,698]

The contributions from ligand field stabilizations are rather small for rare earth metal ions involving $4f$ electrons.

3.4.2. *Entropies of Hydration*

The entropies of complex formation at zero ionic strength have also been used extensively for distinguishing inner and outer sphere complexes, for identifying coordination of water molecules in complexes, and for otherwise characterizing complex ions.

In the case of rare earth metal ions the entropies of hydration as a function of atomic number exhibit an S-shaped curve[50,373] and this type of curve is generally obtained (Fig. 14) for both heats and entropies of complex formation with these metal ions[582] as well as the Jones–Dole B coefficients.[741] On the other hand, the values of $S_{X,II}$ obtained from eqn. (53) depend linearly on the Jones–Dole B coefficients. The molecular interpretation of the S-shaped curves is not known.

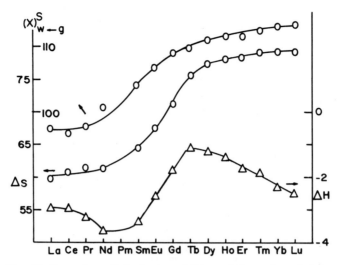

Fig. 14. Entropies of hydration (gibbs mol^{-1}) and enthalpies (kcal mol^{-1}) and entropies (gibbs mol^{-1}) of complexation of lanthanides. $(X)^S_{w \leftarrow g}$ data from Bertha and Choppin.[50] Data for EDTA complexes from Mackey *et al.*[547]

4. OTHER THERMODYNAMIC PROPERTIES OF IONS AT INFINITE DILUTION IN WATER

4.1. Partial Molar Volume of Solute X at Infinite Dilution, $V_{X(aq)}$

4.1.1. *Monatomic Ions*

The partial molar volumes of the solutes in ionic solutions at infinite dilution, usually denoted as $\bar{V}_2^{\ominus}$, have often been used in efforts to elucidate the ion–solvent interactions in molecular terms. A selection of "best" single-ion values of $V_{X(aq)}$, taken mostly from the compilation of Millero,[575] is given in Table VIII. These conventional single-ion values are based on $V_{H^+(aq)} = -5.4$ cm^3 mol^{-1}.

The various methods used to obtain absolute single-ion values for $V_{X(aq)}$ have recently been reviewed by Panckhurst[622] and Millero.[576] An interesting observation is that mutually consistent results are given by very diverse methods: by measurement of ionic vibration potentials,[877] by the assumption that $V_{X(aq)}$ is the same for K$^+$ and F$^-$[839] or for Ph$_4$As$^+$ and Ph$_4$B$^-$,[574] by the $(C_nH_{2n+1})_4$N$^+$ extrapolation procedure,[132] and by making use of the observation that packing densities approach a uniform

TABLE VIII. Partial Molal Volumes (cm^3 mol^{-1}) of Ions in Water Based on $V_{H^+(aq)} = -5.4$ cm^3 mol^{-1} at 25°C^a

Ion	$V_{X(aq)}$	Ion	$V_{X(aq)}$
H$^+$	-5.4	Tb^{3+}	-56.4
Li$^+$	-6.3	Dy^{3+}	-57.0
Na$^+$	-6.6	Ho^{3+}	-58.0
K$^+$	3.6	Er^{3+}	-59.1
Rb$^+$	8.7	Yb^{3+}	-60.4
Cs$^+$	15.9	Th^{4+}	-75.1
Ag$^+$	-6.1	NH$_4^+$	13
Tl$^+$	5.2	Me$_4$N$^+$	84.2
Be^{2+}	-22.8	Et$_4$N$^+$	143.7
Mg^{2+}	-32	n-Pr$_4$N$^+$	209.3
Ca^{2+}	-28.6	n-Bu$_4$N$^+$	270.3
Sr^{2+}	-29.0	n-Pen$_4$N$^+$	333.8
Ba^{2+}	-23.3	(HOEt)$_4$N$^+$	146.6
Pb^{2+}	-26.3	[Bu$_3$N(CH$_2$)$_3$NBu$_3$]$^{2+}$	517.7
Mn^{2+}	-28.5	MeNH$_3^+$	30.7
Fe^{2+}	-35.5	EtNH$_3^+$	47.5
Co^{2+}	-34.8	n-PrNH$_3^+$	64.0
Ni^{2+}	-34.8	n-BuNH$_3^+$	80.1
Cu^{2+}	-35.0	n-PenNH$_3^+$	96.0
Zn^{2+}	-32.4	n-HexNH$_3^+$	111.9
Cd^{2+}	-30.8	n-HepNH$_3^+$	127.8
Hg^{2+}	-30.1	n-OctNH$_3^+$	143.7
Al^{3+}	-58.4	Me$_2$NH$_2^+$	49.2[485]
Fe^{3+}	-59.9	Et$_2$NH$_2^+$	83.4[485]
Cr^{3+}	-55.7	n-Pr$_2$NH$_2^+$	115.4[485]
La^{3+}	-55.3	n-Bu$_2$NH$_2^+$	147.4[485]
Pr^{3+}	-58.7	Me$_3$NH$^+$	67.3[485]
Nd^{3+}	-59.5	Et$_3$NH$^+$	116.1[485]
Sm^{3+}	-58.5	n-Pr$_3$NH$^+$	163.5[485]
Gd^{3+}	-56.6	(HOEt)$_3$NH$^+$	117.2[485]

TABLE VIII. (*Continued*)

Ion[b]	$V_{X(aq)}$	Ion	$V_{X(aq)}$
Pip$^+$	83.5[485]	SO$_3$NH$_2^-$	46.9
1-Methyl pip$^+$	102.3[485]	HCO$_3^-$	28.8
1,1-Dimethyl pip$^+$	117.1[485]	HSO$_4^-$	41.1
Py$^+$	67.8[127]	HSeO$_4^-$	36.5
2-Methyl py$^+$	85.4[127]	H$_2$PO$_4^-$	34.5
2,6-dimethyl py$^+$	102.7[127]	H$_2$AsO$_4^-$	40.6
(N)-Methyl py$^+$	85.2[127]	CO$_3^{2-}$	6.5
1,2,6-Trimethyl py$^+$	116.6[127]	SO$_3^{2-}$	19.7
Ph$_4$As$^+$	295.3	SO$_4^{2-}$	24.8
Ph$_4$P$^+$	286.8[412]	SeO$_4^{2-}$	31.8
F$^-$	4.2	CrO$_4^{2-}$	30.5
Cl$^-$	23.2	MoO$_4^{2-}$	39.7
Br$^-$	30.1	WO$_4^{2-}$	36.5
I$^-$	41.6	C$_2$O$_4^{2-}$	26.8
S^{2-}	2.6	S$_2$O$_3^{2-}$	44.8
OH$^-$	1.4	Cr$_2$O$_7^{2-}$	83.8
NO$_2^-$	31.6	PtCl$_6^{2-}$	160.8
NO$_3^-$	34.4	HPO$_4^{2-}$	18.5
OCN$^-$	31.5	AsO$_4^{3-}$	0.6
SCN$^-$	41.1	Fe(CN)$_6^{3-}$	137.0
SeCN$^-$	55.1	Fe(CN)$_6^{4-}$	95.6
ClO$_3^-$	42.1	CHO$_2^-$	31.7
BrO$_3^-$	40.7	CH$_3$CO$_2^-$	45.9
IO$_3^-$	29.9	CH$_3$CH$_2$CO$_2^-$	59.4
ClO$_4^-$	49.5	CH$_3$(CH$_2$)$_2$CO$_2^-$	75.8
MnO$_4^-$	47.9	PhO$^-$	74.1
ReO$_4^-$	53.6	PhSO$_3^-$	114.3
BF$_4^-$	48.4	p-CH$_3$PhSO$_3^-$	125.0
SO$_3$F$^-$	55.3	Ph$_4$B$^-$	283.0

[a] The data (unless otherwise specified) are taken from the compilation of Millero.[575]
[b] pip$^+$ = piperidinium ion; py$^+$ = pyridinium ion.

value as the size of the solute increases.[447] Since it is known by other measures that the hydration of K^+ differs from that of F^-, the hydration of Ph_4As^+ differs from that of Ph_4B^-, and the $(C_nH_{2n+1})_4N^+$ extrapolation procedure fails when water is the solvent, it must be concluded that in general $V_{X(aq)}$ is not a sensitive measure of the cosphere effects in the hydration of X.

The chemical model suggests the dissection of $V_{X(aq)}$ as follows [cf. eqn. (53)]:

$$V_{X(aq)} = V_{X,self} + V_{X,I} + V_{X,II} + V_{X,Born} \qquad (55)$$

where the first term on the right is the contribution of the actual volume of particle X, the second is the contribution of the change in volume of the water which enters the type I cosphere state, the third is the contribution of the change in volume of water which enters the type II cosphere state, and the last is the Born-model contribution for the volume change associated with the polarization of the water by the field of the ion.

For monatomic ions it is usually assumed that the intrinsic term is given by

$$V_{X,self} = 4\pi (r_X^P)^3/3$$

where r_X^P is the Pauling crystal radius of X.

The Born-model contribution is obtained by differentiating eqn. (25) with respect to pressure, with the result for water at 25°C, with $[\partial(\ln \varepsilon)/\partial P]_T = 47.1 \times 10^{-6}/bar$,

$$V_{X,Born} = -4.175 z_X^2/(r_X^P + \Delta) \quad \text{ml Å mol}^{-1}$$

One can choose $\Delta = 0$, corresponding to the Born-model polarization of the type I cosphere, or $\Delta = 2.76$ Å, corresponding to assessing this term only for the water outside of the innermost cosphere. Sometimes a model with a continuously varying dielectric constant, as discussed in Section 2.3.4, has been used.

Often a particular contribution to $V_{X,I}$ due to void space in a particular model for the packing of the solvent particles in cosphere I is used,[131,278] so that $V_{X,II}$ can be determined by subtracting the rest of the contributions from $V_{X(aq)}$. The various ways of doing this and the conclusions have been summarized by Millero;[576] the conclusions for the cosphere contributions at 25°C are not very clear-cut, as is to be expected because of the observation that $V_{X(aq)}$ is not very sensitive to cosphere effects.

4.1.2. Alkylammonium Ions

Even for these ions, for which the chemical-model effects seem to be very important for other properties, the volumes are partially understood on the basis of simpler considerations.

At 25°C the partial molar volumes at infinite dilution of tetraalkyl-ammonium salts are rather accurately given[132] by the equation

$$V_{(C_nH_{2n+1})_4NX} = V_{X^-(aq)} + 4bn \tag{56}$$

where b is a coefficient determined by the data. However, the analogous linear dependence upon number of methylene groups is not found for other properties such as enthalpies of transfer of alkylammonium ions to propylene carbonate[472] (Section 6.3) and the molar effect of the solutes upon the self-diffusion coefficient of water in the solutions.[357]

Equation (56) can be generalized to apply to other alkylammonium ions[127,485,795] or even alkyl-substituted pyridinium ions, thus defining a coefficient b which has the significance of the increment in $V_{X(aq)}$ per methylene group inserted in the cation. These coefficients have been determined for a number of series by Conway and his co-workers with the results given in Table IX. These contributions per methylene group to the volumes of infinitely dilute aqueous solutions may be compared with 11.7 cm^3 mol^{-1}, the contribution per methylene group to the increment of the molar volume as one passes from liquid pentane to liquid heptane, all at 20°C. For the purpose of this discussion we take 11.7 cm^3 mol^{-1} as $V_{CX_2,\text{self}}$, the self-volume of a mole of methylene particles.

The difference, $b - 11.7$ cm^3 mol^{-1}, comprises a contribution from the hydration of the methylene group, like the second and third terms of the right of eqn. (55), together with a reduction of the Born-charging contribution due to the displacement of water molecules by methylene groups in the field of the charge. These effects are qualitatively evident in the data of Table IX. In the case of R_3NH^+ and $R_2NH_2^+$ ions an effect due to the interference of the methylene groups with the type I_{hb} cosphere at the NH^+ center can also be identified.[485]

TABLE IX. Conway b Coefficients for Volumes of Substituted Ammonium Ions^a

Series	$R_4N^{+[127]}$	$R_3NH^{+[795]}$	$R_2NH_2^{+[795]}$	$RNH_3^{+[795]}$	$PyH^{+[127]}$	N-Me-Py$^{+[127]}$
b, ml mol^{-1} of CH$_2$	15.61	15.98	16.99	17.82	17.18	16.02

^{a} PyH$^+$ = pyridinium ion and N-Me-Py$^+$ = N-methyl pyridinium ion.

4.1.3. Oxyanions

Couture and Laidler[136] proposed the empirical equation for monatomic ions

$$V_{X(aq)} = 16 + 4.9r_X{}^G - 26 \mid z_X \mid \tag{57}$$

where $r_X{}^G$ is the Goldschmidt crystal radius of ions X. This equation was based on $V_{H^+(aq)} = -6$ cm³ mol⁻¹. It has been extended for oxyanions[138]:

$$V_{X(aq)} = 58.8 + 156.25 \times 10^{-4}[n(r_{XO} + 1.4)]^3 - 26 \mid z_X \mid \tag{58}$$

where n is the number of charge-bearing oxygen atoms and r_{XO} is the distance from the center of the ion to the oxygen nucleus.

4.2. Partial Molar Heat Capacity of Solute X at Infinite Dilution, $C_{X(aq)}$

4.2.1. Monatomic Ions

Partial molar heat capacities at infinite dilution, often denoted as $\bar{C}_{P_2}^{\ominus}$, are given for various ions in Table X. These data may be analyzed in the same way as $V_{X(aq)}$ in eqn. (55). Now we have

$$C_{X,\text{Born}} = -12.96 z_X{}^2/(r_X{}^P + \varDelta) \quad \text{cal Å mol}^{-1}\text{ deg}^{-1}$$

as the appropriate derivative of eqn. (25) for water at 25°C, for which $[\partial^2(\ln \varepsilon)/\partial T^2]_P = 54 \times 10^{-8}$ deg⁻². There have been only a few attempts to obtain absolute single-ion values for $C_{X(aq)}$ [149,189] Here we have used the convention $C_{H^+(aq)} = 28$ cal mol⁻¹ deg⁻¹, obtained by Criss and Cobble using the "correspondence principle."[149]

The $C_{X(aq)}$ values for cations become increasingly negative (less positive) with increasing radius. This cannot be explained on the basis of cosphere state I alone. But assuming a state II$_{sb}$ contribution to $C_{X(aq)}$ similar to the one for entropy (Section 3.3) accommodates the data in the chemical model.

4.2.2. Alkylammonium Ions

Since the changes in heat capacity on solutions in water are measured for most of the alkylammonium salts by calorimetry and since the heat capacities of the solids are not available for many of the solutes, one mostly only has data $(X)_{w \leftarrow c}^C = C_{X(aq)} - C_{X(\text{crystal})}$. Most of these values are highly positive, unlike $(X)_{w \leftarrow c}^C$ for simple electrolytes. These ions have contributions to the heat capacity from internal degrees of freedom unlike monatomic ions, but possible changes in internal motions in the w ← c transfer do

TABLE X. Ionic Partial Molal Heat Capacities in Water at 25°C[a]

Ion	$C_{X(aq)}$	Ion	$C_{X(aq)}$	Ion	$C_{X(aq)}$
H^+	28	Zr^{4+}	98	CNS^-	$-35^{(625)}$
Li^+	$42^{(625)}$	U^{4+}	93	ClO_3^-	$-46^{(625)}$
Na^+	37	UO_2^{2+}	56	BrO_3^-	$-52^{(625)}$
K^+	31	$Fe(OH)^{2+}$	59	IO_3^-	$-57^{(625)}$
Rb^+	30	NH_4^+	$45^{(625)}$	NO_3^-	$-56, -47^{(625)}$
Cs^+	26	$MeNH_3^+$	$57^{(703)}$	NO_2^-	$-49^{(625)}$
Ag^+	35	$EtNH_3^+$	$73^{(703)}$	ClO_4^-	$-40, -32^{(559,560)}$
Tl^+	28	Me_4N^+	$85^{(559,560)}$	MnO_4^-	$-45^{(625)}$
Be^{2+}	76		$(-11)^b$	ReO_4^-	-30
Mg^{2+}	62	Et_4N^+	$(30)^b$	SO_4^{2-}	$-115, -121^{(559,560)}$
Ca^{2+}	54	Pr_4N^+	$(106)^b$	SO_3^{2-}	-139
Sr^{2+}	52	Bu_4N^+	$351^{(412)}$	CO_3^{2-}	-151
Ba^{2+}	44		$(176)^b$	PO_4^{3-}	-226
Ra^{2+}	40	$i\text{-}Pen_4N^+$	$(219)^b$	HCO_3^-	-48
Mn^{2+}	57	Ph_4As^+	$325^{(412)}$	HSO_3^-	-43
Cu^{2+}	59	Ph_4P^+	$320^{(412)}$	HSO_4^-	-35
Cd^{2+}	54	F^-	-57	HPO_4^{2-}	-132
Pb^{2+}	44	Cl^-	-59	$H_2PO_4^-$	-55
Cr^{3+}	88	Br^-	$-61, -60^{(625)}$	HCO_2^-	$-42^{(625)}$
Fe^{3+}	86	I^-	$-62, -60^{(625)}$	$CH_3CO_2^-$	$-27^{(625)}$
Al^{3+}	89	OH^-	$-57, -62^{(625)}$	$CH_3CH_2CO_2^-$	$1^{(625)}$
Gd^{3+}	72	SH^-	-60	BPh_4^-	$227,^{(761)} 230^{(412)}$
Sc^{3+}	79				

[a] The convention used here is that $C_{H^+(aq)}$ = 28 cal mol^{-1} deg^{-1}. The heat capacities are given in cal mol^{-1} deg^{-1} = gibbs mol^{-1}. Unless otherwise specified, data are from Ref. 149.

[b] These values are not single-ion values. These are the change in heat capacity of solution, $(R_4NBr)^C_{w \leftarrow c}$ taken from Arnett and Campion.[21]

not account for the positive values of $(X)^C_{w \leftarrow c}$. The data are not additive with respect to chain length, as they are for nonelectrolytes.[22,131] Here again the results are explained on the basis of the chemical model by assuming large positive contributions to $C_{X(aq)}$ from cosphere state II$_{al}$ and negligible contributions from State I. The widely differing heat capacities of Bu_4N^+ and BPh_4^- again suggests further differentiation of type II cosphere states into type II$_{al}$ and type II$_{ar}$ (Table IV). The contributions from state I are apparent only for RNH_3X salts.[703]

4.2.3. *Oxyanions*

There are not enough accurate data for various oxyanions. The $C_{X(aq)}$ data for some of these large ions are positive due to contributions from rotational and vibrational degrees of freedom. This is also evident from the internal heat capacity of $OsO_4(g)$, 14.9 cal mol^{-1} deg^{-1} at 25°C. Criss and Cobble find the correlation

$$C_{X(aq)} = A + B\bar{S}^{\ominus}_{X(aq)} \qquad (59)$$

with the A values for cations, OH^-, oxyanions, and acid oxyanions being 41.6, -56.5, -145, and -136, respectively, and with the corresponding B values -0.523, 0.179, 2.20, and 3.07.[149] This correlation was obtained using $\bar{S}^{\ominus}_{H^+(aq)} = -5.0$ gibbs mol^{-1} and $C_{H^+(aq)} = 28$ cal mol^{-1} deg^{-1}.

4.3. Partial Molar Isothermal Compressibility of Solute X at Infinite Dilution, $KT_{X(aq)}$, and Partial Molar Adiabatic Compressibility of Solute X at Infinite Dilution, $KS_{X(aq)}$ *

4.3.1. *Monatomic Ions*

The partial molar isothermal compressibilities $KT_{X(aq)}$ obtained from bulk compression measurements are well suited for values at high pressures. However, extrapolation to get values at 1 atm ($\sim$zero pressure for this purpose) are generally based upon some modification of the Tait equation and are not very satisfactory. Compressibilities obtained from sound velocities at 1 atm are adiabatic coefficients $KS_{X(aq)}$. The relation of $KS_{X(aq)}$ to $KT_{X(aq)}$ involves the partial molar volume at infinite dilution $V_{X(aq)}$, as well as the concentration dependence of C_P/C_V of the solutions in the limit of infinite dilution. For a few monatomic ions both sets of data are available and the error in assuming $KS_{X(aq)} = KT_{X(aq)}$ is seen to be about 10%. The data are given in Table XI.

We can dissect $KT_{X(aq)}$ in terms of an equation analogous to eqn. (55). In this case we have

$$KT_{X,Born} = -8.307z_X^2/(r_X^P + \varDelta) \quad \text{ml Å bar}^{-1} \text{ mol}^{-1}$$

using $[\partial^2(\ln \varepsilon)/\partial P^2]_T = -71.53 \times 10^{-10}$ bar^{-2} for water at 25°C.

---

* While the isothermal or adiabatic compressibility of a material is defined as $-[\partial(\ln V/\partial P)]_{TorS}$, the partial molar compressibilities are customarily defined as $-(\partial V_i/\partial P)_{TorS}$, where V_i is the partial molar volume of component i. Thus for pure water at 25°C and 1 atm the isothermal compressibility is 4.56×10^{-5} bar^{-1} and, for water as a solute in the same pure water we have $KT_{H_2O(aq)} = 0.822$ μl mol^{-1} bar^{-1}.

TABLE XI. Partial Molal Isothermal and Adiabatic Compressibilities of Electrolytes at Infinite Dilution in Water at 25°C[a]

Electrolyte, X	$KT_{X(aq)}$, μl mol⁻¹ bar⁻¹	$KS_{X(aq)}$, μl mol⁻¹ bar⁻¹
HCl	−0.8	—
LiCl	−4.2	−4.0[257]
	−4.1	
	−3.87[257]	
NaCl	−5.2	−5.08[257]
	−4.8	−5.28[299]
	−4.67[257]	
	−4.63[618]	
KCl	−4.5	−4.26[257]
	−4.4	−4.56[299]
	−3.93[257]	
	−4.04[618]	
NaBr	−4.2	
	−3.77[618]	
KBr	−3.4	
	−3.19[618]	
CsBr	−2.5	
LiI	−1.7	
KI	−1.8	

Electrolyte,[b] X	$KT_{X(aq)}$,[c] μl mol⁻¹ bar⁻¹	$KS_{X(aq)}$, μl mol⁻¹ bar⁻¹
MgSO₄	−15.3	−14.9[257]
CuSO₄	−13.2	—
ZnSO₄	−14.0	—
CdSO₄	−12.7	—
CeCl₃	−17.6	—
Me₄NCl	(−1.41)	−1.66[130]
Me₄NBr	(−0.05)	−0.31,[257] −0.71[130]
Et₄NBr	(−0.16)	−0.47,[257] −0.82[130]
n-Pr₄NBr	(−0.35)	−0.77,[257] −1.18[130]
n-Bu₄NBr	−1.07[257]	−1.82,[257] −2.33[130]
Me₄NI	(+0.66)	+0.33[130]
Et₄NI	(+0.53)	+0.16[130]
Pr₄NI	(+0.26)	−0.23[130]
Bu₄NI	(−0.42)	−1.19[130]
MeNH₃Cl	—	−2.20[130]
Me₂NH₂Cl	—	−2.02[130]

LiOH	-7.8	—	Et$_2$NH$_2$Cl	—	-2.51[485]
NaOH	-8.9	—	n-Pr$_2$NH$_2$Cl	—	-1.79[485]
KOH	-8.1	—	n-Bu$_2$NH$_2$Cl	—	-2.50[485]
Li$_2$SO$_4$	-14.6	-14.1[257]	Me$_3$NHCl	—	-1.81[130]
Na$_2$SO$_4$	-15.4	-16.5[257]	Et$_3$NHBr	—	-0.56[485]
K$_2$SO$_4$	-13.9	—	(EtOH)$_3$NHBr	—	-0.86[485]
Cs$_2$SO$_4$	-11.9	—	Me$_4$NBF$_4$	—	+1.80[485]
(NH$_4$)$_2$SO$_4$	-9.1	—	NH$_4$Cl	—	-2.56[485]
KNO$_3$	-3.0	—	PipCl	—	-2.20[485]
KCNS	-2.2	—	1-Me-PipCl	—	-2.09[485]
KHCO$_3$	-3.5	—	PyCl	—	-1.86[127]
Na$_2$CO$_3$	-17.1	—	2-Me-PyCl	—	-1.64[127]
K$_2$CrO$_4$	-13.9	—	2,6-DiMe-PyCl	—	-1.50[127]
KO$_2$CCH$_3$	-4.7[d]	—	2,6-DiMe-PyBr	—	-0.66[127]
CaCl$_2$	-8.7	—	2,6-DiMe-PyI	—	+0.01[127]
BaCl$_2$	-11.5	—	1(N)Me-PyI	—	+0.20[127]
BeSO$_4$	-9.3	—	1,2,6-TriMe-PyI	—	-0.04[127]

[a] The data (unless otherwise specified) are taken from Owen and Brinkley.[617]

[b] pip = piperidinium ion; py = pyridinium ion.

[c] The values in parentheses were calculated using data from Ref. 130 and Table X of this chapter.

[d] At 30°C and 100 bar.

4.3.2. *Alkylammonium Ions*

Conway and his associates have examined in detail the partial molar adiabatic compressibilities of a number of organic solutes and compared these with $V_{X(aq)}$ data.[127,130,485] They found a linear relationship between $KS_{X(aq)}$ and $V_{X(aq)}$ (or molecular weight of the cations) for R_3NH^+, pyridinium, and piperidinium salts but not for R_4N^+ or $R_2NH_2^+$ salts. Using $KS_{X(aq)}/V_{X(aq)}$ as a function of carbon number of X, some effects were identified as contributions from state I cosphere water, as discussed in Section 4.1.2. The ratio $KS_{X(aq)}/V_{X(aq)}$ was essentially a constant for R_4N^+ and pyridinium iodides. Utilizing the appropriate derivative of eqn. (56), they derived absolute single-ion values for partial molal adiabatic compressibilities. They find

$$KS_{H^+(aq)} = -0.39 \quad \mu l \; mol^{-1} \; bar^{-1}$$

The less negative adiabatic compressibilities of these organic salts were also explained in terms of the chemical model: (a) The contributions from X, self, and X,I terms are negative or close to zero since they involve the compressibility of the ion itself or the void space between the ion and solvent molecules. (b) The contribution from state II is positive. The order $I^- > Br^- > Cl^-$ for $KS_{X(aq)}$ was ascribed to increasing contributions from cosphere type I.

Recent measurements at very high dilutions do indicate that $KT_{X(aq)}$ and $KS_{X(aq)}$ differ widely for these solutes.[257]

4.4. Partial Molal Expansibility of Solute X at Infinite Dilution, $E_{X(aq)}$

The partial molal expansibilities of various electrolytes, often denoted as $\bar{E}_2^{\ominus}$, are given in Table XII.* These data may be analyzed in the same way as $V_{X(aq)}$ in eqn. (55). Now we have

$$E_{X,Born} = -2.742z_X^2/(r_X^P + \Delta) \quad \mu l \; Å \; deg^{-1} \; mol^{-1}$$

using $\partial^2(\ln \varepsilon)/\partial P \, \partial T = 9.27 \times 10^{-8} \; bar^{-1} \; deg^{-1}$ for water at 25°C. Millero and Drost-Hansen utilized the appropriate derivative of eqn. (56) to obtain absolute single-ion values of $E_{X(aq)}$.[578] They find $E_{H^+,(aq)} = -12 \; \mu l \; mol^{-1}$

---

* The partial molal expansibility of a solute X is defined as $E_X = \partial V_X/\partial T$, where V_X is the partial molar volume of the solute.

TABLE XII. Partial Molal Expansibility of Electrolytes at Infinite Dilution in Water at 25°C

Electrolyte, X	LiCl	NaCl	KCl	RbCl	CsCl	NaF	KBr	KI
$E_{X(aq)}$, $\mu l\,mol^{-1}\,deg^{-1}$	25[a]	93[a]	85[a]	70[a]	70[a]	67[577]	93[a]	127[a]
	32[577]	78[577]	69[577]	62[577]	58[577]		60[577]	92[577]

Electrolyte, X	KNO$_3$	HCl	MgCl$_2$	CaCl$_2$	SrCl$_2$	BaCl$_2$	Na$_2$SO$_4$	NH$_4$Cl
$E_{X(aq)}$, $\mu l\,mol^{-1}\,deg^{-1}$	123[577]	34[a]	-13[571]	50[571]	82[571]	80[571]	171[579]	18[578]

Electrolyte, X	Me$_4$NCl	Et$_4$NCl	Pr$_4$NCl	Bu$_4$NCl	NaBPh$_4$	Ph$_4$AsCl
$E_{X(aq)}$, $\mu l\,mol^{-1}\,deg^{-1}$	79[578]	100[578]	141[578]	223[578]	332[572]	407[573]

[a] Quoted in Ref. 577.

$\deg^{-1}$ compared to Noyes's value of -51 μl mol^{-1} $\deg^{-1}$.[601] The coefficient $E_{X(aq)}$ decreases from Na^+ to Cs^+, whereas it increases from F^- to I^- and from Mg^{2+} to Ba^{2+}.

Millero and Drost–Hansen had, in effect, used only Me_4NCl and Et_4NCl data for extrapolation since Pr_4NCl and Bu_4NCl data deviated considerably from the linear function. They interpret the $E_{X(aq)}$ of R_4N^+ in terms of intrinsic contributions and contributions from states I and II.

4.5. Variation of Heat Capacity of Solute X at Infinite Dilution with Temperature, $\partial C_{X(aq)}/\partial T$

4.5.1. *Monatomic Ions*

The temperature dependence of partial molar heat capacities at infinite dilution is shown in Fig. 15. The order of $C_{X(aq)}$ at 25°C, namely more negative with increasing radius of the cation or anion, is no longer obeyed at higher temperatures. Also, in the range 50–80°C the coefficient $\partial C_{X(aq)}/\partial T$ of every electrolyte passes through zero.

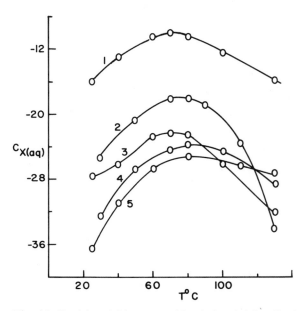

Fig. 15. Partial molal heat capacities (cal mol^{-1} deg^{-1}) of (1) LiCl, (2) KF, (3) KCl, (4) CsCl, and (5) CsI plotted against temperature. Data from Rüterjans *et al.*[703] and 25°C values from Table X.

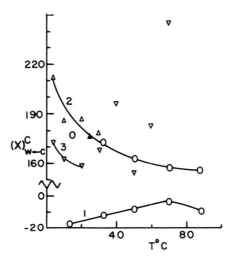

Fig. 16. The change in partial molal heat capacities for dissolution of (1) Me$_4$NBr, (2) Bu$_4$NBr, and (3) NaBPh$_4$ in water plotted against temperature. (○) Data from Mastroianni and Criss.[560] (△, ▽) Data from Subramanian and Ahluwalia[761] and Sarma and Ahluwalia.[715] (▲) Data from Arnett and Campion.[21]

These results have also been explained on the basis of the chemical model.[703] The positive values of $\partial C_X/\partial T$ at 25°C are attributed to the decreasing influence of state II$_{sb}$ with increasing temperature.

4.5.2. Alkylammonium Ions

Even though $\partial C_X/\partial T$ of solutes like MeNH$_3$Cl and EtNH$_3$Cl is positive as for an alkali halide, the contributions from state II$_{al}$ are noticeable with increasing chain length. For R$_4$NX salts $\partial C_X/\partial T$ is negative (with the exception of Me$_4$NBr), as shown in Fig. 16, unlike the behavior of alkali halides but in this respect similar to the alcohols and dialkyl amines in water.[7,237] It is unfortunate that because of the slight solubility of hydrocarbons in water, accurate heat capacity data are not available. Whatever data are available are often confusing, for example, $\partial C_{X(aq)}/\partial T$ for C$_3$H$_8$ or C$_4$H$_{10}$ is reported to be positive in H$_2$O and negative in D$_2$O.[468] The negative $\partial C_{X(aq)}/\partial T$ values for R$_4$NX salts can again be understood on the basis of the chemical model by assuming decreasing contributions from state II$_{al}$ with increasing temperature.

4.6. Variation of Partial Molal Isothermal Compressibility of Solute X at Infinite Dilution with Temperature, $\partial KT_{X(aq)}/\partial T$

Owen and Kronick[618] have measured these for sodium and potassium chlorides and bromides. They found $\partial KT_{X(aq)}/\partial T$ is positive. The next derivative, $\partial^2 KT_{X(aq)}/\partial T^2$, is negative.

4.7. Variation of Partial Molal Expansibility of Solute X at Infinite Dilution with Temperature, $(\partial E_{X(aq)}/\partial T)_P$

A typical set of data is shown in Fig. 17. While $(\partial E_{X(aq)}/\partial T)_P$ is negative for NaCl, Ph₄AsCl, and NaBPh₄, it is positive for Bu₄NCl, which fits in with the classification of II_{al} and II_{ar}, suggesting a differing type of contribution from state II for aliphatic and aromatic groups as included in Table IV. Hepler[338] has used the coefficient

$$(\partial C_{X(aq)}/\partial P)_T = -T(\partial^2 V_{X(aq)}/\partial T^2)_P = -T(\partial E_{X(aq)}/\partial T)_P$$

to classify type II cospheres as II_{sb} or II_{al} depending on whether $(\partial E_{X(aq)}/\partial T)_P$ is negative or positive, respectively.[338] On this basis alkali and alkaline earth chlorides can be classified as dominated by state II_{sb}.

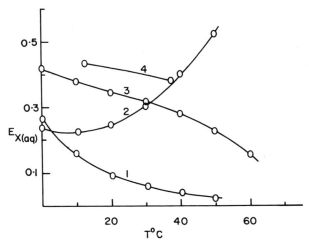

Fig. 17. Partial molal expansibilities (cm³ mol⁻¹ deg⁻¹) of aqueous (1) NaCl, (2) Bu₄NCl, (3) NaBPh₄, and (4) Ph₄AsCl plotted against temperature. Data from Millero.[572,573]

5. SOLVENT-ISOTOPE EFFECT IN HYDRATION

5.1. Introduction

The study of differences in the hydration of solutes in H_2O and D_2O has proved useful in reaching definite conclusions about those hydration effects which naturally fall within the scope of the chemical model discussed in Section 2.5 and outside the scope of the Born and Debye–Pauling models. The reason is that the dielectric properties of H_2O and D_2O are almost identical, as are most of the other macroscopic properties and many of the molecular properties.[593] (See Volume 1, Chapter 10.) Perhaps the largest difference in the properties of the pure liquids is in the viscosity, which is 23% higher in D_2O than in H_2O at 25°C or, compared in another way, is the same for D_2O at 25°C as for H_2O at a temperature 8.5°C lower. The difference in densities is also large, but from the point of view of interpretation of solvent properties it is not important because the molar volumes are almost identical.

Such differences as there are in the thermodynamic properties of liquid H_2O and D_2O can be qualitatively understood in terms of the O—D$\cdots$O hydrogen bond being somewhat stronger than the O—H$\cdots$O hydrogen bond. This in turn is due to the larger moments of inertia of D_2O compared to H_2O; in the liquid this results in a larger reduced mass for the librational motion of the hydrogen-bonded species corresponding to the infrared bands at $700 \, cm^{-1}$ in water and $520 \, cm^{-1}$ in D_2O.[48] Swain and Bader[764] and Van Hook[887] have shown how this effect, together with the isotope effects in the intramolecular frequencies, is adequate to account quantitatively for some of the differences in thermodynamic properties of the liquids, using the well-known simplification of statistical mechanical calculations of chemical equilibrium applied to isotope effects.[372] In a more elaborate calculation Némethy and Scheraga[593] find that the differences in intermolecular and intramolecular frequencies also account for the difference in temperature of maximum density of the liquid and that, in the framework of the model they use, the mean number of hydrogen bonds formed by each water molecule is about 5% larger in D_2O than in H_2O when both are at 25°C.

Remarkably enough, the data for solvent-isotope effects in hydration phenomena are consistent with the assumption that the enthalpy change associated with the process

$$\text{water(bulk)} \rightarrow \text{water(type II cosphere)} \tag{60}$$

is about 5% greater for D_2O than for H_2O for each of the varieties of type II cosphere. Also there is a large measure of enthalpy–entropy compensation (Section 2.5) in the solvent-isotope effect of this process. Of course, not even the sign of the isotope effect in eqn. (60), let alone the magnitude, can be deduced by qualitative reasoning from the difference in hydrogen bonding in the pure liquids. A corresponding difference is also present in each of the final states in the process in eqn. (60), so a quantitative treatment of models seems to be required, especially in view of the very small differences in the mean thermodynamic properties per water molecule in the initial and final states in eqn. (60) (see Fig. 11).

Whatever the interpretation, the measurement of D_2O/H_2O differences in hydration properties is probably the most widely applicable method for identifying contributions of type II cospheres to hydration. The reason is that any hydration property of any solute that can be studied in normal water can be studied in D_2O or, say, 5% H_2O in 95% D_2O if some protons are necessary for the measurement. Then it is "only" a matter of overcoming experimental difficulty to measure the hydration property with sufficient accuracy in the two solvents to obtain the solvent-isotope effect.

Early examples of the study of solvent-isotope effects in hydration properties may be found in the work of Shearman and Menzies[732] and Lange and Martin.[490] The excellent review by Arnett and McKelvey[25] has been extensively used in preparing the present account.

5.2. Free Energies

The solvent-isotope effect upon the hydration free energy of a solute species X is conveniently expressed in terms of the standard free energy of transfer of the species to D_2O from H_2O, $(X)_{d \leftarrow w}^G$, in the notation described in Section 1.4. The available data given in Table XIII are not as concordant as one would like, but "best" values of the single-ion contributions (Table XIV), based on the convention $(Na^+)_{d \leftarrow w}^G = 0$, have been selected and then the additivity checks appiied with the results shown in Table XIII. The single-ion free energies of transfer are shown as a function of ionic radius in Fig. 18.

It happens, as we shall see, that in the process of transfer of ions to D_2O from H_2O one finds a large measure of entropy–enthalpy compensation, with ΔG having the same general dependence on species as ΔH but being smaller as well as more difficult to measure. Therefore it seems reasonable to discuss all these effects in terms of the molecular interpretation of the enthalpy effects (Section 5.3) and the mechanism of compensation (Section 5.4).

TABLE XIII. Standard Free Energies of Transfer of Electrolytes to D_2O from H_2O^a

X	$(X)_{d \leftarrow w}^G$	X	$(X)_{d \leftarrow w}^G$	X	$(X)_{d \leftarrow w}^G$
LiF	−58	KBr	180	$(CH_2)_4MeSBAn_4$	279
	(−140)		(195)		(278)
LiCl	32	KI	240	$(CH_2)_4MeSNPi_2$	279
	110		(245)		(280)
	202[707]	KBPh_4	269	Tol-Me_2SBAn_4 ^{d}	269
	30		(215)		(278)
LiBr	67	$KBAn_4$ ^{b}	377	Tol-Me_2SNPi_2	288
	(50)		(323)		(280)
LiI	100	$RbBPh_4$	269	n-Bu_4NBAn_4	42
	(100)		(265)		(128)
NaF	−28	$RbBAn_4$	369	n-Bu_4NNPi_2	140
	(−60)		(373)		(130)
NaCl	140	CsCl	170	NH_4Cl^e	164
	110		(170)	Me_4NCl^e	134
	116[444]	$CsBPh_4$	212	$MgCl_2$	280
	212[707]		(210)	$CaCl_2$ ^{e}	300
	(110)	$CsBAn_4$	316	$SrCl_2$ ^{e}	300
NaBr	170		(318)	$BaCl_2$ ^{e}	340
	(130)	$CsNPi_2$ ^{c}	316	$CdCl_2$	436
NaI	230		(320)	HCO_2Na^e	−19
	(180)	Me_3SBAn_4	279	$CH_3CO_2Na^e$	−34
KF	−3		(273)	$C_2H_5CO_2Na^e$	−45
	(5)	Me_3SNPi_2	269	$C_3H_7CO_2Na^e$	−46
KCl	130		(275)	$C_4H_9CO_2Na^e$	−39
	180	t-BuMe_2SBAn_4	279	$C_5H_{11}CO_2Na^e$	−39
	225		(268)	$AgBrO_3$	190
	121[444]	t-BuMe_2SNPi_2	260	$Cu(IO_3)_2$	520
	219[707]		(270)		
	(175)				

^{a} Values in parentheses are calculated from single-ion values in Table XIV to test the degree to which the data are made up of additive ionic contributions. Values in cal mol^{-1} at 25°C. Standard states: hyp. 1 M in both solvents, where the molality is taken to be the number of moles of solute per 55.51 mol of solvent in D_2O as well as H_2O, the so-called aquamolality. The electrolyte data (unless otherwise specified) are taken from the compilation given by Arnett and McKelvey.[25]

^{b} BAn_4^- is the tetraanisylborate ion.

^{c} NPi_2 is the dipicrylamide ion.

^{d} Tol = $C_6H_5CH_2$—.

^{e} These data were obtained by an emf method employing an anion exchange membrane to separate the H_2O and D_2O solutions. A correction for solvent transport through the membrane has not been made for these data so the results are subject to uncertainties of the order of 20%. The data for alkaline earths are from Greyson and Snell[293] and carboxylates from Snell and Greyson.[740]

TABLE XIV. Single-Ion Standard Free Energies of Transfer to D_2O from H_2O Based on the Convention that $(Na^+)^G_{d \leftarrow w} = 0^a$

X	Li^+	Na^+	K^+	Rb^+	Cs^+	Me_3S^+	$t\text{-}BuMe_2S^+$	$(CH_2)_4MeS^+$	—
$(X)^G_{d \leftarrow w}$	-80	0	65	115	60	15	10	20	—

X	$TolMe_2S^+$	NH_4^+	Me_4N^+	$n\text{-}Bu_4N^+$	Mg^{2+}	Ca^{2+}	Sr^{2+}	Ba^{2+}	Cd^{2+}
$(X)^G_{d \leftarrow w}$	20	54	24	-130	60	80	80	120	216

X	F^-	Cl^-	Br^-	I^-	BPh_4^-	BAn_4^-	NPi_2^-	HCO_2^-	$CH_3CO_2^-$
$(X)^G_{d \leftarrow w}$	-60	110	130	180	150	258	260	-19	-34

X	$C_2H_5CO_2^-$	$C_3H_7CO_2^-$	$C_4H_9CO_2^-$	$C_5H_{11}CO_2^-$
$(X)^G_{d \leftarrow w}$	-45	-46	-39	-39

^{a} Values in cal mol^{-1}. Standard States: hyp. 1 M in both solvents where the molality is taken to be the number of moles of solute per 55.51. mol of solvent in D_2O as well as H_2O, the so-called aquamolality.

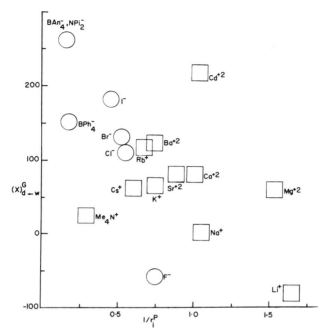

Fig. 18. Single-ion free energies of transfer to D_2O from H_2O. Squares and circles indicate cation and anion experimental data from Tables XIII and XIV. Ionic radius r_i is taken as the Pauling crystal radius for monatomic ions and an estimate from models for the others.

5.3. Enthalpies

5.3.1. *Data*

The enthalpies of transfer to D_2O from H_2O are usually determined as differences in heats of solution in the two solvents. Because of the importance of these measurements the original data are all given in Table XV. Conventional "best" single-ion enthalpies of transfer deduced from these data are given in Table XVI. The convention used there is now known to be unsatisfactory in some respects for solutions in water and methanol (see Section 6.3) but is used here for reasons to be discussed. In many cases the electrolyte transfer enthalpies calculated from the single-ion table agree with all the experimental values within ± 0.05 kcal mol^{-1} but in other cases further experimental work is needed to find the sources of larger inconsistencies.

TABLE XV. Calorimetric Heats of Solution in D_2O and H_2O in kcal mol^{-1} at 25°C[a]

X	$(X)_{d \leftarrow c}^H$	$(X)_{w \leftarrow c}^H$	$(X)_{d \leftarrow w}^H$	Ref.
LiF	—	—	−0.155	([b])
LiCl	−8.42	−8.82	0.40	490
LiBr	—	—	0.535	490
LiI	—	—	0.675	490
LiClO$_4$	−5.95	−6.35	0.40	474
		−6.345		625
LiO$_2$CCF$_3$	−6.19	−6.22	0.03	474
		−6.33		864
NaF	0.36	0.36	0.00	155
			0.01	([b])
NaCl	1.51	0.98	0.53	155
	1.58	1.02	0.56	490
	1.61	1.02	0.59	25
NaBr	—	—	0.695	490
NaI	—	—	0.835	490
NaClO$_4$	3.95	3.35	0.60	474
	4.03	3.46	0.57	490
		3.317		625
NaO$_2$CCF$_3$	−1.88	−1.98	0.10	474
		−1.99		864
NaNO$_3$	5.41	4.92	0.49	490
			0.481	295
NaClO$_3$	5.70	5.17	0.53	474
		5.19		625
NaBPh$_4$	−3.99	−4.77	0.78	474
	−4.00	−4.77	0.77	25
		−4.79		864
MeSO$_3$Na	1.96	1.67	0.29	155
PhSO$_3$Na	2.06	1.78	0.28	155
p-TolSO$_3$Na	2.58	2.29	0.29	155
Na$_2$CO$_3$	—	—	0.418	295
Na$_2$SO$_3$	—	—	0.388	295
Na$_2$SO$_4$	—	—	0.591	295

TABLE XV. (*Continued*)

X	$(X)_{d \leftarrow c}^{H}$	$(X)_{w \leftarrow c}^{H}$	$(X)_{d \leftarrow w}^{H}$	Ref.
HCO_2Na	0.464	0.269	0.195	739
			0.205	739
CH_3CO_2Na	−3.905	−3.973	0.068	739
			0.065	739
$C_2H_5CO_2Na$	−2.999	−3.015	0.016	739
$C_3H_7CO_2Na$	−3.402	−3.385	−0.017	739
i-$C_3H_7CO_2Na$	—	—	−0.036	739
$C_4H_9CO_2Na$	—	—	−0.054	739
$C_5H_{11}CO_2Na$	—	—	−0.099	739
KF	−3.80	−3.86	0.06	490
KCl	4.84	4.23	0.61	490
	4.86	4.13	0.73	25
	4.71	4.10	0.61	474
		4.115		625
			0.75	487
KBr	5.60	4.85	0.75	490
KI	5.82	4.93	0.89	490
KCN	3.30	2.80	0.50	490
KNO_3	8.84	8.33	0.51	474
		8.34		625
$KBrO_3$	10.42	9.90	0.52	490
RbCl	4.69	4.09	0.60	474
		4.13		625
		4.04		864
RbBr	6.08	5.28	0.80	490
RbO_2CCF_3	2.05	1.88	0.17	474
		1.89		864
CsCl	4.79	4.14	0.65	474
		4.25		625
		4.155		864
CsBr	7.06	6.25	0.81	490
CsO_2CCF_3	1.41	1.20	0.21	474
		1.53		864
$AgNO_3$	5.86	5.44	0.42	490

TABLE XV. (*Continued*)

X	$(X)_{d \leftarrow c}^{H}$	$(X)_{w \leftarrow c}^{H}$	$(X)_{d \leftarrow w}^{H}$	Ref.
$AgBrO_3$	—	—	0.80	663
$Hg(CN)_2$	3.70	3.50	0.20	490
$Cu(IO_3)_2$	—	—	0.23	662
$MgCl_2$	−36.16	−37.03	0.87	293
			0.751	294
$CaCl_2$	−18.03	−19.23	1.20	293
			1.05	294
	−18.52	−19.53	1.01	487
$SrCl_2$	−10.72	−11.99	1.27	293
			1.22	294
	−11.0	−12.15	1.15	487
$BaCl_2$	−1.60	−2.96	1.36	293
			1.363	294
	−1.57	−2.85	1.28	487
NH_4Cl	3.82	3.56	0.26	474
		3.53		625
$NH_4O_2CCF_3$	3.69	3.92	−0.23	474
Me_4NCl	1.44	1.06	0.38	474
	1.32	0.94	0.38	25
		0.975		625
Me_4NBr	—	5.92	—	865
	6.29	—	—	472
Me_4NI	10.79	10.08	0.71	474
		10.06		625
		10.08		865
Et_4NCl	−2.90	−3.10	—	474
		−3.07		865
		−3.02		23
Et_4NBr	1.79	—	—	472
	—	1.49	—	865
Et_4NI	7.08	6.67	—	474
		6.67		865
		6.83		23
Pr_4NCl	—	−5.30	—	472
	−5.42	—	—	474

TABLE XV. (*Continued*)

X	$(X)_{d \leftarrow c}^{H}$	$(X)_{w \leftarrow c}^{H}$	$(X)_{d \leftarrow w}^{H}$	Ref.
Pr$_4$NBr	−0.99	−1.03	0.04	472
		−1.10		72
Pr$_4$NI	2.76	—	—	472
	2.765	—	—	54
	—	2.91	—	474
Bu$_4$NCl	−7.40	—	—	474
	—	−7.14	—	472
Bu$_4$NBr	−2.17	−2.02	−0.15	472
Pen$_4$NCl	—	−9.13	—	472
	−9.57	—	—	474
Pen$_4$NBr	0.49	0.77	−0.28	472
Ph$_4$AsCl	−2.47	−2.57	0.10	472
	—	−2.60	—	20
Ph$_4$PCl	−2.07	−2.19	0.12	472
Bu$_3$N(CH$_2$)$_8$NBu$_3$Br$_2$	−7.94	−7.90	−0.04	474
MeNH$_3$Cl	1.40	1.42	−0.02	475
EtNH$_3$Cl	2.03	2.08	−0.05	475
PrNH$_3$Cl	0.22	0.33	−0.11	475
BuNH$_3$Cl	−0.80	−0.64	−0.16	475
PenNH$_3$Cl	−0.43	−0.23	−0.20	475
HexNH$_3$Cl	0.04	0.28	−0.24	475
HepNH$_3$Cl	0.63	0.90	−0.27	475
Oct NH$_3$Cl	2.75	3.04	−0.29	475
MeNH$_3$Br	—	—	0.064	714
PrNH$_3$Br	—	—	−0.026	714
BuNH$_3$Br	—	—	−0.067	714
Oct NH$_3$Br	—	—	−0.214	714
Me$_3$SI	—	—	0.21	25

[a] Heats of transfer are given in those cases in which both heats of solution come from the same laboratory. Data have been corrected to infinite dilution where necessary.
[b] Quoted in Ref. 292.

TABLE XVI. Single-Ion Enthalpies of Transfer of Solutes to D_2O from H_2O[a]

X	$(X)_{d \leftarrow w}^{H}$	X	$(X)_{d \leftarrow w}^{H}$	X	$(X)_{d \leftarrow w}^{H}$	X	$(X)_{d \leftarrow w}^{H}$	X	$(X)_{d \leftarrow w}^{H}$
Li^+	450	NH_4^+	300	$EtNH_3^+$	0	Cl^-	−50	CO_3^{2-}	−800
Na^+	610	Me_4N^+	430	$PrNH_3^+$	−60	Br^-	90	SO_3^{2-}	−830
K^+	660	Et_4N^+	210	$BuNH_3^+$	−110	I^-	230	SO_4^{2-}	−630
Rb^+	680	Pr_4N^+	−50	$PenNH_3^+$	−150	$CF_3CO_2^-$	−510	HCO_2^-	−420
Cs^+	710	Bu_4N^+	−230	$HexNH_3^+$	−190	ClO_4^-	−40	$CH_3CO_2^-$	−540
Ag^+	540	Pen_4N^+	−380	$HepNH_3^+$	−220	NO_3^-	−120	$C_2H_5CO_2^-$	−590
Mg^{2+}	970	Ph_4As^+	160	$OctNH_3^+$	−240	CN^-	−160	$C_3H_7CO_2^-$	−630
Ca^{2+}	1300	Ph_4P^+	170	Hg^{2+}	−120	ClO_3^-	−80	$i\text{-}C_3H_7CO_2^-$	−650
Sr^{2+}	1370	$diBu^{2+}$	−220	Me_3S^+	−20	BrO_3^-	−140	$C_4H_9CO_2^-$	−660
Ba^{2+}	1460	$MeNH_3^+$	30	F^-	−610	BPh_4^-	60	$C_5H_{11}CO_2^-$	−710

[a] Based on the Convention $(AsPh_4^+)_{d \leftarrow w}^{H} = (BPh_4^-)_{d \leftarrow w}^{H}$. Values in cal mol⁻¹ at 25°C.

The dependence of the ionic transfer enthalpies upon the ionic radius is shown for some of these ions in Fig. 19. The isotopic Walden product ratio measured by Kay and Evans,[427]

$$R = (\lambda\eta)_{D_2O}/(\lambda\eta)_{H_2O} \qquad (61)$$

where λ is the limiting ionic mobility of the ion and η is the viscosity of the pure solvent, is also plotted on the same figure. The striking similarity between the two sets of data has been shown by adjusting the R scale so that the enthalpy of transfer datum of Cs^+ lies on the R curve. The R values are absolute ionic quantities since they are determined from experimental conductivities and transference numbers. Then the good agreement of the

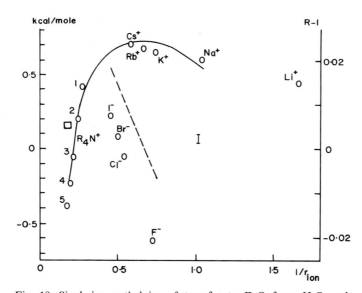

Fig. 19. Single-ion enthalpies of transfer to D_2O from H_2O and solvent-isotope effect on the Walden products for various ions. Ionic radius r_{ion} is taken as the Pauling crystal radius for monatomic ions and an estimate from models for the others. The R_4N^+ circles are the enthalpy data for these ions with R ranging from methyl (1) to n-amyl (5). The other circles are the enthalpy data for alkali ions and the halide ions. These single-ion values are based on the convention that the enthalpies of transfer of $AsPh_4^+$ and BPh_4^- are equal; their common value is given by the square data point. The lines are smooth curves through the Walden product ratios R for the cations (—) and the anions (– –) reported by Kay and Evans.[427] The error bar is an estimate of the uncertainty in both the calorimetric and conductivity data. Enthalpy data from Krishnan and Friedman.[474]

radius dependence for the cations in the two different experiments suggests that the single-ion convention used here is realistic and that the two sets of data have a common underlying physical basis.

5.3.2. Discussion for Cations

The data in Fig. 19 can be interpreted in terms of the chemical model in Section 2.5 if one makes the additional assumption that the enthalpy change in the process in which normal bulk water changes to become cosphere water in any one of the type II states [cf. eqn. (60)] is somewhat larger in magnitude for D_2O than for H_2O.

Consider first the alkali metal ions. The actual properties and the amount of the type I_c cosphere water (cf. Fig. 11) presumably changes considerably in the series Li^+ to Cs^+, while the value of $(X)_{d \leftarrow w}^{H}$ retains a relatively constant positive value. This might be due to contribution from type II_{sb} cospheres, since one does not expect a significant solvent isotope effect in type I_c cospheres for alkali metal ions because (a) hydrogen bonding is only of secondary importance in stabilizing I_c cosphere water and (b) the zero-point energy of the metal ion–water vibration is expected to be nearly the same in H_2O and D_2O (see Section 5.3.4). In type II_{sb} cospheres the stabilization of the water by hydrogen bonding is taken to be less complete than in the bulk liquid. If this so-called structure-breaking effect is larger in D_2O than in H_2O, it accounts for the signs of $R - 1$ as well as the enthalpies of transfer of the alkali ions. As shown in Fig. 20, the solvent-isotope effect $(X)_{d \leftarrow w}^{H}$ also correlates well with the corresponding term $S_{X,II}$ in the hydration entropy, which in turn correlates well with the molar contributions of the ions to the viscosity of the solutions as measured by their Jones–Dole B coefficients (Section 3.3).

Now turning to the tetraalkylammonium ions, the opposite trend with size for these (Fig. 19) shows that a different effect is dominant. It is assumed that as the ion gets larger, from Cs^+ through Me_4N^+ and Et_4N^+, the contribution of type I_c cosphere states to hydration, if present for Cs^+, quickly decays, while the type II cosphere state changes gradually from type II_{sb} to type II_{al} (cf. Fig. 11). In the II_{al} state the water is stabilized by hydrogen bonding more than in the bulk, and again this effect is assumed to be larger for D_2O than for H_2O. This contribution tends to make $(X)_{d \leftarrow w}^{H}$ negative, the more so the greater the amount of water in the II_{al} cosphere state. Apparently the contributions of II_{sb} and II_{al} states in the cospheres balance near Et_4N^+. This point of balance is consistent with some, but not all, measures of this effect, but of course it is not surprising if the relative

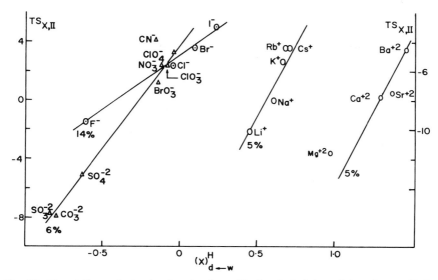

Fig. 20. Correlation of structural entropies with the enthalpies of transfer to H_2O from D_2O. All energies are in kcal mol^{-1}. Lines show the correlation expected in the $S_{X,II}$ data if the enthalpy of transfer to D_2O from H_2O were a definite percentage of the structural enthalpy effect of the aqueous ion.

effects of the II_{sb} and II_{al} states depend somewhat on the probe used to investigate the solutions.

The interpretation of Fig. 19 given here is essentially the same as the one given by Kay and Evans for the radius dependence of their R values.[427] On the other hand, it has been pointed out that the close agreement of the two kinds of data exhibited in Fig. 19 implies that for the alkali metal as well as the tetraalkylammonium ions the ionic cospheres exchange solvent so rapidly with the bulk that even the inner cospheres are exchanged as rapidly as the ions move through the medium rather than being dragged along intact.[354,357] However, very recently Kay and Broadwater found $R \simeq 0$ for Li^+, so this ion may move with its inner cosphere intact.*

The fact that the data point for Ph_4As^+ and Ph_4B^- is off the curve for R_4N^+ in Fig. 19 shows that for aromatic groups the state of water in the cospheres must be different from the II_{al} state; accordingly, a new state II_{ar} is defined, as noted in Table IV. Various studies have indicated that the water in II_{ar} states depends very much upon the substituents on the

---

* R. L. Kay, private communication.

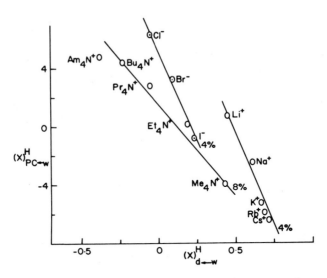

Fig. 21. Correlation of enthalpies of transfer to propylene carbonate from water with enthalpies of transfer to D_2O from H_2O.

aromatic groups[411,471,572,573,884] but a further analysis of this complication has yet to be made.

As shown in Fig. 20, the comparison of $(X)^H_{d \leftarrow w}$ with $S_{X,II}$, the part of the hydration entropy assigned to the outer cospheres for the alkali metal ions, suggests that ΔH for forming the cospheres [eqn. (60)] is about 5% more positive in D_2O than in H_2O. It was found for normal alcohols and tetraalkylammonium ions that ΔH in eqn. (60) for type II_{al} cospheres is about 5% more negative in D_2O than in H_2O.[471,472] The same effect is shown in a cruder but similar way in Fig. 21 if one assumes that for the ions shown there the quantity $(X)^H_{d \leftarrow w}$ is dominated by the terms due to hydration cospheres. This point is discussed in Section 6.3.

There is now extensive evidence that hydration cospheres on one or more alkyl chains connected to the same group do not interfere with each other even though other types of cosphere–cosphere interference have been identified. The data in Fig. 22 suggest that the cosphere effect for alkyl groups in R_4N^+ is just four times that for one alkyl group in RNH_3^+, while in ROH there seems to be a strong interference between the hydration of the R— chain and the —OH group.

From the near parallelism of the curves for RNH_3^+ and $\frac{1}{4}R_4N^+$ an approximate value for the enthalpy of transfer of an NH_3^+ group was deduced, namely -0.05 kcal mol^{-1}.[475] It was not possible to find directly how much of the enthalpy of transfer of NH_3^+ might be due to the transfer without exchange and how much due to the establishment of the isotope exchange equilibrium between NH_3^+ and the D_2O medium, but in the case of an OH group the exchange is the dominant effect in the heat of transfer to D_2O from H_2O. Desnoyers *et al.*[173] also studied this system and arrived at a value of -0.07 kcal mol^{-1} for the exchange reaction. Thus, when considering amino acids, alcohols, and acids as well as ammonium ions, one should allow for the contribution of the exchange reaction before classifying these solutes as structure makers or breakers based on $(X)_{d\leftarrow w}^H$ values.

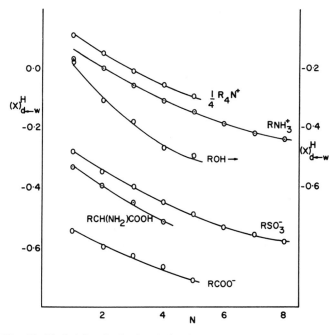

Fig. 22. Enthalpies (in kcal mol^{-1}) of transfer to D_2O from H_2O for various solutes plotted as a function of N, the number of carbon atoms in each alkyl chain R. The R_4N^+ data are from Krishnan and Friedman.[474] The RNH_3^+ data are from Krishnan and Friedman.[475] The RSO_3^- data are from Krishnan and Friedman, unpublished work. The ROH and $RCH(NH_2)COOH$ data are from Kresheck *et al.*[468] The $RCOO^-$ data are from Snell and Greyson.[740]

5.3.3. Discussion for Anions

In the type I_a cosphere on an anion X^- the hydrogen bond

$$X^- \cdots H\!-\!O\!-\!H$$

may make an important contribution, depending on the characteristics of X^-. The librational motion of the hydrogen-bonded water molecule then leads to a large solvent-isotope effect in I_a cospheres, just as in bulk water itself, but in a way which has no counterpart in type I_c cospheres. (See Table IV for nomenclature for cospheres.) This effect accounts for the trend from I^- (weak hydrogen bond) to F^- (strong hydrogen bond) if one also assumes that the X^-–w hydrogen bond is stronger in D_2O than in H_2O.

As shown in Fig. 20, this interpretation is consistent with $S_{X,II}$ for the halide ions if the resulting overall stabilization of the cosphere water is 14% greater in D_2O than in H_2O, judging by the slope of the correlation. This increment is large, but not so large as to be unrealistic in view of the upper limit of 35% for this kind of effect which one may estimate from the model studied by Swain and Bader.[764] On the other hand, a lower figure of about 5% greater stabilization in D_2O than in H_2O may be deduced from the slope of the correlation in Fig. 21. This seems to show that we have still not achieved a quantitative estimate of the relation of the cosphere effects in the D_2O from H_2O transfers to the overall cosphere contributions to hydration.

The slope of the correlation of the oxyanions in Fig. 20 suggests, by contrast with the halide data, that hydrogen bonding is much less important in stabilizing the cospheres of these species.

The alkylsulfonates and alkylcarboxylates provide more ways to investigate the hydration of alkyl chains. The data in Fig. 22 show, first, that the hydrogen-bonding contribution to the solvent isotope effect is much larger for carboxylates than for sulfonates, as one might expect by assuming that the stronger base (CO_2^-) is also the stronger hydrogen-bond acceptor. Figure 22 shows that the increment in $(RX)_{d\leftarrow w}^H$ per methylene group in the longer chains is about -30 cal mol^{-1} in each case, perhaps even for the alcohols. The work with the primary alkylammonium ions and with the alkylsulfonates was undertaken to investigate this point, since the data for the alcohols taken by themselves suggest a "saturation" of the enthalpy of transfer effect with increasing chain length. Owing to experimental difficulties, this has not been investigated further using the alcohols themselves.

5.3.4. *Solvation of Ions in* $H_2^{18}O$ *and* $H_2^{16}O$

It is interesting to compare the D_2O from H_2O transfer of the alkali metal ions with $H_2^{18}O$ from $H_2^{16}O$ transfers. The latter have not been measured directly but were estimated[474] from the $H_2^{16}O-H_2^{18}O$ fractionation between aqueous solutions of alkali halides and the vapor.[214] The results are summarized in Table XVII along with the data for some polyvalent ions for comparison.

As discussed by Feder and Taube,[214] these isotope effects can, in most cases, be understood in terms of the effects on the zero point energy of a vibration in which the motion changes the distance between the metal ion and water molecule in the type I cosphere. If the force constant for this vibration is much weaker than that for stretching an O—H bond within a water molecule, then each water molecule moves as a whole in its vibration against the central ion and the D_2O-H_2O isotope effect ought to include, as one component, the effect reported in Table XVII. The latter is apparently very much smaller than the observed D_2O-H_2O effect. It must be concluded that D_2O-H_2O effect in Fig. 19 for the alkali metal ions can only derive from changes in the hydrogen bonding in the neighborhood of the metal ions, i.e., resembling what is pictured in the chemical model in Section 5.3.2. This confirms that the dominant effect for the alkali metal

TABLE XVII. Free Energies and Enthalpies of Transfer for Some Ions to $H_2^{18}O$ from $H_2^{16}O$ Estimated from Fractionation Data[a]

Ion	$\Delta G^{\ominus}$, kcal mol^{-1}	$\Delta H^{\ominus}$, kcal mol^{-1}
Li$^+$	−0.007	−0.11
Na$^+$	0.000	0.08
K$^+$	0.005	0.04
Cs$^+$	0.007[b]	—
H$^+$	−0.027	0.203
Ag$^+$	−0.004	0.0
Mg^{2+}	−0.036	−0.07
Al^{3+}	−0.089	—
Cr^{3+}	−0.103	—
Co(NH$_3$)$_5$H$_2$O^{3+}	−0.012	—

[a] Krishnan and Friedman.[474] Estimated from the data in Refs. 214, 771, and 213.
[b] This value is at 4°C.

ions in the D_2O from H_2O transfers is a structure-breaking effect which is larger in D_2O than in H_2O.

It is then interesting to consider whether the structure-breaking effect is also reflected in the data of Table XVII. In fact, the simplest interpretation of the trend in free energies shown there is that there is a competition between inner shell binding, reflected in the M—OH_2 zero point energy, and a structure-breaking effect, in which the first dominates for Li^+ and the second for Cs^+, with something of a balance for Na^+. It would be of great interest to have more extensive data of this sort.

5.3.5. Solvation of Ions in CH_3OH and CH_3OD

In this section the solvent-isotope effect on enthalpies of solvation in methanol is compared with the effect in water to see to what extent the effects which seem to require such an elaborate model for interpretation are peculiar to water. It is assumed that O—D$\cdots$O bonds are stronger than O—H$\cdots$O bonds in methanol, just as in water. However, the suggestions which have been made[230,232,753] for the interpretation of the type II cosphere effects in water emphasize the central role of the branched (i.e., three-dimensional) networks of hydrogen bonds in water. Therefore effects corresponding to the type II cospheres may be expected to be absent in methanol, in which only unbranched chains of hydrogen bonds are possible.

The data in Fig. 23 may be discussed from this point of view.[476] To facilitate the discussion, the nomenclature for cosphere states given in Table IV will be used for methanol as well as for water in this section.

The data for hydrocarbons in Fig. 23* show that the type II_{al} cosphere contributions, proposed to explain the hydration of alkyl chains, are absent in methanol. For, in water the solvent-isotope effect $(X)^H_{d\leftarrow w}$ is negative, if X is an alkyl group, the more so the longer the chain, while in methanol the corresponding solvent-isotope effect $(X)^H_{MeOD\leftarrow MeOH}$ is positive, the more so the longer the chain. The latter behavior is readily explained in terms of the somewhat larger energy required to form a cavity in the deuterated solvent; cosphere effects are not needed.

It also appears from Fig. 23 that the marked difference in solvation of alkyl and aryl groups in water, leading to the characterization of type II_{ar} cospheres as distinct from type II_{al} cospheres in water, is completely

---

* The enthalpies of transfer to D_2O from H_2O for C_3H_8 and C_4H_{10} give the expected trend at 4° and 50° but not at 25° and hence the inconsistency in Fig. 23 for these solutes.

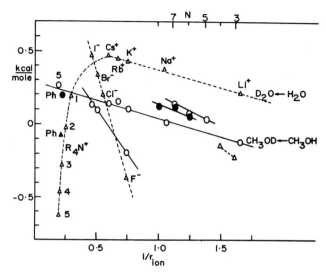

Fig. 23. Comparison of solvent-isotope effect in methanol and water. $(\bigcirc, \bullet)$ CH$_3$OD $\leftarrow$ CH$_3$OH; $(\triangle, \blacktriangle)$ D$_2$O $\leftarrow$ H$_2$O; for ions as a function of reciprocal ionic radius (lower scale) estimated as Pauling crystal radius for monatomic ions and from models for the others. For hydrocarbons $[(\bigcirc)$ aliphatic; $(\bullet)$ aromatic] in methanol and in water $[(\triangle)$ aliphatic] as a function of N (upper scale), the number of carbon atoms in the molecule. The tetraalkylammonium ions are indicated by integers: 1 for Me$_4$N$^+$, 2 for Et$_4$N$^+$, etc. Ph stands for Ph$_4$As$^+$. The single-ion enthalpies are based on the convention that the enthalpies of transfer of Cs$^+$ and I$^-$ are equal. Data from Krishnan and Friedman.[474,476]

absent in methanol; again no type II cospheres at all seem to be needed in methanol.

Because of lack of data for (NaBPh$_4$)$_{\mathrm{MeOD\leftarrow MeOH}}^{H}$ the convention for single-ion enthalpies of transfer used in Fig. 19 cannot be used here and the single-ion values given in Fig. 23 are based on the convention of equal transfer enthalpies for Cs$^+$ and I$^-$.

For the alkali metal ions the trend in the solvent-isotope effect in Fig. 23 is similar in the two solvents. That the effect is about 0.3 kcal mol^{-1} more positive in water than in methanol is attributed to a contribution of type II$_{sb}$ cospheres in water but not in methanol; it would seem that the structure-broken region is absent or at least much smaller in methanol.

More telling on this point is the observation that the solvent isotope effect for the salt CsI is >0.9 kcal mol^{-1} in water, indicative of a large net structure-breaking effect for this neutral combination of ions, while the same quantity is scarcely 0.3 kcal mol^{-1} in methanol. The smaller quantity in methanol is easily accounted for in terms of the cavity energy, as comparison with the hydrocarbon data shows. The trend in alkali ion values in methanol is attributed to an isotope effect in the I_c cosphere material which was not mentioned before. The cosphere material is drawn from a more stable state in the deuterated solvent than in the other, due to the difference in hydrogen bond energies, and this effect is not compensated by significantly greater strength in the $M^+ \cdots O$ interaction for the deuterated species (cf. Section 5.3.4). This effect seems conceptually similar to the cavity-energy effect giving the trend for the solvent-isotope effect of the hydrocarbons in methanol.

In the case of the halide ions in either solvent, the change in stability of the type I_a cosphere solvent accounts for the trend with ion size; the interpretation proposed for water in Section 5.3.3 applies equally to methanol.

Finally, the comparison of the solvent-isotope effect in water and in methanol for $(n\text{-}C_5H_{11})_4N^+$ is just what one would expect on the basis of the hydrocarbon data already discussed and the comparison for Ph_4As^+ is just what one would expect on the basis of the data for the aromatic hydrocarbons.

It is concluded that the data in Fig. 23 provide definite evidence that the type II cosphere effect depends on the branched chains of hydrogen bonds in water.

5.4. Entropies

The solvent-isotope effect in the single-ion entropies of hydration is readily calculated from the single-ion free energies in Table XIV and enthalpies in Table XVI after adjusting the single-ion conventions to a consistent basis, for example, zero for Na^+ transfers in both cases. The results given in Table XVIII, however, are based on a different single-ion convention for the single-ion entropies of transfer which was chosen after observing that, for neutral salts X, one finds

$$(X)^H_{d \leftarrow w} \simeq T(X)^S_{d \leftarrow w} \tag{62}$$

where $T = 298°K$. Thus there is a high degree of entropy–enthalpy compensation in the neutral salt data. The single-ion convention for entropies in the second column of Table XVIII is chosen to preserve the relation

TABLE XVIII. Quantities Related to the Solvent-Isotope Effect in Single-Ion Hydration Entropies[a]

Species, X	$T(X)_{d \leftarrow w}^{S}$ Experimental[b]	$T(X)_{d \leftarrow w}^{S}$ Experimental[c]	$T(X)_{d \leftarrow w}^{S}$ Calculated
Li^+	0.49	0.05	0.006
Na^+	0.57	0.13	0.068
K^+	0.56	0.11	0.092
Rb^+	0.53	0.07	0.105
Cs^+	0.61	0.17	0.115
F^-	−0.51	−0.07	−0.065
Cl^-	−0.12	0.33	0.148
Br^-	0.00	0.45	0.205
I^-	0.09	0.54	0.262
Mg^{2+}	0.84	—	—
Ca^{2+}	1.13	—	—
Sr^{2+}	1.19	—	—
Ba^{2+}	1.25	—	—
Me_4N^+	0.37	—	—
HCO_2^-	−0.35	—	—
$CH_3CO_2^-$	−0.47	—	—
$C_2H_5CO_2^-$	−0.51	—	—
$C_3H_7CO_2^-$	−0.54	—	—
$C_4H_9CO_2^-$	−0.59	—	—
$C_5H_{11}CO_2^-$	−0.63	—	—

[a] All quantities in kcal mol^{-1} at 25°C.
[b] Single-ion values based on $T(Na^+)_{d \leftarrow w}^{S} = 0.57$ kcal mol^{-1}.
[c] Single-ion values based on $T(F^-)_{d \leftarrow w}^{S} = -0.07$ kcal mol^{-1}.

in eqn. (62) when these entropies are compared with single-ion enthalpies from Table XVI.

The resulting Barclay–Butler correlation, shown in Fig. 24, exhibits a very high degree of compensation! However, it must be noted that choosing $T(Na^+)_{d \leftarrow w}^{S} = 0.47$ kcal mol^{-1} gives a correlation which is almost as good, but with

$$\Delta H = 20 + 420 \, \Delta S \quad \text{cal mol}^{-1}$$

Still it can be concluded on the basis of these data that, in the notation of eqn. (42), $T^* = (1.2 \pm 0.2) \times 298$ deg for the process of transfer to D_2O from H_2O. It has already been noted that in most other cases of compensation one finds T^* to be somewhat smaller than T (Section 2.5).

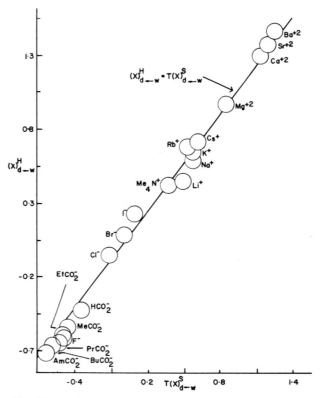

Fig. 24. Barclay–Butler correlation for the solvent-isotope effect in ionic hydration. Enthalpy values are from Table XVI while entropy values are from the second column of Table XVIII. All energies in kcal mol⁻¹.

Swain and Bader[764] calculated $(X)_{d \leftarrow w}^{S}$ for alkali and halide ions from a particular model using experimental data for neutral salt solvent-isotope effects in the enthalpy together with certain spectral data. Their calculation also yields a definite assignment for $(X)_{d \leftarrow w}^{H}$ single-ion values and, taken together, the entropies and enthalpies may be found accurately to fit a Barclay–Bulter plot with the equation

$$(X)_{d \leftarrow w}^{H} = 15 + 730(X)_{d \leftarrow w}^{S} \quad \text{cal mol}^{-1}$$

That this model gives the Barclay–Butler relation accurately can be no surprise because the essential ingredient of the model is the representation of the fluid as a collection of independent harmonic oscillators, other major contributions to the partition functions of the individual solutions being assumed to cancel in calculating the solvent-isotope effect. Frank already showed that such models give Barclay–Butler entropy–enthalpy correlations.[224] The surprising and promising new development is that the Swain–Bader calculation comes close to reproducing the experimentally observed compensation; their T^*/T is only about 2.5, whereas in Barclay–Butler plots for processes without compensation one may find that T^* is an order of magnitude or more larger than T.

Another way of comparing the Swain–Bader results with experiment is shown in Table XVIII where, in the third column, the experimental entropies are based on a single-ion convention to make them agree with the Swain–Bader value for F^-, which is the key one in their calculations. Their results are in the fourth column. It is apparent that the calculated trends are about a factor of two too small.

5.5. Volumes

The solvent-isotope effect in the partial molar volumes at infinite dilution is shown for several solutes in Table XIX. Single-ion values based on these data and the assumption that extrapolating the partial molar volumes of $(C_nH_{2n+1})_4NBr$ to $n = 0$ gives the single-ion value for $Br^{-[129]}$ (see Section 1.7) are given in Table XX and compared with the corresponding enthalpy effects. Since volume changes for a series of solutes in a given process most often show the same trends as the entropy changes, and since the entropy changes are found to be closely parallel to the enthalpy changes for this process (Section 5.4), the results in Table XX for the R_4N^+ series seem highly anomalous, for the trends in volumes are in the opposite direction to those in the enthalpies. Conway and Laliberté[129] find that $(X)_{d \leftarrow w}^V$ is linearly correlated with the field acting on a water molecule in the first hydration layer of ion X, calculated according to the equation

$$\text{field} = \text{const} \cdot z_X/(r_X + \Delta)^2$$

where z_X is the charge and r_X the radius of ion X, and Δ is taken as 1.38 Å for monatomic ions and zero for tetraalkylammonium ions. However, this finding scarcely removes the anomaly.

It seems interesting to recall that a similar anomaly for excess enthalpies and volumes for mixtures of spherical molecules proved to be the downfall of the conformal solution theory but is accounted for by more powerful theories for such mixtures.[504,557,650]

TABLE XIX. Partial Molal Volumes at Infinite Dilution for Several Solutes in H_2O and D_2O at 25°C and the Solvent-Isotope Effect in the Partial Molal Volumes[129]

Solute, X	$\bar{V}^{\ominus}_{X(w)}$, cm³ mol⁻¹	$\bar{V}^{\ominus}_{X(d)}$, cm³ mol⁻¹	$(X)^V_{d \leftarrow w}$, cm³ mol⁻¹
NaF	−2.37	−3.79	−1.42
	−1.19	−3.12	−1.93[685]
NaCl	16.61	15.78	−0.83
NaBr	23.64	23.30	−0.34
NaI	35.19	33.84	−1.35[685]
$C_6H_5SO_3Na$	107.73	101.77	−5.96[685]
KCl	26.80	26.43	−0.37
KBr	33.67	33.48	−0.19
Me_4NBr	114.26	114.14	−0.12
Et_4NBr	173.70	173.90	0.20
	174.28	174.53	0.25[128]
$n\text{-}Pr_4NBr$	239.4	239.82	0.42
$n\text{-}Bu_4NBr$	300.96	301.90	0.94
$(HOC_2H_4)_3NHBr$	147.28	—	—
$(DOC_2H_4)_3NHBr$	—	144.58	—
HOD	18.10	18.66	0.56
Pyridine	77.70	77.90	0.20

TABLE XX. The Solvent-Isotope Effect in Ionic Volumes and Enthalpies

Ion	$(X)^V_{d \leftarrow w}$,[129] cm³ mol⁻¹	$(X)^H_{d \leftarrow w}$, kcal mol⁻¹
F^-	−1.3	−0.61
Cl^-	−0.7	−0.05
Br^-	−0.5	0.09
Na^+	−0.1	0.61
K^+	0.3	0.66
Me_4N^+	0.4	0.43
Et_4N^+	0.7	0.21
$n\text{-}Pr_4N^+$	0.9	−0.05
$n\text{-}Bu_4N^+$	1.4	−0.23

5.6. Heat Capacities

The only data found for the solvent-isotope effect in the partial molal heat capacities (C_P) at infinite dilution, $(X)^C_{d \leftarrow w}$, are -6.8 for NaCl and -8.8 for KCl, both in gibbs mol^{-1}.[155,487] This could be due to the fact that ΔH for forming type II$_{sb}$ cospheres on the cations is positive, so that the cospheres form to a smaller extent at higher temperatures, with the effect being larger in D$_2$O, or to an actual ΔC_P in the process of forming the cospheres, but the former effect must dominate in view of the conclusions already reached regarding ΔH and ΔS for the formation of type II$_{sb}$ cospheres in H$_2$O and D$_2$O. Of course, other interpretations also suggest themselves; more data are needed to reduce the number of possibilities.[44]

6. REFERENCE SOLVENTS

6.1. Free Energies in Nonaqueous Solvents

For a better understanding of electrolyte solutions in water it is helpful to have similar data in a variety of other solvents for comparison. Even though the thermodynamics of transfer is more complicated to interpret than solvation thermodynamics, the transfer data can be obtained with much greater accuracy, so that even subtle effects can be noticed. However, it is hard to obtain the free energy data in nonaqueous solvents because of complications from ion–ion interactions and also because of the difficulty of finding reversible electrodes that operate in these solvents. Also, solubility problems often restrict the study of a series of similar ions. In spite of these difficulties, data have been obtained in a few solvents and these are tabulated as $(X)^G_{s \leftarrow w}/2.303RT = \log m\gamma_i$ in a recent review.[641]

In general, transfer free energies are not as sensitive to specific solvation effects as are enthalpies and entropies.

6.2. Entropies in Nonaqueous Solvents

Criss et al.[150] compared $\bar{S}^{\ominus}_{i(a)}$, the limiting ionic entropies in solvent a, with the entropies in water $\bar{S}^{\ominus}_{i(w)}$ and found that the variation with ionic species i is governed by the equation

$$\bar{S}^{\ominus}_{i(a)} = S_{a,w} + \lambda_{a,w}\bar{S}^{\ominus}_{i(w)} \tag{63}$$

where the coefficients $S_{a,w}$ and $\lambda_{a,w}$ do not depend upon the ionic species

TABLE XXI. Entropy Correlation in Eqn. (63)[a]

Solvent, a	$S_{a,w}$, gibbs mol^{-1}	$\lambda_{a,w}$	T_a, °K
D$_2$O	0.8	1.04	173
H$_2$O	0.0	1.00	174
Formamide	-1.6	0.64	162
N-methylformamide	-5.7	0.72	128
Methanol	-10.9	0.82	117
Ethanol	-16.0	0.79	97
Dimethylformamide	-15.9	0.79	98
NH$_3$	-22.4	0.82	57

[a] Data taken from Ref. 150.

but may be expected to depend upon the species of the two solvents as shown. In their paper the experimental data for a given solvent pair a–w give the same $\lambda_{a,w}$ whether i is a cation or an anion, so, for a given single-ion convention for $\bar{S}_{i(w)}^{\ominus}$, it is possible to choose a single-ion convention for $\bar{S}_{i(a)}^{\ominus}$ such that both cations and anions lie on the same line. The coefficients obtained in this way are given in Table XXI. They are based upon data for the ions H$^+$, Li$^+$, Na$^+$, K$^+$, Rb$^+$, Cs$^+$, Cl$^-$, Br$^-$, and I$^-$.

The correlation expressed by eqn. (63) is quite accurate, mostly better than 2 gibbs mol^{-1}, so it may be the basis of a useful procedure for estimating the entropies of some aqueous ions from data in other solvents in which the ions are more readily investigated, or viceversa.

Although the concept of absolute single-ion entropies may be used in the discussion of eqn. (63), it is easily verified that one does not require absolute single-ion entropies to determine $S_{a,w}$ and $\lambda_{a,w}$. Thus, suppose

$$\bar{S}_{i(w)}^{\ominus} = \bar{S}_{i(w)}^{\ominus\ominus} + z_i K_w$$

is only some conventional single-ion entropy and that $\bar{S}_{i(w)}^{\ominus\ominus}$ is another conventional value, whether or not it is closer to the absolute value. Suppose also that for the conventions $\bar{S}_{i(a)}^{\ominus}$ and $\bar{S}_{i(w)}^{\ominus}$ one finds that eqn. (63) holds for ions of both signs. Then we have

$$\bar{S}_{i(a)}^{\ominus} = S_{a,w} + z_i \lambda_{a,w} K_w + \lambda_{a,w} \bar{S}_{i(w)}^{\ominus\ominus} \tag{64}$$

or

$$\bar{S}_{i(a)}^{\ominus\ominus} \equiv \bar{S}_{i(a)}^{\ominus} - z_i \lambda_{a,w} K_w = S_{a,w} + \lambda_{a,w} \bar{S}_{i(w)}^{\ominus\ominus} \tag{65}$$

Thus the correlation must work equally well for aqueous single-ion entropies based on an arbitrary convention, for example, $\bar{S}^{\ominus}_{\text{H}^+(\text{w})} = 0$. On the other hand, the correlation establishes a relation between the single-ion conventions in different solvents: For example, by choosing $\bar{S}^{\ominus}_{\text{H}^+(\text{w})} = 0$, one will not in general find $\bar{S}^{\ominus}_{\text{H}^+(a)} = 0$ in a second solvent a by the use of eqn. (64).

It is striking that in Table XXI one finds $\lambda_{a,\text{w}} = 0.73 \pm 0.09$ except, of course, for $\lambda_{\text{w,w}} = 1$. In this respect the correlation shows that ionic solvation by water is qualitatively different from solvation by the other media.*

On the other hand, $S_{a,\text{w}}$ is strongly correlated with the quantity

$$T_a \equiv \text{boiling point of } a$$
$$- \text{ boiling point of model for nonpolar } a$$

where the model is a real or hypothetical species having the molecular weight and structure of a but lacking strong dipoles or the ability to form hydrogen bonds. Criss and co-workers have estimated the T_a values in Table XXI and have shown that the single-ion entropies correlate well with them. It is clear from Table XXI that this is due to the behavior of $S_{a,\text{w}}$ rather than $\lambda_{a,\text{w}}$ and we find

$$S_{a,\text{w}} = -50.8 + 0.483T_a - 1.10 \times 10^{-3}T_a^2 \quad \text{gibbs mol}^{-1} \qquad (66)$$

with an accuracy of about ± 1 gibbs mol^{-1} except for $a = \text{NH}_3$. The coefficient $S_{a,\text{w}}$ cannot reflect what goes on in the cospheres since it is just an additive term whose effect is independent of the species. Thus it seems to reflect some global contribution to the ionic entropy, perhaps from the Born-model charging through $d\varepsilon^{-1}/dT$, perhaps from a free-volume or communal entropy effect.

With the observed regular behavior for both $S_{a,\text{w}}$ and $\lambda_{a,\text{w}}$ it seems possible to use this correlation to get $\bar{S}^{\ominus}_{i(\text{w})}$ from $\bar{S}^{\ominus}_{i(a)}$ or vice versa even when no other ionic entropies have been determined in solvent a. However, using the data calculated from the available free energies and heats[708] and applying the correlation to entropies in propylene carbonate (PC) and water, one finds $S_{\text{PC,w}} = -2$ gibbs mol^{-1} and $\lambda_{\text{PC,w}} = 1.57$ for the best correlation line, which, however, misses some of the data by as much as 10 gibbs mol^{-1}.

---

* It is also interesting to note that 1.04 for $\lambda_{\text{d,w}}$ in Table XXI is close to what one would expect from the evidence (Section 5) that the structural effects in solvation are 5% larger in D_2O than in H_2O for many of these ions.

TABLE XXII. Entropy Correlation Applied to Water–Methanol Mixtures

Wt% methanol	0	10	20	43.1	68.3	100
$S_{a,w}$, gibbs mol^{-1}	0	0	0	-3	-8	-15
$\lambda_{a,w}$	1	1.00	1.00	0.95	0.89	0.79

In another application Franks and Reid[235] applied the correlation in eqn. (63) to the entropies of alkali halides in water–methanol mixtures of varying composition. The correlation is reasonably accurate, giving the parameters shown in Table XXII, which gives a very economical presentation of the variation in ionic entropies of all the alkali halides with the composition of the solvent mixture. The nonlinear behavior is striking, with the maximum in the nonlinearity coming at the composition expected on the basis of some other properties of water–methanol mixtures.

To the extent that the correlations provided by eqn. (63) are accurate, they indicate that the cosphere properties change smoothly as one changes ionic species in the series made up of alkali and halide ions. This is perhaps not surprising in view of the smallness of the cosphere effects for these species in water, when the effects are expressed in terms of contributions per molecule of cosphere water, as in Fig. 11. To interpret these correlations more completely in terms of cosphere effects it is necessary explicitly to allow for the Born-model effects (cf. Frank and Evans[230] and Section 3.3).

6.3. Enthalpies in Nonaqueous Solvents

6.3.1. *Monatomic Ions*

A study of enthalpies in dipolar aprotic solvents is especially interesting since it helps to identify the factors that control the enthalpy in the absence of hydrogen bonding. We can then turn our attention to simple hydrogen-bonding solvents (Section 5.3.5) and finally to the unique case of water. Enthalpies in nonaqueous solvents can be more easily and accurately determined than free energies. Typical data for monatomic cations are shown in Fig. 25. As discussed in Section 2.2, these results cannot be accounted for in terms of the Born model or the Debye–Pauling model. The results for cation transfers between two nonaqueous solvents can be qualitatively understood in terms of an interaction of the ions as Lewis

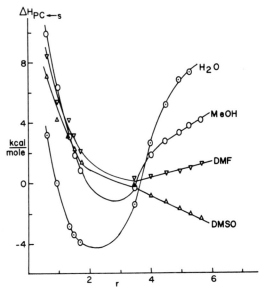

Fig. 25. Single-ion enthalpies of transfer to propylene carbonate from various solvents. For PC ← DMSO and PC ← DMF the values are based on the convention that the enthalpies of transfer of Ph_4As^+ and Ph_4B^- are the same. For PC ← CH_3OH the values are obtained by the $n\text{-}C_nH_{2n+1}OH$ extrapolation procedure. For PC ← H_2O the values are based on the assumption $(Cs^+)_{PC\leftarrow w} = (I^-)_{PC\leftarrow w} - 0.7$ kcal mol^{-1}. Data from Krishnan and Friedman.[473]

acids with the solvents as Lewis bases. However, the order of basicities of the solvents depends on the probe used.[477]

In Fig. 25 the $(X)^H_{PC\leftarrow H_2O}$ transfers of alkali metal ions are much more negative than the $(X)^H_{PC\leftarrow s}$ transfers, where s is a nonaqueous solvent. In order to explain this and also to explain the trend from Li$^+$ to Cs$^+$, contributions from the cosphere state II$_{sb}$, present only in water (Section 5.3.5), seem to be required.[472,473,865]

6.3.2. Alkylammonium Ions

The enthalpies of transfer of various alcohols, hydrocarbons, and alkylammonium ions to PC from other nonaqueous solvents show very regular behavior. Thus there is no evidence of chain–chain interactions, methylene contributions within a chain are additive, and the chain contributions are the same whether they are in primary, secondary, tertiary, or quaternary ammonium ions.[472,475] Also, the interference of the alkyl chains with the formation of hydrogen bonds to solvents is negligible even though interferences between multiple hydrogen bonds are observed.[475]

This is strong evidence that many of the complications in hydration which may be discussed in terms of interference of neighboring cospheres are associated with the branched network of hydrogen bonds in water.

The single-ion values obtained on the basis of the convention $Ph_4As^+ = Ph_4B^-$ are nearly the same as those obtained by an extrapolation procedure similar to eqn. (56) when two aprotic solvents are involved. However, when a hydrogen-bonded solvent such as methanol or water is involved there is no agreement between the conventions.[473] In the case of methanol this disparity was traced to the peculiar solvation of Ph_4B^-, while Ph_4As^+ and Ph_4P^+ behave more as would be expected from the solvation of benzene. Thus the use of reference solvents has helped in identifying the influence of the details of the charge distribution in aromatic groups on solvation properties.

Now turning to the study of $PC \leftarrow H_2O$ transfers as a method of studying hydration of polyatomic ions, we notice a remarkable feature of the $(X)_{PC \leftarrow H_2O}^{H}$ curve as a function of ion size included in Fig. 25: The data depend nonlinearly on the number of methylene groups in the chains, just as in the case of alcohols (see Fig. 1). This nonadditive effect is not always clear in other measurements, for example, in the data for $V_{X(aq)}$ (see Section 4.1.2). However, it is exhibited in other properties, such as the self-diffusion coefficient of water in the presence of these salts (Section 4.1.2), and in heat capacities (Section 4.2.2).

The results can again be understood on the basis of the chemical model. It indicates that the contributions from state II_{sb} are dominant for Me_4N^+, something of a balance between state II_{sb} and state II_{al} for Et_4N^+, while state II_{al} is dominant for Pr_4N^+, Bu_4N^+, and Am_4N^+ (cf. Section 5.3).

In the case of partially substituted alkylammonium ions the data are less complete in PC and hence we turn to $DMSO \leftarrow H_2O$ transfers to study their hydration. If it is assumed that $(X)_{DMSO \leftarrow H_2O}^{H}$ results from independent action of the alkyl chains, then the hydrogen bond contribution to $(X)_{DMSO \leftarrow H_2O}^{H}$ can be obtained.[475] The data thus obtained are plotted as a function of the number of N—H bonds for various chain lengths in Fig. 26. If the N—H$\cdots$O bonds were independent of each other and of alkyl groups and if the N—H$\cdots$O bond strength is given by R_3NH^+ data, then all the data would fall on line I. On the other hand, if the N—H$\cdots$O bond strength is given by NH_4^+ datum, then all the data would fall on line II. Although

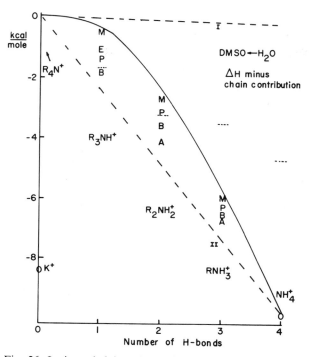

Fig. 26. Ionic enthalpies of transfer to DMSO from water deduced on the assumption that the alkyl chains make independent contributions. Data points: M = methyl, E = ethyl, P = propyl, B = butyl, A = amyl. (———) Rough extrapolation to zero chain length. (I) data would fall on this line if H bonds were independent of each other and of R groups, with H-bond strength given by R_3NH^+ data. (II) Same as I but with H-bond strength given by NH_4^+ datum. (· · · ·) DMSO ← PC transfers extrapolated to zero chain length. Data from Krishnan and Friedman.[475]

the data do not lie on line II, it certainly represents the dominant effect in the transfer.

Evidence that the enthalpy effect of chain–chain interactions on their cospheres is small is evident from Table XXIII.

The interference of the cospheres of alkyl chains with the cospheres of the N—H· · ·O bonds is manifested by the spread of data points for various chains but fixed number of N—H· · ·O bonds. The interference of the cospheres of alkyl groups and polar end groups, which contributes to the spread of data points, would be expected to be largest when there is

TABLE XXIII. Evidence for the Negligible Influence of Chain–Chain Interactions in the Hydration of Alkylammonium Ions[475]

Ion, X	$(X)^H_{PC \leftarrow H_2O}$		$(X)^H_{PC \leftarrow DMSO}$	
	Obs.[a]	Calc.[b]	Obs.[a]	Calc.[b]
Et_3MeN^+	−0.70	−0.85	−0.77	−0.69
Et_3PrN^+	0.81	0.83	−1.04	−0.94
Bu_3MeN^+	2.39	2.32	−1.39	−1.30
Bu_3PrN^+	4.14	3.99	−1.51	−1.55

[a] Observed enthalpy of transfer.
[b] Enthalpy of transfer calculated from data for symmetrical tetraalkylammonium ions, assuming additive contributions of alkyl chains.

only one N—H $\cdots$ O bond and to change uniformly with chain length. Most of the data are consistent with this and also suggest that increasing chain length either strengthens N—H $\cdots$ O hydrogen bonds to DMSO more than those to water or weakens the hydrogen bonds to water more than those to DMSO.

The data further suggest that the mutual interaction of hydrogen bonds is more destructive in water as solvent than in DMSO. The magnitude of this enormous effect is shown by considering the dotted lines in Fig. 26. These represent the data for PC $\leftarrow$ DMSO transfers after extrapolation to zero chain length. They suggest an average interference of about 2 kcal mol^{-1} per hydrogen bond. Interference effects of this order have been found in recent quantum mechanical calculations with three water molecules.[310]

Similar effects have been observed in the partial molal volumes at infinite dilution discussed in Section 4.1.2. Also the rate constant k_H for breaking the N—H $\cdots$ O hydrogen bond in aqueous ammonium salts decreases in the sequence RHN_3^+, $R_2NH_2^+$, R_3NH^+, as is consistent with the finding that the N—H $\cdots$ O hydrogen bond gets stronger in this order.[296,298] Evidence for chain–hydrogen bond interference effects are also found in salting out[175] and $^{79}Br^-$ relaxation[521] studies.

It should be emphasized that the selection of a suitable reference solvent has made it possible to identify many of these peculiar effects in the aqueous solutions. Similar solvation studies of biological compounds might well be helpful in understanding more about the role of hydration in biological processes.

6.4. Heat Capacities in Nonaqueous Solvents

There are practically no data available for solute partial molal heat capacities at infinite dilution in nonaqueous solvents except in methanol.[181,182,559,560] The data for $C_{X(\text{MeOH})}$ are given in Table XXIV. The influence of solvents on the three-dimensional hydrogen-bonded network structure of water discussed in Section 5.3.5 is also evident in the heat capacity data. The correlation between $C_{X(\text{MeOH})}$ and $C_{X(\text{H}_2\text{O})}$ for alkali halides is not clear. However, the $C_{X(\text{MeOH})}$ data for R_4NX salts are much more negative than $C_{X(\text{H}_2\text{O})}$ and the $(X)^C_{\text{MeOH}\leftarrow\text{crystal}}$ values are negligible. This is comparable with Arnett and Campion's[21] value for

$$(\text{Bu}_4\text{NBr})^C_{\text{EtOH}\leftarrow\text{crystal}} = 10 \text{ cal mol}^{-1} \text{ deg}^{-1}.$$

Mastroianni and Criss[559,560] have studied $C_{X(\text{MeOH})}$ for $X = \text{NaClO}_4$ and Me_4NBr in methanol as a function of temperature and find $\partial C_{X(\text{MeOH})}/\partial T$ to be negative. The corresponding functions in water are positive. The difference may be attributed to the cosphere state II_{sb} being present in water but not in methanol.

7. IONIC HYDRATION AND EXCESS PROPERTIES

The basis for the relation between ionic hydration and the excess thermodynamic functions of the solutions has been outlined in Section 1.5. The underlying ideas were developed and applied in a qualitative way by many authors, beginning with Frank and Robinson[231] and represented in recent times by Wood and Anderson[851] and Desnoyers et al.[172] The usual interpretation is that the effect of overlap of cospheres is destructive: Part of the water in the cospheres is changed to normal bulk water as two ions approach each other (cf. Fig. 4a).

TABLE XXIV. Partial Molal Heat Capacities in Methanol

Solute	LiCl	LiBr	LiI	NaBr	NaI	KI
$C_{X(\text{MeOH})}$, cal mol^{-1} deg^{-1}	-27.7[181]	-24.1[181]	26.4[181]	-54.3[181]	-2.9[181]	-28.2[181]

Solute	NaClO$_4$	Me$_4$NBr	Bu$_4$NBr	Ph$_4$PCl	Ph$_4$AsCl	NaBPh$_4$
$C_{X(\text{MeOH})}$, cal mol^{-1} deg^{-1}	23.0[559]	14.0[560]	119.5[412]	105.0[412]	121.0[412]	132.5[412]

For nonelectrolytes such as alcohols the situation is quite clear. The deviation of the alcohol partial pressure from Henry's law in aqueous solutions, at least at low alcohol concentrations and excepting methanol, tends to be negative, as though two alcohol molecules dissolved in water attract each other.[459]* This effect may be attributed to the mutual destruc-

---

* The analysis of these data is rather subtle because they provide an example of the situation mentioned as a possibility in Section 1.3, namely the qualitative interpretation of the data depends upon the choice of concentration scales. In this case we are concerned with excess functions, while in Section 1.3 the concern was with solvation functions, but otherwise the problem is the same. The present example has several interesting aspects and seems worth detailed discussion.

The excess free energies of the aqueous alcohol solutions are positive[234] and at low mole fraction x of the alcohol the partial pressure of the alcohol exceeds the Raoult's law value xP_{Alc}^{pure}. This, of course, is a manifestation of a solvation effect; when the liquid alcohol is dissolved in a large excess of water the dominant contribution to the excess free energy is due to the reverse of the process in eqn. (67), for type II_{al} cospheres. It is assumed that hydrogen-bonding contributions in the initial and final states nearly cancel.

The same data show that at small x and excepting methanol the partial pressure of the liquid alcohol shows negative deviations from Henry's law. (See, for example, the graph for water–ethanol and the discussion in Guggenheim's text, Fig. 6.1.[300]) This negative deviation implies that alcohol molecules attract each other in solution: hydrophobic bonding, since it becomes larger the longer the alkyl chain. However, the conclusion may be misleading. For, by calculating the osmotic pressure π from the excess free energy data and expressing the result in the form, where c is the alcohol concentration in molecules per unit volume,

$$\pi/ckT = 1 + B_2 c + B_3 c^2 + \cdots$$

thus showing the deviations from van't Hoff's law, one finds that the coefficient B_2 is positive for ethanol through butanol. Positive deviations from van't Hoff's law imply that the solute molecules repel each other, just as positive deviations from the ideal gas law imply that the repulsive part of the interaction between pairs of molecules is dominant.

The paradox is entirely due to the fact that the sign of the deviations from Henry's law or from the van't Hoff equation depend on the concentration scale when, as in these cases, the deviations are relatively small. If one believes that concentrations in units of moles per liter or particles per unit volume are somehow more fundamental than concentrations in units of mole fraction, then the negative deviations from Henry's law on a mole-fraction scale may be attributed to the fact that the mean molar volume of the solutions increases with x, at least when x is small. With this philosophy the hydrophobic bonding is gone.

However, an explicit formula for B_2 in terms of the solute–solute forces in the dilute solutions is given in eqn. (20). Using this equation and the data for the aqueous alcohols, Kozak et al.[459] were able to identify two contributions to B_2, one due to the actual size of the solute molecule and a remainder. The first part is positive and may

tion of the type II_{al} cospheres on the alkyl chains when the alcohol molecules come so close together that their cospheres overlap. For these cospheres ΔG for the following process is negative:

$$\text{water in cosphere state} \rightarrow \text{water as pure bulk liquid} \qquad (67)$$

as one may deduce from consideration of the solvation data for solutes with alkyl groups.[234,420]

Corresponding to this free energy effect is the concentration dependence of the partial molal enthalpy of the alcohols in water, which is positive,* reflecting the fact that ΔH for eqn. (67) is positive for type II_{al} cospheres.[471] Of course, the assumed consistency of the excess and solvation properties depends entirely on the model in which overlap tends to destroy the cospheres, as in Fig. 4(a).

The same relation of excess properties to solvation properties seems to apply to ionic solutions, but now the interpretation of the data is greatly complicated by various factors as emphasized in Section 1.5. By suitably accurate statistical mechanical methods, all of the complicating factors may be taken into account, as described briefly in Section 1.5. It seems appropriate to summarize some of the main results of the first investigations of this kind.[244,660,661,670]

The first conclusion is that one cannot find a model which fits the excess-function data uniquely well. This reflects a general rule. For a simpler example, one cannot establish the "correct" pair potential for the interaction of argon atoms by comparing the experimental thermodynamic data with the calculated behavior of various models. On the other hand, in the calculations with ionic solution models having a term in the pair potential to represent the overlap effect it is found that some aspects of the

---

be estimated on the basis of chemical bond lengths or, as these authors did, from the partial molal volumes of the solutes at infinite dilution. The result is that the remainder is negative, and more so the longer the alkyl chain, thus reflecting the hydrophobic bonding of pairs of alcohol molecules in water.

In this case the same final conclusion is reached by qualitative analysis of the deviation from Henry's law on the mole fraction scale and the quantitative analysis of B_2. In fact, the repulsive or excluded volume contribution to B_2 seems to be introduced by the change from x to c as the concentration scale! Whether this is generally the case has not been established.

Finally it may be noted that the pitfalls in qualitative interpretation of the thermodynamic data are, of course, much greater in ionic systems.

* In the notation of eqn. (4), where now c is taken to be the molality, the coefficient B_2 is negative for the alcohols (excepting methanol) in water, while the corresponding coefficient $\partial B_2/\partial(1/T)$ in the expression for the partial molal enthalpy H_A is positive.

parameters of the overlap effect are independent of the more arbitrary features of the model.[660] In the following discussion only these aspects are emphasized.

The parameters of the overlap effect have been obtained for models which fit the data for the excess functions and have an overlap term corresponding to the effect pictured in Fig. 4(a), assuming that the various distance parameters in the potential correspond to what may be estimated from crystal radii and that the cospheres are all one water molecule in thickness. Then the overlap effect may be expressed in terms of the thermodynamic coefficients for eqn. (67) expressed per mole of water changing state. The coefficients are A_{ij}, the increase in Helmholtz free energy, and E_{ij}, the increase in energy [cf. eqn. (23)].

In the case of the models which have been studied there is only one cosphere about each ion. In view of the evidence that it would be more realistic to allow for two cospheres about each ion, as in Fig. 11, it is clear that the model parameters A_{ij} and E_{ij} for the overlap of ions of species i and j each expresses a superposition of several effects. Nevertheless, the trends in the parameters can be understood in terms of the chemical model of hydration described in Section 2.5.

Typical results of the calculations are given in Table XXV. The entropy changes are not independent parameters but are determined by the equation

$$A_{ij} = E_{ij} - TS_{ij} \tag{68}$$

and are presented in Table XXV because the entropy changes are emphasized in Section 2.5 and in other discussions.[231]

The A_{+-} values for the chlorides become increasingly negative from LiCl to CsCl, apparently reflecting the diminishing positive contribution of type I_c in the presence of a negative contribution from type II_{sb} cospheres. It is remarkable that the corresponding S_{+-} parameters are found to be negative; this apparently shows that the dominant contribution is due to the overlapping of the II_{sb} cospheres because the opposite sign would be expected from the overlap of I_c cospheres.

The data for the potassium halides seem amenable to a similar interpretation; they reflect a diminishing positive contribution of I_a cospheres in the presence of a negative contribution of type II_{sb} cospheres.

The A_{+-} values for the tetraalkylammonium chlorides must be interpreted in terms of the increasing contribution of type II_{al} cospheres with increasing ion size, as established by the solvation properties. To interpret the trend in the parameters in terms of destruction of the II_{al} cospheres

TABLE XXV. Gurney Cosphere Parameters for Some Aqueous 1-1 Electrolytes Calculated from a Particular Model[a]

	LiCl	NaCl	KCl	RbCl	CsCl	Me$_4$NCl	Et$_4$NCl	Pr$_4$NCl	Bu$_4$NCl
A_{+-}	50	−47	−80	−90	−110	−135	−65	−20	0
E_{+-}	−100	−180	−152	—	−183	−277	−373	−249	−195
TS_{+-}	−150	−133	−72	—	−73	−142	−308	−229	−195
A_{++}	0	−50	−100	−100	−100	−135	−190	−194	−182
E_{++}	1500	0	0	0	0	0	0	0	0

	KF	KCl	KBr	KI	Pr$_4$NF	Pr$_4$NCl	Pr$_4$NBr	Pr$_4$NI
A_{+-}	−19	−80	−86	−95	235	−20	−70	−125
E_{+-}	−80	−152	−178	−196	809	−249	−307	−283
TS_{+-}	−60	−72	−92	−101	574	−229	−237	−158

[a] For Aqueous Solutions at 25°C. Values in cal mol^{-1} of water. From Refs. 660, 661; $A_{--} = 0$ assumed throughout. In the calculations with models for the tetraalkylammonium halides it was found possible to fit the heat of dilution data by adjusting only E_{+-}, giving the results shown here. Of course, it would he more physically reasonable to have a contribution from E_{++} as well, in view of the importance of A_{++} in fitting the free energies, but this was not done in the initial study.

due to overlap would be inconsistent with the results for A_{++} parameters, which show a strongly negative contribution due to overlap of II_{al} cospheres. In the case of the A_{+-} parameters, then, one is forced to assume something more complicated than mutual destruction of the cospheres. Perhaps this reflects a structure in which the anion is incorporated in the II_{al} cosphere much in the same way it may be incorporated in the water structure in the clathrate hydrates.[408] Such structures have been assumed in solution before, for example, by Lindman *et al.*[520]

The E_{+-} and TS_{+-} parameters for the tetraalkylammonium chlorides also do not show the behavior expected if they are dominated by the effect of disrupting type II_{al} cospheres. (One would have E_{ij} positive and somewhat smaller than TS_{ij}, see Section 2.5.) On the other hand, there is not much basis for predicting the sign of the heat and entropy effect in A_{+-} if the anion is incorporated in the II_{al} cosphere.

Now turning to the A_{++} values, one finds a uniform trend with ion size. For the smaller ions this is expected on the basis of a decreasing positive contribution from I_c overlap together with a large negative contribution from II_{sb} overlap. Passing to the tetraalkylammonium ion series, one finds A_{++} continuing to decrease; this may be interpreted as due to increasing contribution from II_{sb} overlap, i.e., from the so-called hydrophobic bonding.[420]

Thus it seems quite possible to formulate a consistent picture of the overlap effects in terms of the chemical model for hydration, although to do so becomes more of a challenge as the parameters for a wider variety of solutions are considered. It is also important, just as in learning about solvation properties, to consider the phenomena in solvents besides water. Thus Wood and his students[465,855] have recently shown that some of the excess thermodynamic function phenomena in aqueous solutions which have been interpreted in terms of hydration phenomena are mimicked in solutions of the same electrolytes in N-methylacetamide. Nevertheless, it seems clear that the cosphere-overlap phenomena play a principal role in determining the species-specific aspects of thermodynamic excess functions in moderately dilute solutions.

CHAPTER 2

Thermodynamics of Aqueous Mixed Electrolytes*

Henry L. Anderson[†]

University of North Carolina at Greensboro
Greensboro, North Carolina

and

Robert H. Wood

University of Delaware
Newark, Delaware

1. INTRODUCTION

Studies of the thermodynamic properties of aqueous mixed electrolyte solutions have been very useful in increasing our understanding of the nature of specific ion interactions in solution. From both an experimental and theoretical point of view it is convenient to treat these mixtures in terms of the way in which the properties of the mixture differ from the properties of the component pure electrolyte solutions. Thus one studies the change on mixing of the desired thermodynamic property (e.g., $\Delta_m G^E$, $\Delta_m H^E$, $\Delta_m V^E$, etc.) at constant total molal ionic strength.

For the purposes of this discussion it is convenient to treat a binary electrolyte mixture made by mixing $y_A m$ moles of MX (molality $= m$) with $y_B m$ moles of NX (molality $= m$) to get m moles of mixture (molality $= m$) in 1 kg of water:

$$y_A m \mathrm{MX}(m) + y_B m \mathrm{NX}(m) \rightarrow m \cdot \text{mixture}(m) \tag{1}$$

* This study was aided by grants from the Office of Saline Water, U.S. Department of the Interior.
† Prof. Anderson was killed in a tragic automobile accident on May 7, 1972.

119

In this mixing scheme y_A and y_B are the mole fractions of the component electrolytes in the final mixture. This example represents a mixture with a common anion, but the mixing scheme is general and does not depend on a common ion.

One can now write the excess free energy change during the mixing as

$$\varDelta_m G^E / W = [G^E(\text{mixture}) / W] - y_A [G^E(\text{MX}) / W] - y_B [G^E(\text{NX}) / W] \quad (2)$$

where $G^E(\text{mixture})$ is the excess free energy of the mixture, $G^E(\text{MX})$ and $G^E(\text{NX})$ are the excess free energies of the component pure electrolytes solutions, and W is the weight of solvent.* It is important to note that all of the free energies in eqn. (2) are at the same molal ionic strength.

Similar equations can be written for the heat and excess volume of mixing,[†]

$$\varDelta_m H^E / W = m\phi_L(\text{mixture}) - y_A m\phi_L(\text{MX}) - y_B m\phi_L(\text{NX}) \quad (3)$$

$$\varDelta_m V^E / W = m\phi_V(\text{mixture}) - y_A m\phi_V(\text{MX}) - y_B m\phi_V(\text{NX}) \quad (4)$$

Numerous experimental determinations of $\varDelta_m G^E$, $\varDelta_m H^E$, and $\varDelta_m V^E$ at constant total molal ionic strength have been made for aqueous binary salt mixtures (i.e., MX–NX–H_2O) and Fig. 1 gives the heat of mixing of NaCl with KCl as measured by Young and Smith.[875] Almost all of the experimental results behave similarly to those shown in Fig. 1; the heat of mixing data very closely conform to a parabola when plotted versus the mole fraction of one salt component and the data are conveniently represented by an expression of the form[876]

$$\varDelta_m H^E / W = y_A y_B [A + (y_A - y_B)B] \quad (5)$$

where A and B are dependent on the total molal ionic strength but independent of mole fraction. When $B = 0$ the equation represents a perfect parabola.

---

* There has been considerable confusion in the literature in recent years with respect to the units in mixed electrolyte solutions. The problem has mainly been with the units of calories per kilogram of solvent (used most frequently) and calories per mole. To attempt to alleviate this problem in this work, G and H will be in calories and W will represent the weight of solvent in kilograms. Thus, G/W is in calories per kilogram of solvent.

† The units of ϕ_L are calories per mole and those of V^E are milliliters per mole.

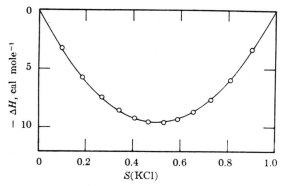

Fig. 1. Heat of Mixing of NaCl–KCl–H$_2$O at $m = 1.0$ as a function of mole fraction. From Ref. 875. Reproduced with the permission of *J. Phys. Chem.*

In terms of a modification of the nomenclature of Friedman[240,242] the excess properties are given by

$$\Delta_m G^E / W = RTI^2 y_A\, y_B [g_0 + (y_A - y_B)g_1] \tag{6}$$

$$\Delta_m H^E / W = RTI^2 y_A\, y_B [h_0 + (y_A - y_B)h_1] \tag{7}$$

$$\Delta_m V^E / W = I^2 y_A\, y_B [v_0 + (y_A - y_B)v_1] \tag{8}$$

where I is the molal ionic strength. In terms of Fig. 1, RTh_0 would be a measure of the height of the parabola at $y = 0.5$ and RTh_1 would be a measure of the skew or asymmetry of the curve. The coefficients in eqns. (6) and (8) have similar meanings.

Normally the experimental data are accurately described by the parabolic and first-order skew terms, with the skew term seldom contributing more than 10% to the overall effect.

2. THEORETICAL FRAMEWORK

2.1. Symmetric, Common-Ion Mixtures

Theoretically, the simplest mixed electrolyte solution to treat is that of a mixture containing electrolytes of the same charge type (e.g., MX–NX; MX$_2$–NX$_2$; M$_2$X–N$_2$X; etc.) and this discussion will be limited to this type of mixture.

Friedman[240,242] has written extensively on the properties of mixed electrolytes and the results of his studies have provided an excellent starting point for considering the types of interactions to expect in an electrolyte mixture. His treatment is based on Mayer's ionic cluster expansion,[562] which basically treats the interactions of ions with their ion atmospheres. These interactions are evaluated and represented in the form of cluster integrals (denoted by the symbol B in this work). The theory assumes a mixing at constant molar ionic strength and constant water activity; thus for a comparison with experimental results (normally obtained at constant molal ionic strength and 1 atm) corrections are needed for the different concentration scales. Friedman[240] has shown that for 1–1 mixtures these correction terms are small and can be neglected for a first approximation. Our comparison of theory and experiment will be focused primarily on the concentration dependence of the various interactions so that the neglect of the correction is justified.

The coefficient describing the excess free energy of mixing for the MX–NX–H_2O mixture can be expressed in the form

$$g_0 = (B_{MN} - B_{MM} - B_{NN}) + I(B_{MNX} - P_{MMX} - B_{NNX}) \qquad (9)$$

where g_0 is the interaction parameter in eqn. (6) and the B's are the cluster integrals corrected for the change in concentration and pressure effects. The subscripts of the cluster integrals reflect the various ion interactions and depend on the composition only through the ion atmosphere (i.e., the ionic strength). As a natural result of his calculations using a primitive model, Friedman found that for common-ion mixtures the leading term in calculating g_0 was the like-charged pair interaction term and that the second and smaller term was for triplet interactions involving two ions of like charge with the common ion [e.g., for the mixture MX–NX see eqn. (9)].

Recently, Robinson et al.[695] treated the excess free energy of mixing in an alternative way. They used an equilibrium constant technique for calculating the various ionic interaction contributions to $\Delta_m G^E$. For example, the specific interaction of M with N is represented as

$$M + N \rightleftharpoons MN$$

where MN denotes the ion pair formed by M and N. The equilibrium constant for this reaction is

$$K_{MN} = \left(\frac{m_{MN}}{m_M m_N} \right) \gamma_{MN} / \gamma_M \gamma_N \qquad (10)$$

Considering only pair and triplet interactions and excluding interactions of three ions of the same sign, they derived an expression for $\Delta_m G^E$,

$$
\begin{aligned}
\Delta_m G^E/(WRT) = y_A\,y_B m^2\{&[K_{MM}(\gamma_M{}^2/\gamma_{MM}) + K_{NN}(\gamma_N{}^2/\gamma_{NN}) \\
&- K_{MN}(\gamma_M\gamma_N/\gamma_{MN})] \\
&+ m[K_{MMX}(\gamma_M{}^2\gamma_X/\gamma_{MMX}) + K_{NNX}(\gamma_N{}^2\gamma_X/\gamma_{NNX}) \\
&- K_{MNX}(\gamma_M\gamma_N\gamma_X/\gamma_{MNX})]\}
\end{aligned}
\tag{11}
$$

It should be noted that this treatment accounts for the pair interactions with their ion atmospheres *via* an equilibrium constant with activity coefficients. In terms of Friedman's nomenclature, g_0 is represented as

$$
\begin{aligned}
g_0 = &[K_{MM}(\gamma_M{}^2/\gamma_{MM}) + K_{NN}(\gamma_N{}^2/\gamma_{NN}) - K_{MN}(\gamma_M\gamma_N/\gamma_{MN})] \\
&+ m[K_{MMX}(\gamma_M{}^2\gamma_X/\gamma_{MMX}) + K_{NNX}(\gamma_N{}^2\gamma_X/\gamma_{NNX}) \\
&- K_{MNX}(\gamma_M\gamma_N\gamma_X/\gamma_{MNX})]
\end{aligned}
\tag{12}
$$

This shows that each cluster integral in Friedman's nomenclature is replaced by an equilibrium constant and an activity coefficient ratio with the sign changed ($-K_{MN}\gamma_M\gamma_N/\gamma_{MN}$ replaces B_{MN}, etc.) in the alternative derivation. In terms of the ionic interactions present upon mixing the results are identical with Friedman's treatment.

One would intuitively expect the results given in eqns. (9) and (12) since upon mixing MX with NX at constant total molal ionic strength, the anion concentration does not change and therefore a given cation would continue to see the same concentration of anion around it. However, the M$\cdots$M and N$\cdots$N interactions should decrease by a simple dilution factor and the M$\cdots$N interactions should correspondingly appear. These are precisely the pair interaction terms represented in eqns. (9) and (12). The same reasoning can be applied to the triplet interactions. In addition, one would intuitively expect triplet interactions between three like-charged ions to be unimportant and these terms have been neglected in the present treatment. The inclusion of such interactions in the theoretical expression produces asymmetry in mole fraction (i.e., cannot be collected as a simple factor of $y_A\,y_B$). The experimental evidence has justified the omission of these interactions in accounting for g_0 since they contribute roughly the same amount to both g_0 and g_1 and the g_1 terms are experimentally small.

An interesting result of the theoretical calculation of g_0 is the limiting law for the excess free energy of mixing at vanishing total molality. First, at $m = 0$, g_0 must be accounted for by pair interactions only [see eqns. (9) and (12)] and thus is a direct measure of these interactions. Another way of expressing this is to look at the limiting law predicted by eqn. (11). If the activity coefficients are calculated by an extended Debye–Hückel expression, eqn. (11) can be rewritten as

$$\Delta_m G^E / (WRTy_A\, y_B m^2)] = (K_{MM} + K_{NN} - K_{MN})\exp[2Am^{1/2}/(1 + Bam^{1/2})] + \cdots$$

(13)

Then

$$\lim_{m \to 0} \frac{d\ln[\Delta_m G^E/(WRTy_A\, y_B m^2)]}{dm^{1/2}} = \frac{d(\ln g_0)}{dm^{1/2}} = 2A$$

(14)

This limiting law is identical to that derived by Friedman[242] from the Mayer cluster expansion method.

Experimentally, a test of the relation in eqn. (14) is beyond current experimental accessibility in that the excess thermodynamic properties fall off as the square of the total ionic strength, and below 0.1 m the best data are not accurate enough to test the limiting law.

2.2. Symmetric, Multicomponent Mixtures

The theoretical results for common-ion mixtures can be readily extended to include non-common-ion mixtures provided the electrolytes are all of the same charge type (charge-symmetric mixtures). In fact, the results have been used by Wood and Anderson[849] for the prediction of the properties of charge-symmetric mixtures and by Reilly and Wood[679] for the prediction of the properties of charge-asymmetric mixtures. For charge-symmetric mixtures the component pure electrolytes can be chosen in such a way that there is no change in the interactions of two ions of opposite sign. In addition, all of the like-charged pair and triplet interactions (except for three ions of the same sign) can be measured by common-ion mixings and the results of these measurements can be used to predict the properties of a many-component mixture. For the excess free energy of a mixture, Wood and Anderson derived an expression of the form

$$[G^E/W](\text{mixture}) = \sum_{M,X} y_M\, y_X [G^E(MX)/W] + [\Delta_m G^E/W]$$

(15)

where

$$\Delta_m G^E/W = 4 \sum_{M>N,X} y_M\, y_N\, y_X [\Delta_m(G^E(MX - NX)/W)] +$$
$$4 \sum_{M,X>Y} y_M\, y_X\, y_Y [\Delta_m(G^E(MX - MY)/W)]$$

(16)

where $\Delta_m(G^E(MX - MY)/W)$ is measured at $y_A = y_B = \frac{1}{2}$. This equation indicates that to predict the behavior of mixed electrolytes, it is only necessary to measure the properties of pure electrolyte solutions and the properties of two-component common-ion mixtures. This is because it is possible to prepare any mixture of electrolytes, no matter how complex, by a series of common-ion mixtures. This is most easily seen by considering the mixing scheme given below.

Mixing scheme: MNOXYZ. Cations, M, N, O; anions X, Y, Z.

$$y_M + y_N + y_O = 1; \qquad y_X + y_Y + y_Z = 1$$

Step 1: Mix $y_M y_X m$ moles of MX, $y_N y_X m$ moles of NX, and $y_O y_X m$ moles of OX to give $y_X m$ moles of MNOX:

$$y_M y_X m MX + y_N y_X m NX + y_O y_X m OX \rightarrow y_X m MNOX$$

Step 2: Similarly mix MY, NY, and OY to give MNOY:

$$y_M y_Y m MY + y_N y_Y m NY + y_O y_Y m \rightarrow y_Y m MNOY$$

Step 3: Mix MZ, NZ, and OZ to give MNOZ:

$$y_M y_Z m MZ + y_N y_Z m NZ + y_O y_Z m \rightarrow y_Z m MNOZ$$

Step 4: Treat solutions in steps 1, 2, and 3 as common-cation solutions and mix them (the common cation is a constant mixture of M, N, and O):

$$y_X m MNOX + y_Y m MNOY + y_Z m MNOZ \rightarrow m MNOXYZ$$

In this scheme an anion is chosen and all of the cations in the mixture are mixed with this common anion. This is repeated in turn for each anion. The resulting mixtures contain the same cations, all at the same concentrations, but different anions. If these are then mixed, the mixing process is a mixing in which there is a common-cation mixture. The amounts of the various components in this scheme are the same as those in the equation for the prediction of charge-symmetric mixtures [eqn. (16)]. It is easy to see from this mixing scheme why the interactions between a single cation and a single anion do not change in any of the mixings. In a common-anion mixing the concentration of the anion does not change during the mixing. Thus the concentration of anions around a given cation is constant and therefore its interaction with the anions is constant. Similar reasoning holds for common-cation mixings, even when the common cation is in reality a constant mixture of cations. This reasoning shows that the first term in eqn. (16) accounts for all of the oppositely charged pairwise interactions in the mixture, and that all of the new interactions in the mixture are the result of common-ion mixings.

2.3. Nomenclature

Many different sets of nomenclature have been used in expressing the excess thermodynamic properties of mixed electrolyte solutions. The Harned coefficients have been used extensively, especially in reporting the results of electromotive force studies and some of the early isopiestic studies.* In terms of the Harned coefficients the excess free energy of mixing is given by the relation[†]

$$\Delta_m G^E/W = -2.303\,RTm_2m_3[\alpha_{23} + \alpha_{32} + 2(m_3\beta_{23} + m_2\beta_{32})$$
$$+ \tfrac{2}{3}(\beta_{23} - \beta_{32})(m_2 - m_3) + \cdots] \tag{17}$$

for 1–1 mixtures. The corresponding expressions for $\Delta_m H^E$ and $\Delta_m V^E$ can be written by taking the proper derivative of the Harned coefficients in eqn. (17).

Scatchard and Prentiss[701,721] introduced an alternative method of treating mixed electrolyte solutions, where the excess free energy of mixing is given by[‡]

$$\Delta_m G^E = RTy_A\,y_B I^2[(B^{(0)}_{AB}/I) + (y_A - y_B)(B^{(1)}_{AB}/I)] \tag{18}$$

where

$$B^{(0)}_{AB}/I = b^{(0)}_{AB} + \tfrac{1}{2}b^{0,1}_{AB}I + \cdots \tag{19}$$

The present authors have adopted the Friedman nomenclature, which is used throughout this work. The reason for adopting the Friedman notation is that it allows for a ready comparison of experiment with the theoretically predicted ionic interactions.

It is useful to make the comparison between the different sets of nomenclature. It is only necessary to make this distinction for the excess free energy of mixing and the other excess properties can be compared by analogy.

By comparing eqns. (6) and (17), one can write

$$g_0 = -2.303[\alpha_{23} + \alpha_{32} + I(\beta_{23} + \beta_{32})] \tag{20}$$

$$g_1 = -\tfrac{1}{3}2.303I(\beta_{23} - \beta_{32}) \tag{21}$$

---

* See, for example, the reviews in Refs. 317, 318, and 694.
† For a derivation see Ref. 510.
‡ For a review see Ref. 718. These equations, slightly modified and recast in modern nomenclature, are given by Scatchard.[717a]

For mixtures obeying Harned's rule (i.e., $\beta_{23} = \beta_{32} = 0$) then, we have

$$g_0 = -2.303(\alpha_{12} + \alpha_{21})$$

This would mean that Harned's rule implies that like-charged pair interactions account for the excess free energy of mixing and RTg_0 would be independent of ionic strength.

By comparing eqns. (6) and (18), one can write

$$g_0 = B_{AB}^{(0)}/I \quad \text{and} \quad g_1 = B_{AB}^{(1)}/I \tag{22}$$

As indicated earlier, g_1 is a skew term and for most mixtures can be neglected. Thus in terms of eqn. (19),

$$g_0 = b_{AB}^{(0)} + \tfrac{1}{2}b_{AB}^{0,1}I + \cdots \tag{23}$$

For many mixtures only the first two terms of eqn. (23) are needed to fit the experimental data; thus $b_{AB}^{(0)}$ is analogous to the like-charged pair cluster integrals and $b_{AB}^{0,1}$ to the triplet cluster integrals of eqn. (9).

3. EXPERIMENTAL TECHNIQUES

Most determinations of the excess free energy of mixing have been calculated from isopiestic,[11,142,518,544,621,637,688-690,700-702] freezing point lowering,[719-721] or emf cell measurements[314,315] of mixed electrolyte solutions. Each experimental method requires considerable treatment of the data and this has been reviewed in several standard reference works.[317,318,694] The experimental methods essentially involve measurements of the partial molal free energy of one component in the entire mixture (i.e., not the excess free energy of mixing). To obtain the partial molal free energy of the other component, some form of the Gibbs–Duhem equation is generally used. The procedure then requires some method of subtracting out the pure component electrolyte contribution to obtain the excess free energy change due to mixing. To appreciate the problems involved, Rush and Robinson's reevaluation of the NaCl–KCl–H_2O system is worth reading.[702] It is difficult to evaluate the reliability of the $\Delta_m G^E$ calculations but the results at $I = 1.0$ are usually good to about 2–10%. Below this concentration the best data probably derive from the freezing point measurements of Scatchard and Prentiss.[721]

The heat of mixing has an advantage over the free energy in that it can be measured directly, using a mixing scheme as shown in eqn. (3). A number of investigators have measured heats of mixing with a variety of standard solution calorimeters using either thermocouples[849,854,866,876] or thermistors[749,750] having a sensitivity in the 1×10^{-6} to 5×10^{-5}°C sensitivity range. More recently a flow calorimeter[413] has been used to measure heats of mixing at constant molar concentration. The errors are directly related to the calorimetric capability and usually range from about 0.1–1.0% at $I = 1.0$ to about 15–30% at $I = 0.1$.

The excess volume of mixing can also be measured directly. Several investigators[829,830,842] have studied mixtures of electrolytes using the dilatometer described by Wirth et al.[842] The sensitivity is about 1×10^{-4} ml, resulting in an error in $\Delta_m V^E$ at 1 m of about 1% and at 4 m of about 0.2%. The excess volume of mixing has also been calculated by making very accurate measurements of the densities of the initial and final solutions.[838,841]

The change in heat capacity on mixing, $\Delta_m C_P$, has been determined for a number of electrolyte mixtures from heats of mixing at several temperatures.[13,14,748–750]

4. EXPERIMENTAL RESULTS AND DISCUSSION

4.1. Concentration and Common-Ion Dependence

Many of the data on mixtures of electrolytes with a common ion conform to two generalizations. The first, due to Young,[876] is that "heats of interaction between cations are, as a rough approximation, independent of the nature of the anions present." This can now be extended to include free energies and volumes. The second generalization is that for many systems the concentration dependences of RTg_0, RTh_0, and v_0 are small. Wood and Smith[854] have shown that both these generalizations are expected if the pairwise interactions of like-charged ions are large compared to the interactions between three ions. This is most easily seen by referring to eqn. (12). If like-charged pair interactions predominate over triplet interactions, then the first term on the right-hand side of this equation is the most important one and the only concentration dependence in this term is due to the activity coefficient ratios, so that a rather small concentration dependence will be expected for this term except at very low concentrations, where the Debye–Hückel limiting law applies, and at very high concentrations, where activity coefficients tend to turn up rather sharply. Since the first term on the right-hand side of eqn. (12) is the same no matter what the nature of the common ion in the mixing, it is expected that if this term predominates, the experimental results will be independent of the particular anion chosen for the measurement. Reference to eqn. (12) also shows that the first generalization, i.e., the thermodynamic property is independent of the common ion, must hold exactly for all charge-symmetric common-ion mixtures as the ionic strength is reduced to zero. This is because the contribution of the second term on the right-hand side of this equation becomes zero and the activity coefficients are all unity. A second point to be noticed from eqn. (12) is that because of the presence of the activity coefficient ratio in this equation and the presence of at least some triplet interactions in most mixtures, both generalizations are only rough approximations. The equation also indicates that if one of the two generalizations is found experimentally to be true, then it is expected that the other generalization will also hold.

Enough data have been collected covering a wide variety of systems so that a third rough generalization can now be tentatively advanced: A marked concentration dependence of RTg_0, RTh_0, or v_0 is not usually observed unless one of the salts in the common-ion mixture is suspected of existing to some extent as ion pairs or at least has a lower than usual osmotic coefficient. This is just what is expected since a lower osmotic coefficient indicates stronger attractive forces between the anion and the cation and this should also lead to more frequent contacts of an anion with two cations or of a cation with two anions. That is, if the forces of attraction between an anion and a cation are larger than normal, not only will there be more pairwise contacts, but also more three-ion contacts.

A number of heats of mixing at constant molal ionic strength are recorded in Table I. These results reflect the direct measurement of the heat of mixing at the molal ionic strength indicated and cover a concentration range of $I = 1.0$–3. In addition, the results of the measurements of Jolicoeur et al.[413] at constant molar ionic strength are recorded in Table II. An examination of these tables show that most of the results conform to the original observation of Wood and Smith that the concentration dependence of RTh_0 is relatively small (see Fig. 2). The mixtures with the highest percentage change in RTh_0 are HCl–KCl, KCl–CsCl, and NaCl–NaNO$_3$. Of these, two mixtures contain one salt that is thought to have strong cation–anion interactions (CsCl and NaNO$_3$).

Partial molal free energy data of mixed electrolyte solutions have been investigated extensively; however, most of them are not of sufficiently high accuracy for calculations of $\Delta_m G^E$. For this reason the authors have

TABLE I. Heat of Mixing: Concentration Dependence[a]

1-1 Electrolytes

Mixture	$I = 3$	$I = 2$	$I = 1$	$I = 0.5$	$I = 0.2$	$I = 0.1$
HCl–NaCl	83[749,750]	112[749,750]	130[866]	—	—	—
HCl–KCl	−53.1[748]	−42.2[748]	−15[866]	−2[748]b	—	—
LiCl–NaCl	58[749,750]	70[749,750]	84.6[866]	87[749,750,854]	79±5[854]	67±15[854]
LiCl–KCl	−93[507]	−80.4[507]	−64.2[866]	−54.1[854]	−50±4[854]	−39±8[854]
LiCl–CsCl	−229[507]	−215[507]	−192[507]	−172[507]	—	—
NaCl–KCl	−33.5[749,750,802]	−35.4[749,750,802]	−38.3[866]	−37.0[854]	−39±5[854]	−38±10[854]
KCl–CsCl	−1.7[507]	2.8[507]	8.0[507]	—	—	—
NaCl–NaBr	—	—	3.2[866]	2.8[854]	—	—
NaCl–NaNO₃	—	—	12.4[866]	16.7[854]	—	—
NaBr–KBr	—	−34.5[801]	−38[801,866]	—	—	—

2-1 Electrolytes[849]

Mixture	$I = 3$	$I = 2$	$I = 1$	$I = 0.6$	$I = 0.3$
MgCl₂–SrCl₂	3.84	3.63	3.3	3.1	—
MgCl₂–BaCl₂	8.26	8.00	6.7	4.5	5±1
MgCl₂–MgBr₂	1.81	1.57	—	—	—

[a] $RT h_0 = \Delta_m H^E / W y_A y_B I^2$ at $y_A = 0.5$.
[b] Author's estimate.

TABLE II. $\Delta_m H^E / V y_A y_B c^2 = A + Bc^a$

Mixture[413]	A,[b] cal dm³ mol⁻²	B, cal dm⁶ mol⁻³
NaF–KF	−34.4	−0.4
NaCl–KCl	−38.4	−1.2
NaBr–KBr	−38.0	−2.8
NaI–KI	−39.2	−3.6
NaNO₃–KNO₃	−54.5	+1.2
NaOAc–KOAc	−33.2	−4.4
Na₂CO₃–K₂CO₃	−35.6	+6.4

[a] c = moles of cation per dm³ of solution.
[b] Based on 0.1–1.0 extrapolation except NaNO₃–KNO₃, 0.4–1.0c.

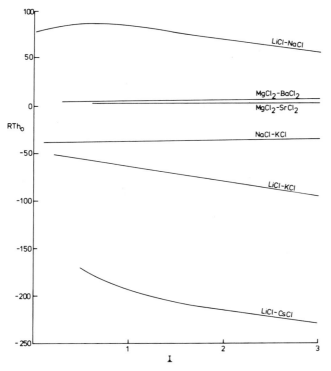

Fig. 2. Ionic strength dependence of RTh_0 for some cation mixtures having a common anion.

TABLE III. Excess Free Energy of Mixing: Concentration Dependence[a]

1–1 Electrolytes

Mixture	$I = 0.5$	$I = 1$	$I = 2$	$I = 3$	$I = 4$	Ref.
NaCl–KCl	—	−16	−17	−18	−19	689
NaNO$_3$–KNO$_3$	—	−14	−15	−16	—	689
NaCl–NaNO$_3$	—	18	14	11	7	689
KCl–KNO$_3$	—	16	12	8	—	689
NaBr–KBr	—	−16	—16	−18	−19	142
NaCl–NaBr	—	2	2	2	2	142
KCl–KBr	—	0	2	1	1	142
NaCl–LiCl	11	12	11	10	8	695
NaNO$_3$–LiNO$_3$	9	8	4	0	−5	695
NaOAc–LiOAc	12	12	10	7	4	695
HClO$_4$–LiClO$_4$	—	16	14	13	11	701
HClO$_4$–NaClO$_4$	—	27	17	8	−5	701
LiClO$_4$–NaClO$_4$	—	4	−2	−7	−11	701
LiBr–Pr$_4$NBr	−8.9	−32	−92	−134	−163	828
NaBr–Pr$_4$NBr	187	157	103	67	40	828
CsBr–Pr$_4$NBr	180	153	105	75	59	828
Me$_4$NBr–Pr$_4$NBr	−27	−31	−53	−59	−65	828
Et$_4$NBr–Pr$_4$NBr	−12	−19	−21	−18	−10	828
Me$_4$NBr–Et$_4$NBr	−23	−14	−15	−15	−18	828
Me$_4$NBr–Bu$_4$NBr	−64	−36	+1.8	+34	+49	828

2–1 Electrolytes

Mixture	$I = 1$	$I = 3$	$I = 6$	$I = 9$	Ref.
CaCl$_2$–Ca(NO$_3$)$_2$	14	9	3	−3	637
Ca(NO$_3$)$_2$–Mg(NO$_3$)$_2$	10	5	−0.1	−5	637
Mg(NO$_3$)$_2$–MgCl$_2$	12	11	9	7	637
CaCl$_2$–MgCl$_2$	1.2	0.9	0.7	0.3	688

[a] $RTg_0 = \Delta_m G^E / (W y_A y_B I^2)$.

arbitrarily selected more recent data where the investigator himself has been concerned with the actual calculation of $\Delta_m G^E$. The results of a number of these studies are given in Table III. Very few studies of $\Delta_m G^E$ have been conducted below $1.0I$, but an examination of Figs. 3 and 4 shows that even in the high concentration ranges, mixtures of alkali metal halides exhibit a rather low concentration dependence of RTg_0 while mixtures with sodium perchlorate, calcium nitrate, or the larger tetraalkylammonium bromides have a rather high concentration dependence. Rush[700] has treated 22 electrolyte mixtures, tabulating the results in terms of the coefficients of eqns. (18) and (19). In every case $b_{AB}^{(0)}$ is the most important term and in most cases no term beyond $b_{AB}^{(0,1)}$ is very important. These results are consistent with the conclusion that pair and triplet interactions are predominant in these mixtures. The original results of Scatchard and Prentice[721] for KNO_3–KCl, $LiCl$–$LiNO_3$, KNO_3–$LiNO_3$, and KCl–$LiCl$ provide notable exceptions to these generalizations. Scatchard and Prentice set $b_{AB}^{(0)}$ equal to zero since it was not necessary to fit their data. These results are the only ones of sufficient accuracy which do not indicate that like-charged pair interactions constitute the leading term.

The excess volume of mixing has been investigated by Wirth and his students and more recently by Wen *et al.*, and the results of their work are

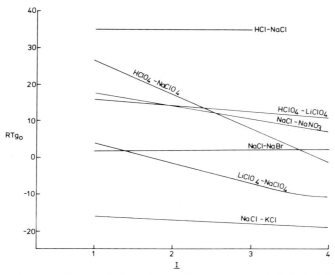

Fig. 3. Ionic strength dependence of RTg_0 for some 1–1 electrolyte mixtures.

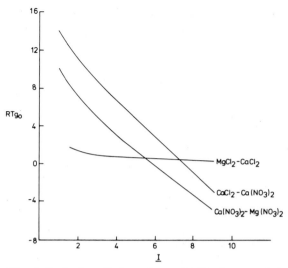

Fig. 4. Ionic strength dependence of RTg_0 for some 2–1 electrolyte mixtures.

given in Table IV. The results for $(CH_3)_4NBr$–KBr, HCl–HClO$_4$ and HCl–NaCl show fairly small concentration dependence and this indicates the relatively small contribution of triplet interactions. However, the higher tetraalkylammonium bromides when mixed with KBr show a large concentration dependence of v_0 as well as a large value of v_0 at low concentrations. This indicates that both pair and triplet interactions are important

TABLE IV. Excess Volume of Mixing: Concentration Dependence[a]

Mixture	$I = 0.2$	$I = 0.5$	$I = 1.0$	$I = 4.17$	Ref.
Me$_4$NBr–KBr	—	0.39	0.40	—	829, 830
Et$_4$NBr–KBr	—	2.54	1.96	—	829, 830
Pr$_4$NBr–KBr	—	4.56	3.60	—	829, 830
Bu$_4$NBr–KBr	7.22	5.44	3.78	—	829, 830
HCl–HClO$_4$	—	—	−0.0744	−0.0934	841, 842
HCl–NaCl	—	—	−0.129	−0.1042	841, 842
HClO$_4$–NaClO$_4$	—	—	−0.0686	−0.0196	841, 842
NaCl–NaClO$_4$	—	—	−0.1087	−0.1422	841, 842

[a] $v_0 = \Delta_m V^E/(W y_A y_B I^2)$.

TABLE V. Heats of Mixing with Varying Common Ions[a]

	$\Delta_m H^E / W y_A y_B m^2$	I
Cl⁻–Br⁻ Mixings		
Li⁺	$3.2^{(866)}$	1.0
Na⁺	$3.2^{(866)}$	1.0
K⁺	$3.2^{(866)}$	1.0
Mg²⁺	$3.5^{(849)}$	2.0
Na⁺–K⁺ Mixings		
Cl⁻	$-38^{(854,866)}$	0.2–1.0
Br⁻	$-38^{(866)}$	1.0
MgCl₂	$-52 \pm 8^{(853)}$	0.5
BaCl₂	$-46 \pm 3^{(853)}$	0.5
NO₃⁻	$-55 \pm 5^{(854,866)}$	0.2–1.0
Li⁺–Na⁺ Mixings		
Cl⁻	$82 \pm 3^{(854,866)}$	0.2–1.0
Br⁻	$83^{(866)}$	1.0
MgCl₂	$68 \pm 12^{(853)}$	0.5
BaCl₂	$72 \pm 20^{(853)}$	0.5
Li⁺–K⁺ Mixings		
Cl⁻	$-57 \pm 7^{(854,866)}$	0.2–1.0
Br⁻	$-68^{(866)}$	1.0
MgCl₂	$-60^{(853)}$	0.5
BaCl₂	$-48^{(853)}$	0.5

[a] m is the molality of the ions to be mixed.

in these mixtures. Mixings of $NaClO_4$ with both $HClO_4$ and $NaCl$ also show a large concentration dependence of v_0. These results are consistent with the third generalization since for both $NaClO_4$ and the larger tetraalkylammonium bromides the osmotic coefficients are unusually low. Tables V–VII show the results of a wide variety of tests of Young's rule. Young's rule was originally proposed on the basis of the results of the heat of mixing of chloride with bromide in the presence of lithium, sodium, and potassium as common ions and the mixing of sodium with potassium in the presence of chloride and bromide. Table V shows that for these mixtures the rule is very accurate. For the majority of cases the rule holds to within 20–30%.

TABLE VI. Excess Free Energies of Mixing with Varying Common Ion[a]

	$\Delta_m G^E / W y_A y_B m^2$	Ref.
Na^+–K^+ Mixings		
Cl^-	−16	689
Br^-	−16	142, 801
NO_3^-	−14	689
Na^+–Li^+ Mixings		
Cl^-	5.6	695
NO_3^-	5.2	695
$C_2H_3O_2^-$	6.4	695
Cl^-–Br^- Mixings		
Na^+	2	142
K^+	0	544
Cl–NO_3^- Mixings		
Na^+	18	689
K^+	16	689

[a] $I = 1.0$.

TABLE VII. Excess Volume of Mixing with Varying Common Ion

	$\Delta_m V^E / W y_A y_B I^2$	I
Cl^-–ClO_4^- Mixings[841,842]		
H^+	−0.0744	1.0
Na^+	−0.1087	1.0
Cl^-–SO_4^{2-} Mixings[840]		
Li^+	0.1698	4.0
	0.2770	1.0
Na^+	0.2189	4.0
	0.4067	1.0
Li^+–Na^+ Mixings[840]		
Cl^-	−0.0639	4.0
	−0.0960	1.0
SO_4^{2-}	−0.0279	4.0
	−0.0687	1.0

When the comparison in Table V is made for mixtures with different charge types or where a third salt is present at a constant concentration in the two solutions to be mixed the results must be corrected so that the comparison is made for mixtures in which the same concentrations of the two ions are mixed. That is, the quantity to be compared is $\Delta_m H^E / y_A\, y_B m^2$, where m is the molality of the two ions being mixed.

The most notable exception to Young's rule is illustrated in Fig. 4, which shows that at low concentrations, where the rule is expected to be most accurate, the free energy of mixing of magnesium and calcium in the presence of a common chloride ion is quite different from the free energy of mixing of calcium with magnesium in the presence of the nitrate ion. Calcium nitrate has a very low osmotic coefficient, so that the large slope of the curve for the calcium nitrate mixtures in Fig. 4 is not surprising. However, Young's rule must hold as the ionic strength is reduced and Fig. 4 shows no evidence of this for the mixtures of calcium with magnesium.

The isopiestic studies of some chloride–bromide mixtures in the presence of the common tetramethylammonium, benzyltrimethylammonium, and (β-hydroxyethyl)benzyldimethylammonium ions in the $I = 1$–9 concentration range reported by Lindenbaum and Boyd[518] provide another exception to the present generalizations. Within the experimental error RTg_0 was found to be practically independent of concentration for all three mixtures, having values of -5.2 ± 2.3, -8.6 ± 1.9, and ~ -2, respectively. This is consistent with like-charged pair interactions because of the concentration independence. However, as seen in Table III, the results are opposite in sign to those of the KCl–KBr and NaCl–NaBr studies. This cannot be easily explained by invoking triplet interactions because the concentration dependence of RTg_0 is too small in both studies. It may be that because of the considerable complexity of the common cation in Lindenbaum and Boyd's work the activity coefficient ratio has a very high concentration dependence at low molalities.

It is easy to show that the same generalizations should hold for mixings at constant molarity as well as constant molality. The measurements by Jolicoeur et al.[413] of the heat of mixing of sodium with potassium in the presence of fluoride, chloride, bromide, iodide, nitrate, acetate, and carbonate ions are shown in Fig. 5. The two generalizations discussed above hold quite well for these data since the slope of the curves are generally small and the values are the same within 20–30%. However, eqn. (12) shows that the intercept of all these curves at $c = 0$ should be identical. A common intercept for all these curves is certainly possible but it must be admitted that the curve for the nitrates must bend down rather sharply

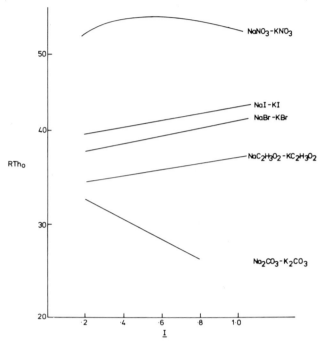

Fig. 5. Ionic strength dependence of RTh_0 for Na^+–K^+ mixtures in the presence of common anions.

below 0.2 M. This implies that either the activity coefficient ratio or triplet interactions change considerably below 0.2 M. It is interesting that this behavior is associated with the presence of a salt, KNO_3, which has a very low osmotic coefficient. Jolicoeur *et al.* concluded from their measurements that the intercepts of the curves in Fig. 5 are not equal and that this was due to the effect of the common ion on the interactions between the two cations. However, the effect of a common ion on a pair of cations is in effect a triplet interaction and eqn. (12) shows that this triplet does not effect the value of RTh_0 in the limit of low concentrations.

Recent results* on the heats of mixing of Bu_4NCl with KCl and $LiCl$ with $CsCl$ in the 0.5–0.04 m concentration range show that RTh_0 does decrease below 0.1 m as predicted by eqn. (14). The results are also compatible with the limiting slope of eqn. (14) but unfortunately are not accurate enough to experimentally verify the limiting law.

* R. H. Wood, J. A. Falcone, and A. L. Levine, unpublished results.

4.2. Young's Sign Rule

Young and co-workers[866,876] observed that the sign of the heat of mixing of common-ion mixtures could be correlated with an arbitrary classification of the ions. They suggested that the classification for the alkali metal cations could be based on ion size or structural properties. For alkali metal and alkaline earth cations no clear distinction can be made between these two classification methods since small ions (H^+, Li^+, Na^+, Mg^{2+}, Ca^{2+}, Sr^{2+}, and Ba^{2+}) are structure-making ions and large ions (K^+, Rb^+, Cs^+) are structure-breaking ions. The structural classification scheme most commonly adopted in mixed electrolyte solution studies has been that of Frank, Evans, and Wen,[230,232] with the sodium ion classified as a structure-maker.

Young's sign rule predicts that at constant molal ionic strenth the mixing of two ions of like classification in the presence of a common ion gives rise to a positive heat of mixing. Similarly, mixing ions of unlike classification results in a negative heat of mixing. For simple alkali metal and alkaline earth charge-symmetric mixtures the only exception to this rule is the mixing of KCl with CsCl at $3\,m$ concentration.[507] At lower concentrations the KCl–CsCl mixture obeys Young's sign rule.

Wood and Anderson[850] studied the heats of mixing of a variety of anions in the presence of the common potassium cation and the results are given in Table VIII. The anions chosen were such that a distinction between size and structural properties could be made. That is, the F^- and $C_2H_3O_2^-$ ions are both structure-making ions but the F^- ion is small and the $C_2H_3O_2^-$ ion is large. Both the Cl^- and Br^- ions are large, structure-breaking anions.

TABLE VIII. Heats of Mixing: Common Cation^a

Mixture	RTh_0
$KF–KC_2H_3O_2$	$+9.5\pm0.3$
$KCl–KBr$	$+3.2$
$KF–KCl$	-22.6 ± 0.2
$KF–KBr$	-16.7 ± 0.2
$KCl–KC_2H_3O_2$	-34.3 ± 0.3
$KBr–KC_2H_3O_2$	-32.4 ± 0.3

^{a} $RTh_0 = \Delta_m H / W y_A y_B I^2$ at $y = 0.5$, $I = 1.0$. From Ref. 850.

The heats of mixing agree with Young's sign rule only if the classification is based on the structural classification, for, if size was the determining factor, one would predict the heat of mixing of $KF-KC_2H_3O_2$ to be exothermic and the heat of mixing of $KCl-KC_2H_3O_2$ to be endothermic.

Examination of Table III shows that the excess free energy of mixing obeys Young's sign rule also. There are several exceptions in that $\Delta_m G^E$ for the $LiClO_4-NaClO_4$ mixture at $I = 1$ is positive as predicted, but at higher concentration changes sign. Similar results are observed for the $Ca(NO_3)_2-Mg(NO_3)_2$ mixture. Young's sign rule apparently is valid for both heats of mixing and excess free energies of mixing at low concentrations for simple salt mixtures.

It is not surprising to find that the heat and excess free energy are largely due to structural properties but is surprising that such a simple classification will predict the sign of the heat and excess free energy of mixing as well as it does.

From eqn. (9), considering only the like-charged pair interactions, one can write

$$RTg_0 = RT(B_{MN}-B_{MM}-B_{NN}) \quad \text{and} \quad RTh_0 = RT^2 \, \partial(B_{MN}-B_{MM}-B_{NN})/\partial T$$

This equation shows that the sign rule is predicting the difference in the free energy and heat when two MN interactions are changed to an MM and an NN interaction. It is not too surprising that the magnitude of RTg_0 or RTh_0 does not follow a simple rule. The deviation from Young's rule at higher concentrations indicates that the rule applies to the pair interactions and is not always applicable to triplet or higher interactions.

4.3. Temperature Dependence of $\Delta_m H^E$

The Frank–Evans–Wen structure model assumes that an ion has three regions of water around it. The first region (A) is highly structured. The second (B) is in a state of disorder. The rest of the water is in the region (C) of normal bulk water. Normal water structure is considerably temperature-dependent[66,812] in the 25–80°C range. Recently the heats of mixing of a number of electrolytes have been measured in this temperature range.[13,14, 748–750] The temperature dependence of RTh_0 for selected mixtures is plotted in Fig. 6. The majority of the mixings are remarkably temperature-independent. This led Anderson and Petree[13] to suggest that the heat of mixing is a direct reflection of the interaction of the heterions in the interface between region A and B. Their reasoning was that the water in this region should be thermally stable and therefore reflect no temperature dependence.

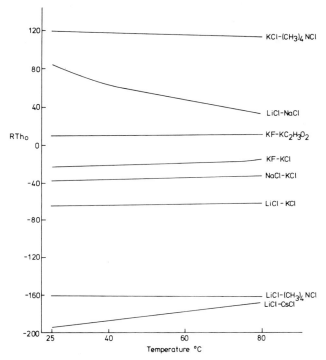

Fig. 6. Temperature dependence of RTh_0 for 1–1 electrolyte mixtures.

The only mixtures showing a definite temperature dependence in the heat of mixing are those containing NaCl or CsCl. Anderson and Petree concluded that the sodium ion is a borderline structure-maker in the classification scheme and that water structure in the A–B region most probably is temperature-sensitive. The cesium ion shows very poor hydration,[670] but it is suspected that as temperature increases it becomes more hydrated; thus mixtures containing CsCl should display a temperature dependence. It is worthy of note that of all the mixtures that display a temperature dependence, the NaCl–CsCl is the only example of a heat of mixing increasing in absolute magnitude as the temperature is increased.

4.4. Tetraalkylammonium Electrolytes

The tetraalkylammonium electrolytes have received considerable attention in recent years. They are of interest because of the cation size and because the larger electrolytes usually show extraordinarily large thermodynamic effects. Frank and Wen[232] proposed that these large cations should be regarded as structure-makers as the size of the alkyl chain increases. Systematic measurements of many physical properties of solutions of alkylammonium ions have tended to confirm this view.[233]

Wood and Anderson[851] measured the heats of mixing of a series of tetraalkylammonium chloride mixtures and the results are recorded in Table IX. The heats of mixing are consistent with Young's sign rule by classifying the $(CH_3)_4N^+$ ion as a structure-breaker and the $(C_2H_5)_4N^+$ ion as a borderline structure-maker [i.e., structure-maker when mixed with $(CH_3)_4N^+$ but a structure-breaker when mixed with Li^+, K^+, or Cs^+]. Using conductivity measurements, Kay and co-workers[428,433] (see Chapter 4) arrived at similar conclusions concerning the structural classification. Partial molal volume data[236,831] suggest a similar classification except that the $(C_2H_5)_4N^+$ ion should be regarded as just a slight hydrophobic structure-maker.

The Pr_4N^+ ion has been classified as a hydrophobic structure-maker and thus Young's sign rule does not apply. Wood and Anderson noted that the large negative heats of mixing ($RTh_0 = -300$ to -600) were consistent with the results of the heats of dilution of the tetrapropylammonium ions.

TABLE IX. Heat of Mixing Tetraalkylammonium Chlorides[a]

Mixture	RTh_0
$LiCl-(CH_3)_4NCl$	-160.8 ± 0.9
$KCl-(CH_3)_4NCl$	118.9 ± 0.8
$CsCl-(CH_3)_4NCl$	82.0 ± 1.3
$LiCl-(C_2H_5)_4NCl$	-172.4 ± 1.0
$KCl-(C_2H_5)_4NCl$	117.5 ± 0.9
$CsCl-(C_2H_5)_4NCl$	73.1 ± 1.3
$LiCl-(n-C_3H_7)_4NCl$	-693.3 ± 3.8
$KCl-(n-C_3H_7)_4NCl$	$-348 \ \pm 4$
$CsCl-(n-C_3H_7)_4NCl$	-438.6 ± 1.5
$(CH_3)_4NCl-(C_2H_5)_4NCl$	-49.8 ± 0.7
$(CH_3)_4NCl-(n-C_3H_7)_4NCl$	-613.5 ± 2.3
$(C_2H_5)_4NCl-(n-C_3H_7)_4NCl$	-308.4 ± 2.2

[a] $RTh_0 = \Delta_m H / W y_A y_B I^2$ at $y = 0.5$, $I = 0.5$. From Ref. 851.

Using eqn. (24) as a guide, it is easily seen that in a Pr_4NX-MX mixture three like-charged pair interactions occur: (1) the heat of diluting the Pr_4N^+ ions and therefore the reduction in the overlap of the Pr_3N^+ hydration sheaths, (2) a similar dilution effect for the M^+-M^+ interactions, and (3) the formation of some $Pr_4N^+-M^+$ overlap. In this mixing it is assumed that the first effect is by far the largest since the size of the ion is so large ($\bar{V}$ = 213.5 cm^3 mol^{-1}).[236] Thus when tetrapropylammonium ions are mixed with the other smaller cations the overlap of the tetrapropylammonium hydration sheaths is reduced, allowing for these ions to more fully complete their hydration sphere, thus liberating heat. This is consistent with the heat of dilution data[428,433] in that the tetrapropylammonium halides exhibit a much larger heat of dilution than the smaller tetramethylammonium or tetraethylammonium halides. In fact, Lindenbaum found that the entropies of dilution of the chlorides and bromides of the tetrapropylammonium ions were almost identical up to 1.5 m and concluded that the dilution was primarily the result of interactions of the tetrapropylammonium ions. This result allowed Wood and Anderson to estimate the heat of mixing of tetrapropylammonium chloride mixtures by taking the slope of the linear portion of the heat of dilution plot. This method gives a value of RTh_0 = -370 at $I = 0.5$ as compared with the experimental values of -300 to -600 (see Table IX). The agreement is excellent and suggests that the thermodynamic properties of the larger tetraalkylammonium halide can be considered to be primarily dependent on the cation.

Recently, Wen and co-workers[828] have determined the excess free energies of mixing of some tetraalkylammonium halides. The results are recorded in Table III. In general the excess free energies are considerably smaller than the heats of mixing. Wen and co-workers point out that this reflects a small g_0 interaction parameter for the large tetraalkylammonium ions, supporting the conclusion of Wood and Anderson that the large tetraalkylammonium ion interactions were predominantly a structural (i.e., entropy) effect.

CHAPTER 3

Hydration Effects and Acid-Base Equilibria

Loren G. Hepler
Department of Chemistry
University of Lethbridge
Lethbridge, Alberta, Canada

and

Earl M. Woolley
Department of Chemistry
Brigham Young University
Provo, Utah

1. IONIZATION OF LIQUID WATER

Water is the most common solvent for all of chemistry, including acid–base chemistry. For this reason it is appropriate to begin our discussion of acids and bases in aqueous systems with a consideration of the thermodynamics of self-ionization of water.

As a result of the thorough electrochemical investigations of Harned and others,[317,694] we now have equilibrium constants for ionization of water from 0 to 60°C and 1.0 atm as listed in Table I, with standard states discussed below. These equilibrium constants all refer to the reaction that is conveniently represented by

$$H_2O(liq) = H^+(aq) + OH^-(aq) \tag{1}$$

By $H^+(aq)$ and $OH^-(aq)$ we mean hydrogen ions and hydroxide ions as they exist in the aqueous solution under consideration, whether or not we know anything about the details of hydration of these species. The equilibrium constant, denoted by K_w, is therefore of the form

$$K_w = a_{H^+}a_{OH^-}/a_{H_2O} \tag{2}$$

Activities of hydrogen ions, hydroxide ions, and water may be based on any standard states, but only a few choices have actually been used. Except where specifically stated otherwise, we always use the "hypothetical 1 m standard state" for solutes, as described by Klotz.[454] For the solvent it is customary and usually most convenient to choose the pure liquid to be the standard state. In this case the denominator of (2) approaches unity for dilute solutions, so that we have the familiar

$$K_w = a_{H^+}a_{OH^-} \tag{3}$$

The K_w and pK_w values listed in Table I are for the equilibrium constants defined in (3) on the "molal scale."

TABLE I. Ionization Constants of Water[317,694]^a

t, °C	pK_w	$K_w \times 10^{14}$
0	14.943_5	0.1139
5	14.733_8	0.1846
10	14.534_6	0.2920
15	14.346_3	0.4505
20	14.166_9	0.6809
25	13.996_5	1.008
30	13.833_0	1.469
35	13.680_1	2.089
40	13.534_8	2.919
45	13.396_0	4.018
50	13.261_7	5.474
55	13.136_9	7.297
60	13.017_1	9.614

^{a} Note discussion in the text in which reference is made to a slightly different K_w value at 25°C. We also present evidence that the temperature dependence of the K_w values given in this table is in error by a small amount.

**TABLE II. Thermodynamics of Ioniza-
tion of Water at 298°K as Derived from
K_w Values at Several Temperatures[143]a**

$\Delta G° = 19{,}095$ cal mol^{-1}

$\Delta H° = 13{,}526$ cal mol^{-1}

$\Delta S° = -18.68$ cal deg^{-1} mol^{-1}

$\Delta C_P° = -46.6$ cal deg^{-1} mol^{-1}

---

a See Table IV for "best" values, partly based
on results of calorimetric investigations.

The K_w (or pK_w) values in Table I can be combined with familiar thermodynamic equations[454] ($\Delta G° = -RT \ln K$, $d(\ln K)/dT = \Delta H°/RT^2$, $d(\Delta H°)/dT = \Delta C_P°$, and $\Delta G° = \Delta H° - T \Delta S°$) to yield values of $\Delta G°$, $\Delta H°$, $\Delta S°$, and $\Delta C_P°$ to be associated with reaction (1). Values for these quantities as calculated recently by Covington et al.[143] are listed in Table II.

About the time Mayer, Joule, and Helmholtz were establishing what we now call the first law of thermodynamics, Hess, Graham, Andrews, Favre, and Silbermann were making calorimetric measurements of heats of neutralization. Improved measurements were made by Berthelot and especially by Thomsen in the last third of the nineteenth century. Further calorimetric determinations of the heats of neutralization of strong acids by strong bases were reported by Wörmann[863] in 1905. His results at several temperatures were represented by Lewis and Randall[509] with the equation

$$\Delta H = 29{,}210 - 53T \tag{4}$$

From this equation we calculate $\Delta H = 13{,}420$ cal mol^{-1} and $\Delta C_P = -53$ cal deg^{-1} mol^{-1} at 298°K for the ionization reaction (1). Allowing for uncertainties in both the measured heats of neutralization and the heats of dilution that must be used to obtain standard heats of reaction, we see that agreement between these calorimetric results and values in Table II is satisfactory.

More recent support for the values in Table II was provided by $\Delta H° = 13.50$ kcal mol^{-1} for reaction (1) from the microcalorimetric measurements reported by Papee et al.[623] In spite of the impressive agreement between the $\Delta H°$ values from Table II and microcalorimetry, several other calorimetric investigations have provided convincing support for a $\Delta H°$

value significantly smaller than the apparently established 13.50 or 13.53 kcal mol^{-1}. The discrepancy is both large enough and important enough to deserve detailed attention.

Before turning to discussion of modern calorimetric measurements leading to $\Delta H°$ for ionization of water as in reaction (1), it is useful to consider $\Delta C_P°$ for this reaction in order to make full use of results obtained at several temperatures near 298°K. Parker[625] has carefully reviewed and compiled the results of many investigations of heat capacities of various aqueous solutions. She has concluded that $\Delta C_P° = -53.5$ cal deg^{-1} mol^{-1} is the best value for ionization of water at 298°K with $d(\Delta C_P°)/dT = 0.9$ cal deg^{-2} mol^{-1} in the range 15–30°C. We shall first make use of calculations based on these quantities and return later to consideration of other $\Delta C_P°$ values. Parker[625] has also reviewed and tabulated heats of dilution that have been useful in evaluating $\Delta H°$ values for reaction (1).

In Table III we summarize the results of a large number of calorimetric investigations leading to $\Delta H°$ for ionization of water at 298°K. All but one of these values has been obtained either by extrapolation of measured heats of neutralization to zero concentration of solute or by combination of measured heats of neutralization with measured heats of dilution, which have themselves included an extrapolation to zero concentration. In some cases the measurements have been carried out at some temperature near 25°C, so that it has been necessary to make use of information about $\Delta C_P°$ to obtain the desired $\Delta H°$ at 25°C. Reference numbers refer to the original publication in which calorimetric results were reported and to subsequent publications in which results of improved calculations of $\Delta H°$ values at 25°C have been reported.

The $\Delta H°$ value attributed to Larson[491] does not involve heats of neutralization but is largely based on calorimetric data. Larson[491] calculated $\bar{S}_2° = -2.7_0$ cal deg^{-1} mol^{-1} for the standard partial molal entropy of OH$^-$(aq) from the thermodynamic properties of NaOH · H$_2$O(c), NaCl(c), and solutions of these compounds. It is important to recognize that this entropy is totally independent of both heat of neutralization measurements and the ionization constant of water. Use of the above $\bar{S}_2°$ for OH$^-$(aq) with $\bar{S}_2° = 0$ for H$^+$(aq) and $S° = 16.71$ cal deg^{-1} mol^{-1} for H$_2$O(liq) at 298°K from NBS Tech. Note 270-3[805] leads to $\Delta S° = -19.4_1$ cal deg^{-1} mol^{-1} for ionization of water at 298°K as in eqn. (1). Taking $\Delta G°$ from Table II and combining it with this $\Delta S°$ gives us $\Delta H° = 13.31$ kcal mol^{-1} for ionization of water at 298°K. This calculation, or the earlier calculation by Larson[491] involving comparison of $\bar{S}_2°$ values from various data sources, provides convincing independent support for

TABLE III. Calorimetric Determinations of $\Delta H°$ of Ionization of Water at 298°K[a]

$\Delta H°$, kcal mol^{-1}	Investigator	Date of publication
13.42	Wörmann[863]	1905
13.35	Richards et al.[683,684] (see also Refs. 493, 625)	1922, 1929
13.36	Gillespie et al.[272] (see also Ref. 625)	1930
13.36	Pitzer[636]	1937
13.33	Bender and Biermann[43] (see also Ref. 625)	1952
13.36	Biermann and Weber[56] (see also Ref. 787)	1954
13.36	Davies et al.[156]	1954
13.17–13.33	Bidinosti and Biermann[55] (see also Refs. 493, 625)	1956
13.50	Papee et al.[623]	1956
13.34	Sacconi et al.[704]	1959
13.34	Hale et al.[306]	1963
13.34	Vanderzee and Swanson[787]	1963
13.33	Gerding et al.[269]	1963
13.33	Wood and Smith[854]	1965
13.33	Anderson et al.[18]	1966
13.35	Vasil'ev and Lobanov[790]	1967
13.36	Wood et al.[852]	1967
13.34	Christensen et al.[105]	1968
13.35	Goldberg and Hepler[284]	1968
13.33	Leung and Grunwald[506]	1970
13.35	Grenthe et al.[291]	1970
13.31	Larson[491] (see discussion)	1970
13.33	Hansen and Lewis[313]	1971

[a] All $\Delta H°$ values are rounded to the nearest 0.01 kcal mol^{-1}.

a $\Delta H°$ of ionization of water smaller than the $\Delta H° = 13.50$ kcal mol^{-1} from calorimetric measurements by Papee et al.[623] or $\Delta H° = 13.53$ kcal mol^{-1} from Table II.

Consideration of the $\Delta H°$ values summarized in Table III and our estimates of uncertainties to be associated with each of these values leads us to adopt $\Delta H° = 13.34$ kcal mol^{-1} for the "best" heat of ionization of H_2O(liq) at 298°K. Parker[625] selected $\Delta H° = 13,345 \pm 25$ cal mol^{-1} as the "best" value on the basis of her review published in 1965, while Larson and Hepler[493] selected $\Delta H° = 13.34$ kcal mol^{-1} in 1969. The $\Delta H_l°$ values listed in NBS Tech. Note 270-3[805] lead to $\Delta H° = 13.345$ kcal mol^{-1}.

It is impossible to state definitely the source of the disagreement between the calorimetrically based $\Delta H°$ we have adopted above and the $\Delta H°$ value in Table II that was derived from K_w values at several temperatures. Because the electrochemical measurements leading to the K_w values were carried out carefully by experienced and highly capable researchers, the discrepancy cannot be casually attributed to gross experimental error in the equilibrium work. Rather, we believe that the difficulty lies in some *small* temperature-dependent error and in uncertainties associated with activity coefficients and extrapolations to zero concentration.

Guggenheim and Turgeon[301] have carried out what appears to be the best treatment of the equilibrium data for ionization of water at 25°C and have obtained $K_w = 1.002 \times 10^{-14}$ ($pK_w = 13.9991$) in contrast to $K_w = 1.008 \times 10^{-14}$ ($pK_w = 13.9965$) obtained originally and listed in Table II. The difference between these values is equivalent to only 4 cal mol^{-1} in $\Delta G°$ and therefore is of minor importance in this connection. But, as illustrated below, small errors in $\Delta G°$ values can lead to relatively large errors in $\Delta H°$ values derived by differentiation.

King[445] has described a thorough and rigorous analysis of the relationship between *random* errors in equilibrium constants and the thermodynamic quantities derived by differentiation. In the case of equilibrium constants determined at 5-deg intervals from 5–50°C with *random* errors in pK of ± 0.02 or ± 0.001, the standard deviations in the derived $\Delta H°$ values amount to ± 190 or ± 10 cal mol^{-1}, respectively. Because the statistical analysis excluded systematic (possibly temperature-dependent) errors in measurement or in extrapolation to zero concentration, these "mathematical uncertainties" are only a lower limit on the "total uncertainty" which is of greater interest.

To demonstrate in simple fashion how small errors in pK values can lead to much larger errors in $\Delta H°$, we write the van't Hoff equation as

$$\Delta pK_w \cong -(\Delta H°)(\Delta T)/2.3RT^2 \qquad (5)$$

for the case of two temperatures reasonably close together. To indicate changes or errors in ΔpK_w and $\Delta H°$ we then have

$$\delta(\Delta pK_w) = -[\delta(\Delta H°)](\Delta T)/2.3RT^2 \qquad (6)$$

For $\Delta T = 10°$ at $T \cong 300°K$ we find that $\delta(\Delta H°) = 180$ cal mol^{-1} (the discrepancy in $\Delta H°$ values) corresponds to $\delta(\Delta pK_w) = 0.004$. It is thus easy to see that small temperature-dependent errors in measurement or evaluation of K_w from the experimental data can cause considerable error in the derived $\Delta H°$ value. It is unfortunate that data are not available for extension of the calculations of Guggenheim and Turgeon[301] to other temperatures.

We now turn to further consideration of the temperature dependence of the various thermodynamic functions of ionization of water, all of which may be performed in terms of $\Delta C_P°$.

As our starting point we take $\Delta C_P° = -53.5$ cal deg^{-1} mol^{-1} at $298°K$ and $d(\Delta C_P°)/dT = 0.9$ cal deg^{-2} mol^{-1} from 15 to 30°C, as reported by Parker[625] on the basis of her careful analysis of heat capacity data and derived partial molal heat capacities. These values lead us to

$$\Delta C_P° = -321.8 + 0.9T = -53.5 + 0.9(t - 25) \tag{7}$$

in which T and t represent degrees kelvin and degrees centigrade, respectively.

We now combine (7) with $d(\Delta H°)/dT = \Delta C_P°$ and our "best" $\Delta H°$ $= 13{,}340$ cal mol^{-1} at $298°K$ to obtain

$$\Delta H_T° = 69{,}283 - 321.8T + 0.45T^2 \tag{8}$$

and

$$\Delta H_t° = 13{,}340 - 53.5(t - 25) + 0.45(t - 25)^2 \tag{9}$$

We may also obtain further information about $\Delta C_P°$ from calorimetric $\Delta H°$ values determined at different temperatures. As a start, we take $\Delta C_P°$ to be a linear function of temperature similar to (7) but with unspecified constants. Thence we obtain a general version of (9), which we rearrange to

$$(\Delta H_t° - \Delta H_{25}°)/(t - 25) = A + B(t - 25) \tag{10}$$

to obtain an equation that is suitable for graphical display and evaluation of parameters A and B.

Calorimetric $\Delta H°$ values for ionization of water at several temperatures from several investigations are displayed in Fig. 1 as suggested by eqn. (10). Before turning to detailed consideration of these results we point out that differentiation of $\Delta H°$ values to obtain $\Delta C_P°$ leads to the same kind of magnification of small errors as does differentiation of pK values to obtain $\Delta H°$.

The two points (hollow squares) in Fig. 1 from the work of Anderson et al.[18] are almost certainly in error and can reasonably be regarded as superseded by more recent results[107,108] (solid squares) from the same laboratory. In every case it appears that a substantial fraction of possible

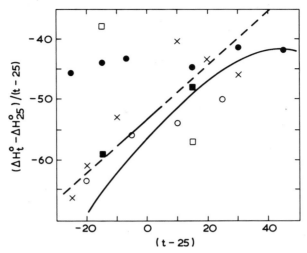

Fig. 1. Display of calorimetrically determined heats of ionization of water, as suggested by eqn. (10). Data points are from following sources: (□) from Anderson et al.;[18] (■) from Christensen et al.;[107,108] (×) from Leung and Grunwald;[506] (○) from Grenthe et al.;[291] (●) from Vasil'ev and Lobanov.[790] The straight line is based on eqns. (7)–(9), which were originally intended to apply only for 15–30°C. The curved line is based on heat capacity data at 20-deg intervals from 10° upward as reported by Ackermann.[2]

errors should be associated with the heats of dilution used in obtaining $\Delta H°$ values from measured heats of neutralization. For example, Vasil'ev and Lobanov[790] obtained $\Delta H° = 13,349$ cal mol^{-1} at 25°C by combination of their calorimetric heat of neutralization with heats of dilution from the literature. In the absence of appropriate heats of dilution at other temperatures they used a semiempirical heat of dilution correction based on the Debye–Hückel theory. This approach led to $\Delta H° = 13,480$ cal mol^{-1} at 25°C, which differs by 131 cal mol^{-1} from their more reliable value cited above. These same two approaches led to $\Delta H°$ values for 18°C that differ by 85 cal mol^{-1}. Although these differences are not very large, the fact that they are temperature-dependent can cause large errors in derived $\Delta C_P°$ values.

In spite of the difficulties, it appears that most of the calorimetric results are in fair agreement with eqns. (7)–(9) from Parker[625] (represented by the straight line in Fig. 1) from about 0°C to about 50°C. Before turning to further calculations based on eqn. (7) we point out that above room temperature there are substantial discrepancies between partial molal heat capacities reported by various investigators. For example, $\bar{C}_P°$ values for HCl(aq) from Ackermann[2] and from Ahluwalia and Cobble[4] differ by 2.2, 10.0, and 13.5 cal deg^{-1} mol^{-1} at 50, 70, and 90°C. It will require further investigations, which should include pK_w, calorimetric $\Delta H°$, and calorimetric $\bar{C}_P°$ values, to provide adequate characterization of the thermodynamics of ionization of hot water.

Our "best" values for K_w, pK_w, $\Delta G°$, $\Delta H°$, $\Delta S°$, $\Delta C_P°$, $d(\Delta C_P°)/dT$, and $\Delta V°$ for ionization of water at 298°K and 1.0 atm are listed in Table IV. The K_w, pK_w, and $\Delta G°$ values are from Guggenheim and Turgeon.[301] The $\Delta S°$ value is from the $\Delta G°$ just cited and our previously discussed "best" $\Delta H°$. The $\Delta C_P°$ and $d(\Delta C_P°)/dT$ values are from Parker,[625] as partly confirmed by results of other investigators discussed in preceding paragraphs and displayed in Fig. 1. The $\Delta V°$ value is from partial molal volumes that are discussed later in this chapter.

Combination of the "best" values from Table IV has already led to eqns. (7) and (8) for $\Delta C_P°$ and $\Delta H°$ as functions of temperature. Combination of our "best" $\Delta S°$ with eqn. (7) and $d(\Delta S°)/dT = \Delta C_P°/T$ leads to

$$\Delta S° = 1545.84 - 321.8 \ln T + 0.9T \qquad (11)$$

Further combination of (8) and (11) gives

$$\Delta G° = 69{,}283 - 1867.64T - 0.45T^2 + 321.8T \ln T \qquad (12)$$

TABLE IV. "Best" Values for Thermodynamics of Ionization of Water at 298°K

$$K_w = 1.002 \times 10^{-14}$$
$$pK_w = 13.9991$$
$$\Delta G° = 19.098 \text{ kcal mol}^{-1}$$
$$\Delta H° = 13.34 \text{ kcal mol}^{-1}$$
$$\Delta S° = -19.31 \text{ cal deg}^{-1} \text{ mol}^{-1}$$
$$\Delta C_P° = -53.5 \text{ cal deg}^{-1} \text{ mol}^{-1}$$
$$d(\Delta C_P°)/dT = 0.9 \text{ cal deg}^{-2} \text{ mol}^{-1}$$
$$\Delta V° = -22.13 \text{ cm}^3 \text{ mol}^{-1}$$

and then from $\Delta G^\circ = -RT \ln K$ we obtain

$$\ln K_w = -(34.865/T) + 939.8563 + 0.22645T - 161.94 \ln T \quad (13)$$

These equations are limited to the range of satisfactory accuracy of (7), which probably includes temperatures from $\sim$5 to $\sim$45°C.

We also note that there have been several determinations of pK_w values at temperatures above 100°C, with some investigations extending over wide ranges of pressure. Recent papers by Mesmer et al.[567] and Quist[657] report their excellent results and cite references to earlier publications. As Quist[657] has pointed out, we now have pK_w values extending from 0 to 1000°C and from 1 to 133,000 bar. In principle we can use pK_w values at various high temperatures to add to our knowledge of ΔC_P° as already discussed. But in practice the experimental difficulties have been sufficiently formidable that uncertainties in published pK_w values largely limit us to agreeing with Ackermann[2] that ΔC_P° probably passes through a maximum in the neighborhood of 70°C. It also appears that K_w (at pressures near the equilibrium vapor pressure) passes through a maximum between 200 and 300°C.

Combination of $\Delta G^\circ = -RT \ln K$ with $[\partial(\Delta G^\circ)/\partial P]_T = \Delta V^\circ$ gives

$$[\partial(\ln K)/\partial P]_T = -(\Delta V^\circ)/RT \quad (14)$$

Thus the standard-state volume change for ionization of water can be evaluated from values of K_w at several pressures, or we can calculate K_w at several pressures from prior knowledge of the ΔV° of ionization of water.

We first turn to evaluation of ΔV° from volumetric data. This problem can be approached by way of apparent molar volumes of strong acids, strong bases, and salts, and also by way of direct measurement of the volume change associated with neutralization reactions involving strong acids and bases.

Bodanszky and Kauzmann[62] have cited earlier dilatometric measurements leading to $\Delta V = 21.0$ cm³ mol⁻¹ for neutralization of 0.1 N HNO₃ by 0.1 N NaOH and $\Delta V = 20.7$ cm³ mol⁻¹ when 0.1 N HCl is added to 0.24 N NaOH. Similar measurements by Kauzmann[62] gave $\Delta V = 20.8$ cm³ mol⁻¹ on mixing equal volumes of 0.1 N HCl and 0.1 N NaOH. All of these quantities refer to reactions represented by

$$\text{HX(finite conc.)} + \text{NaOH(finite conc.)} = \text{H}_2\text{O(liq)} + \text{NaX(finite conc.)} \quad (15)$$

and must be adjusted to zero concentrations of solutes to yield the desired standard-state volume change. Because we can do these calculations more accurately for single electrolyte systems as discussed in the following paragraphs, we only note that the results cited above lead to $\Delta V^\circ \cong 21$ cm^3 mol^{-1} for the neutralization of H$^+$(aq) by OH$^-$(aq) and thence to $\Delta V^\circ \cong -21$ cm^3 mol^{-1} for the ionization of water at 25°C. We thus see from eqn. (14) that K_w is expected to increase with increasing pressure.

The apparent molar volume of solute is defined by

$$\phi V_2 = (V - n_1 v_1{}^\circ)/n_2 \tag{16}$$

and the partial molar volume by

$$\bar{V}^2 = (\partial V/\partial n_2)_{n_1, P, T} \tag{17}$$

in which V and $v_1{}^\circ$ represent the volume of solution and the molar volume of pure solvent, while n_1 and n_2 represent numbers of moles of solvent and solute in the solution. Since it may be shown[454] that $\bar{V}_2 = \phi V_2$ at $n_2/n_1 = 0$, our problem is reduced to finding limiting values of ϕV_2 as the concentration of solute tends to zero. This requires very accurate density data for dilute solutions and proper extrapolation to infinite dilution.

Because the Debye–Hückel theory provides information about the free energy of dilution of electrolytes in very dilute solutions, it is possible to use $(\partial \bar{G}_2/\partial P)_T = \bar{V}_2$ to obtain information about the concentration dependence of $\bar{V}_2$ (or ϕV_2) as reviewed by Redlich and Meyer[677] and Millero.[575] The result is that

$$\phi V_2 = \phi V_2{}^\circ + k(M)^{1/2} + \cdots \tag{18}$$

in which $\phi V_2{}^\circ = \bar{V}_2{}^\circ$ is the apparent or partial molar volume at infinite dilution. The Debye–Hückel constant k is a function of temperature, dielectric constant and density of the solvent, and the charge type of the solute. For 1:1 electrolytes in water at 25°C, it is now clear[575,677] that the physical data for pure water suggest a value of k between 1.86 and 1.87. Millero[575] has reviewed a considerable number of experimental investigations that provide convincing support for the theoretical value of k.

Now we proceed to use ϕV_2 data for aqueous HCl, NaOH, and NaCl with the molar volume of H$_2$O(liq) to obtain the desired ΔV° for ionization of water [represented by eqn. (1)] from

$$\Delta V^\circ = \bar{V}^\circ_{\text{HCl}} + \bar{V}^\circ_{\text{NaOH}} - \bar{V}^\circ_{\text{NaCl}} - V^\circ_{\text{H}_2\text{O}} \tag{19}$$

Review of ϕV_2 data and extrapolation to zero concentration to obtain $\bar{V}_2{}^\circ$ values led Owen and Brinkley[617] to report $\Delta V^\circ = -23.4$ cm^3 mol^{-1}

for ionization of water, calculated by way of eqn. (19). It has subsequently been pointed out by several authors that the $\bar{V}_2{}^\circ$ values used by Owen and Brinkley[617] were based on incorrect extrapolations, with their $\bar{V}_2{}^\circ = -6.8$ cm³ mol⁻¹ for NaOH(aq) apparently contributing the largest error to their calculated ΔV°.

More recently, Bodanszky and Kauzmann[62] have reviewed earlier data and made dilatometric measurements on aqueous sodium hydroxide at 30°C with concentrations down to 0.06 M. Their resulting $\phi V_2{}^\circ = \bar{V}_2{}^\circ = -4.60$ for NaOH(aq) at 25°C led them to $\Delta V^\circ = -21.28$ cm³ mol⁻¹ for ionization of water. Redlich and Meyer[677] have pointed out that this result can be improved by making use of a better $\bar{V}_2{}^\circ$ value for NaCl(aq) and thereby obtained $\Delta V^\circ = -21.45$ cm³ mol⁻¹ for ionization of water at 25°C and 1.0 atm.

Subsequent to the work of Bodanszky and Kauzmann,[62] Hepler *et al.*[341] developed an improved dilatometric method for determination of ϕV_2 values for dilute solutions and applied this method to NaOH solutions down to 0.001757 M. Extrapolation of resulting ϕV_2 values to zero concentration by way of eqn. (18) led to $\phi V_2{}^\circ = \bar{V}_2{}^\circ = -5.25$ cm³ mol⁻¹ for NaOH(aq). Then $\Delta V^\circ = -21.86$ cm³ mol⁻¹ was calculated for ionization of water at 25°C. Later dilatometric measurements[186] on dilute solutions of HCl confirmed Redlich's observation that the frequently quoted[317] $\phi V_2{}^\circ$ values (18.20 and 18.07 cm³ mol⁻¹) for HCl(aq) are incorrect, largely due to an incorrect extrapolation to zero concentration. The new measurements[186] led to $\phi V_2{}^\circ = 17.82$ cm³ mol⁻¹ for HCl(aq), in good agreement with the earlier work of Redlich and Bigeleisen.[674] Dunn *et al.*[186] combined this $\bar{V}_2{}^\circ = 17.82$ cm³ mol⁻¹ for HCl(aq) with $\bar{V}_2{}^\circ = -5.25$ cm³ mol⁻¹ for NaOH(aq),[341] $\bar{V}_2{}^\circ = 16.61$ cm³ mol⁻¹ for NaCl(aq) from the work of Kruis[478] (obtained by extrapolation based on the correct Debye–Hückel limiting slope), and $V^\circ = 18.07$ cm³ mol⁻¹ for H₂O(liq) to obtain $\Delta V^\circ = -22.11$ cm³ mol⁻¹ for ionization of water. Since more recent investigations[236,792] lead to $\bar{V}_2{}^\circ = 16.65$ and 16.63 cm³ mol⁻¹ for NaCl(aq), we adopt $\Delta V^\circ = -22.13$ cm³ mol⁻¹ as the "best" value for the standard volume change for ionization of water at 25°C and 1 atm. Unfortunately, available data are inadequate for accurate evaluation of ΔV° at other temperatures.

As mentioned earlier, eqn. (14) provides a connection between ΔV° of ionization and the pressure dependence of the equilibrium constant for ionization. One way to evaluate ΔV° from ionization constants at several pressures is to obtain the limiting slope of pK_w against pressure as $P \to 1$ atm. A better approach is that developed by Lown *et al.*,[527] who allowed for the pressure dependence of ΔV° by introducing an isothermal change in compressibility defined by

$$[\partial(\Delta V^\circ)/\partial P]_T = -\Delta \varkappa^\circ \tag{20}$$

We take $\Delta \varkappa^\circ$ to be independent of pressure and combine it with eqns. (14) and (20) to yield

$$[RT/(P-1)]\ln(K_P/K_1) = -\Delta V_1^\circ + \tfrac{1}{2}\Delta \varkappa^\circ(P-1) \tag{21}$$

in which subscripts P and 1 indicate values of K and ΔV° that refer to pressure P and to 1 atm, respectively.

Lown et al.[527] have applied their eqn. (21) to the pK_w values determined by Hamann[308] at pressures from 1 to 2000 atm and have obtained $\Delta V^\circ = -20.5$ cm³ mol⁻¹ and $\Delta \varkappa^\circ = -3.4 \times 10^{-3}$ cm³ mol⁻¹ bar⁻¹ from the intercept and slope of their graph. In spite of the ingenuity of this treatment and the good quality of the experimental K_w values, the above ΔV° is probably not as "good" as the "best" value cited earlier because this calculation (analogous to calculation of ΔH° from K values at several temperatures with ΔC_P° taken to be independent of temperature) involves differentiation of the K_w values and thus effectively magnifies small experimental errors.

Several investigators[143,282,628] have reported ionization constants for D_2O and reviewed results of earlier investigations. The investigation by Covington et al.[143] included determination of equilibrium constants at 5-deg intervals from 5 to 50°C and calculation of $\Delta H^\circ = 14,311$ cal mol⁻¹ for the ionization reaction at 25°C. Subsequent calorimetric measurements by Goldberg and Hepler[284] led to $\Delta H^\circ = 14,488$ cal mol⁻¹ for the ionization reaction at 25°C. It is presently impossible to account for the discrepancy between these two ΔH° values.

When comparing the self-ionization of D_2O with that of H_2O it is necessary to give careful consideration to the standard states on which equilibrium constants are based, as discussed previously.[143,284] On the basis of reasonable choices based on molarity, molality, or "aquamolality," we can say that H_2O is more ionized than is D_2O at 25°C and can also say that the difference in enthalpies of ionization contributes more than does the difference in $T\Delta S^\circ$ of ionization.

We also note that the equilibrium constant for self-ionization of T_2O has recently been determined by Goldblatt and Jones,[285] whose results indicate that T_2O is less ionized than D_2O or H_2O.

There have been a considerable number of experimental and theoretical investigations of the equilibrium

$$H_2O + D_2O = 2HDO \tag{22}$$

and of the self-ionization of H_2O–D_2O mixtures. Several of these investigations have been concerned with equilibria between the various products of self-ionization that can be described by H^+(solvated), D^+(solvated), OH^-(solvated), and OD^-(solvated) in H_2O–D_2O mixtures. Because of limitations of space, we can only refer the interested reader to some of the published work.[245,280–282,331,332,388,466,467,546,605,606,628,655,656] We also call attention to calorimetric measurements on H_2O–D_2O mixtures.[185] Two reviews[499,777] that are partly concerned with isotope effects are of interest.

1.1. Ionization of Water in Aqueous Electrolyte Solutions

The most extensive studies of the ionization of water in aqueous electrolyte solutions are those of Harned et al.[317] mostly involving measurements on cells of type

$$H_2 \mid HX, MX \mid AgX, Ag \tag{23}$$

and

$$H_2 \mid MOH, MX \mid AgX, Ag \tag{24}$$

From the results they have calculated values of the ion product of water, $Q_w = m_{H^+}m_{OH^-}$, for solutions with electrolyte concentrations up to about 5 m at several temperatures. Knowledge of the value of K_w (zero ionic strength) has allowed calculation of activity coefficient ratios $(\gamma_{H^+}\gamma_{OH^-}/\gamma_{H_2O})$ at each electrolyte concentration and each temperature.

We also call attention to investigations[513,514] with glass electrodes of ionization of water in moderately concentrated electrolyte solutions, and to several investigations cited by Sillen and Martell.[733]

The investigation by Mesmer et al.[567] of the ionization of water in 1.0 m KCl is of particular interest because of the wide range of temperature (50–292°C).

1.2. Ionization of Water in Aqueous Organic Mixtures

Until recently there have been few data on the ionization of water in aqueous organic mixtures. Harned and Fallon[316] made potentiometric measurements with hydrogen electrodes and have calculated values of K_w

for ionization of water in 20, 40, and 75 mass per cent dioxan–water mixtures at several temperatures.

Gutbezahl and Grunwald[303] have also made potentiometric measurements with hydrogen electrodes and have calculated K_w values for ionization of water in 20, 35, 50, 65, and 80 mass per cent ethanol–water mixtures at 25°C.

Banerjee et al.[34] have used the method of Harned and Fallon[316] to obtain K_w values for ionization of water in 10, 30, 50, 70, and 90% ethylene glycol–water mixtures.

More recently, Woolley et al.[859]* have devised a rapid and convenient method making use of a glass electrode and pH meter for investigation of the ionization of water in aqueous organic mixtures. Results have been reported for binary mixtures of water with ethanol, 1-propanol, 2-propanol, 2-methyl-2-propanol, ethylene glycol, acetone, and p-dioxan at 25°C. Various choices of standard states have been discussed in some detail.

Reported results for ionization of water in aqueous dioxan,[316,859] aqueous ethanol,[303,859] and aqueous ethylene glycol[34,859] obtained with hydrogen electrodes[34,303,316] and glass electrodes[859] are in excellent agreement, thus confirming that these different electrodes respond to the same solvated H⁺ species in these solutions.

One striking feature of the previously reported results[34,303,316,859] is the qualitative difference between pK_w values for ionization of water in aqueous ethylene glycol and in all the other systems mentioned above. It is generally expected that these pK_w values will increase with increasing organic content of solvent, as observed for most systems.[303,316,859] But two independent investigations[34,859] have led to reported pK_w values that decrease with increasing ethylene glycol content. Because of the excellent agreement between results obtained by way of different electrode systems, it appears certain that neither the "normal" behavior of water in most systems nor the "abnormal" behavior in ethylene glycol can be attributed to experimental errors.

Reported pK_w values have been implicitly based on the assumption that the organic cosolvent does not ionize to contribute H⁺(solvated) to the solution, or at least that such ionization is negligible in comparison with ionization of water. This assumption is obviously justified in the case of an ideal "inert" cosolvent and is equally obviously incorrect for distinctly acidic cosolvents such as acetic acid or even weaker acids such as phenol. For still less acidic cosolvents (such as ethylene glycol) an ex-

* Note that eqn. (4) should read $K_{a/c} = (K_{a/1})/C_w$.

tensive analysis of experimental data is required, as reported by Woolley *et al.*[860] This analysis has led to accurate ionization constants for such weak acids as glucose, sucrose, glycerol, and ethylene glycol (all as dilute solutes in water) and accounts for the apparent decrease in pK_w for ionization of water in mixtures with these substances. Further results obtained by way of measurements with glass electrodes in other solvent systems have also been reported.[267,857,860]

The glass electrode–pH meter method[859] with calculations that allow for the acidity or basicity of the cosolvent[860] has been applied[857] to mixtures of water with various slightly basic compounds. This same method has also been applied[267] to several D_2O organic mixtures.

Calorimetrically determined values of $\Delta H°$ for ionization of water in several water–ethanol mixtures are available.[30,52] These results show that both enthalpies and entropies must be considered in attempts to understand these ionization reactions. Further measurements of heats of neutralization in other water–organic mixtures will be of considerable interest.

2. HYDRATION OF H⁺ AND OH⁻

Any reasonably complete understanding of acids and bases in aqueous solution must be extensively concerned with interactions of H^+, OH^-, and solute species (such as acetic acid, acetate ion, aniline, and anilinium ion) with water. Because H^+ and OH^- are common to all of these acid–base reactions, we focus our attention on them in the following discussions.

2.1. Gas-Phase Hydration of H⁺ and OH⁻

The most extensive and reliable thermodynamic data for the gas-phase hydration of H^+ and OH^- are from the recent mass spectrometric investigations of Kebarle *et al.*[27,436,437,439,440] and De Paz *et al.*[167–170]

Kebarle's group has measured ion intensities of the hydrated species $H^+(H_2O)_n$, where, $1 \leq n \leq 8$ and $OD^-(D_2O)_n$, where $0 \leq n \leq 4$, at several temperatures. From these measured intensities they calculated gas phase equilibrium constants for reactions of types

$$H^+(H_2O)_n(g) + H_2O(g) = H^+(H_2O)_{n+1}(g) \qquad (25)$$

and

$$OD^-(D_2O)_n(g) + D_2O(g) = OD^-(D_2O)_{n+1}(g) \qquad (26)$$

They then applied the van't Hoff equation to their equilibrium constants to obtain $\Delta H°$ and $\Delta S°$ values.

Equilibrium constants for reaction (25) with $n = 3$ and $n = 4$ derived from the kinetic data of Puckett and Teague[652] are in good agreement with corresponding values reported by Kebarle et al.[440]

De Paz et al.[168–170] have used a tandem mass-spectrometric method to measure kinetic energy thresholds for dissociation of ions of types $D^+(D_2O)_n$, where $1 \le n \le 5$, and $OH^-(H_2O)_n$, where $1 \le n \le 3$. From their data they calculated $\Delta H°$ values for the hydration reactions analogous to reactions (25) and (26), mostly in good agreement with results from Kebarle's group.

For the important reaction of type (25) with $n = 0$ there are several reported $\Delta H°$ values,[167,169,170,309,439,440,526] with the most reliable value apparently about -165 kcal mol^{-1}.

Several papers have described quantum mechanical calculations of the structures and properties of hydrates of H^+ and OH^-.[153,167,458]* In one of the most extensive of these papers De Paz et al.[167] have used the CNDO/2 method in calculating geometries, charge distributions, and energies of the species $H^+(H_2O)_n$, where $0 \le n \le 4$, and $OH^-(H_2O)_n$, where $0 \le n \le 3$. They have calculated $\Delta H°$ values for reactions of types (25) and (26) that are in reasonable agreement with experimental values discussed above. Others have employed CNDO/2,[153] LCAO-MO-SCF,[170,458] valence bond,[170] and Slater orbital[170] methods in calculating structures and properties of various hydrated H^+ and OH^- species.

We list in Table V our estimates of "best" values for the thermodynamics of reactions of types (25) and (26) at 298°K.

2.2. Hydration of H^+ and OH^- in Aqueous Solution

In this section we are primarily concerned with $\Delta G°$, $\Delta H°$, and $\Delta S°$ values for the hydration of the proton represented by

$$H^+(g) = H^+(aq) \qquad (27)$$

Unfortunately, results of thermodynamic measurements and rigorous classical thermodynamics yield only sums of these quantities for like-charged ions and differences between these quantities for oppositely charged ions. We are therefore forced to rely upon various theoretical estimates of

---

* Summarized in Ref. 170, Table III.

TABLE V. Thermodynamics of Hydration of Gaseous H⁺ and OH⁻ at 298°K[a]

n	$-\Delta H^\circ$	$-\Delta G^\circ$	$-\Delta S^\circ$
	$H^+(H_2O)_n(g) + H_2O(g) = H^+(H_2O)_{n+1}(g)$		
0	165	—	—
1	34	25	30
2	23	13.6	32
3	17	8.5	29
4	15	5.5	32
5	13	3.9	31
6	12	2.8	32
7	10	2.2	26
	$OH^-(H_2O)_n(g) + H_2O(g) = OH^-(H_2O)_{n+1}(g)$		
0	24	16.9	23
1	18	10.7	24
2	16	7.7	28
3	14	5.5	28
4	14	4.2	33

[a] Standard states are based on ideal gas behavior at $P = 1.0$ atm. Units for ΔH° and ΔG° are kcal mol⁻¹ while units for ΔS° are cal deg⁻¹ mol⁻¹.

the quantities of interest or upon combinations of thermodynamic data with some extrathermodynamic assumption (more or less well justified) as to how the sums and differences mentioned above should be divided.

Because so many papers have been published on the thermodynamics of hydration of ions, we can cite only a modest number of representative efforts. Although results of no single investigation can be regarded as entirely convincing, the net result of several investigations is that we can cite approximate "best" values with a reasonable degree of confidence.

We begin with consideration of the "absolute" partial molal entropy of H⁺(aq) in the usual hypothetical 1 m standard state. Reported[131,148, 166,600,601,697] values for the absolute $\bar{S}^\circ$ for H⁺(aq) range from -1.1 to -6.3 cal deg⁻¹ mol⁻¹ and we choose -4 ± 2 cal deg⁻¹ mol⁻¹ as the "best" value. Combination of this value with $S^\circ_{298} = 26.0$ cal deg⁻¹ mol⁻¹ for H⁺(g) permits us to calculate that $\Delta G^\circ = \Delta H^\circ + 9$ (expressed in kcal mol⁻¹) for the hydration reaction represented by eqn. (27).

On the basis of a large number of investigations and reviews[131,166, 483,600,601,697,708,756] that yield $\Delta G°$ of hydration values ranging from -235 to -292 kcal mol^{-1} we adopt $\Delta G° = -260 \,(\pm?)$ kcal mol^{-1} and $\Delta H° = -269$ kcal mol^{-1} as "best" values for the hydration process represented by eqn. (27).

We now find that combination of $\Delta H°$ values in Table V with the above $\Delta H°$ of hydration of H$^+$ and the $\Delta H°$ of vaporization of water leads to $\Delta H° = -64$ kcal mol^{-1} for

$$H^+(H_2O)_8(g) = H^+(aq) \tag{28}$$

Although there is no direct experimental or theoretical check on this value, it seems to be in reasonable accord with values for "large" ions from papers cited earlier.

A large number of investigations leading to an absolute $\bar{V}°$ for H$^+$(aq) have been reviewed by Millero,[575] King,[447] and Laliberté and Conway.[485] On the basis of values cited in these papers we select $\bar{V}° = -5 \pm 1$ cm^3 mol^{-1} as the "best" value for H$^+$(aq).

3. ORGANIC ACIDS AND BASES IN AQUEOUS SOLUTION

Strengths of organic acids and bases in water are commonly and usefully discussed in terms of equilibrium constants and related thermo-dynamic functions for certain reactions. For example, we consider the acidity of phenol in terms of the reaction represented by

$$C_6H_5OH(aq) = H^+(aq) + C_6H_5O^-(aq) \tag{29}$$

and the corresponding equilibrium constant

$$K = a_H a_P / a_{HP} \tag{30}$$

in which activities of the solute species are indicated by subscripts (charges omitted for convenience). The numerical value for K (also $\Delta G°$ and $\Delta S°$) depends on the choice of standard states. The most common standard states for thermodynamic investigations are the hypothetical 1 m solution for solutes and the pure liquid for solvent, as already discussed in connection with the self-ionization of water. In some investigations the molar scale is used either for experimental convenience or to permit comparison of K values based on equal volumes of solution for different solvents. In this case it is necessary to allow for the thermal expansion of the solvent in calculating ΔH° and ΔS° values from equilibrium constants at several temperatures.[445]

We may similarly represent the basicity of aqueous aniline in terms of

$$C_6H_5NH_2(aq) + H_2O(liq) = C_6H_5NH_3^+(aq) + OH^-(aq) \qquad (31)$$

and

$$K_{31} = a_{AnH}a_{OH}/a_{An} \qquad (32)$$

Or we can represent the related acidity of anilinium ions in terms of

$$AnH^+(aq) = H^+(aq) + An(aq) \qquad (33)$$

and

$$K_{33} = a_Ha_{An}/a_{AnH} \qquad (34)$$

Because the equilibrium constants for reactions represented by (31) and (33) are related by

$$K_{31}K_{33} = K_w \qquad (35)$$

we may choose to treat the acidity–basicity of the anilinium–aniline conjugate pair in terms of either reaction (33) or reaction (31).

The equilibrium constants and other thermodynamic functions for reactions such as (29), (31), and (33) depend on the properties of the "solvent" (also a "reactant") as well as on the properties of the various solute species. In fact, the acidity–basicity of a solvent limits the range of acid and base strengths that can be studied in that solvent. Because of the dependence of acidity or basicity on solvent, it is often advantageous to use solvents that are themselves more acidic or basic than is water. For example, some acids that appear to be of equal strength because they are (nearly) completely ionized in aqueous solution may be differentiated in some more acidic solvent such as glacial acetic acid. Another reason for using nonaqueous or partly aqueous solvents rather than pure water is that many organic acids and bases are only sparingly soluble in water but dissolve readily in various mixed solvent systems such as water–ethanol or water–cellosolve. Further, it appears from results of several recent investigations that study of acid–base reactions in aqueous organic solvents can lead to useful information about molecular interactions in solution. In this connection we have space only to cite the excellent book by Bates[37] and a number of his and Robinson's recent publications.[38,619,620,686,687, 726,745,856]

Because of limitations of space, we can only cite a few references and point out here that studies of isotope effects on acid–base properties have led to considerable information about both solutes and solvent in aqueous systems. Laughton and Robertson[499] have reviewed a variety of solvent-isotope effects and have prepared an extensive table of pK values for acids in H_2O and their deuterated counterparts in D_2O. King[445] has given a clear account of several theoretical investigations. We also call attention to an analysis[336] of combined isotope-substituent effects that is related to a treatment of substituent effects that is described later in this chapter.

3.1. Thermodynamics of Ionization Reactions

King[445] and Robinson and Stokes[694] have reviewed and described in detail most methods for determining equilibrium constants for acid and base ionization reactions, with emphasis on methods that are capable of high accuracy.

Several procedures based on the glass electrode–pH meter combination have been usefully applied to many problems, although the accuracy is often poorer than might be attained by other more difficult methods.[445,694] For example, O'Hara et al.[607] have described one such convenient method and applied it to determination of ionization constants for o-allylphenol and o-propylphenol in a study of intramolecular hydrogen bonding. Liotta et al.[523] have subsequently applied this same method over a range of temperature and have reported ionization constants (also $\Delta H°$ and $\Delta S°$ values) for p- and m-hydroxybenzaldehyde. A glass electrode–pH meter method previously cited[267,857,860] has proven useful for investigation of very weak acids and bases. While these methods may be applied with some confidence to determinations of K and $\Delta G°$, caution is called for in assessing $\Delta H°$ and $\Delta S°$ derived from the temperature dependence of K determined with the glass electrode (or any other electrode system with a liquid junction).

Christensen, Izatt, and their students have described and applied[104,106,109,110,312] a calorimetric method for determination of equilibrium constants for proton ionizations. Their method offers the practical advantage of yielding K, $\Delta G°$, $\Delta H°$, and $\Delta S°$ values from a single calorimetric run.

The equilibrium constant for an acid or base ionization reaction leads directly to the standard free energy change for that reaction. When equilibrium constants have been determined at several temperatures it is also possible to calculate $\Delta H°$, $\Delta S°$, and $\Delta C_P°$ values. King[445] has discussed methods of carrying out these calculations and assessing errors in derived thermodynamic quantities. More recently, Ives and his students[399–401] have given careful attention to these questions in relation to their own very accurate K values determined by an improved conductance method. The Clarke and Glew[111] method, which involves expression of ΔC_P° as a Taylor series expansion, is easily adapted to computer calculations and is now widely used. Bolton[64] has recently reviewed these calculations, with particular emphasis on the Clarke and Glew method.

Calorimetry offers the most direct method for determination of ΔH° values for acid and base ionization reactions, and many of the "best" ΔH° values presently available are based on calorimetric measurements. Both titration calorimetry and the more traditional "ampoule" calorimetry have been extensively applied to various acid–base reactions. At least for imidazole and presumably for many other monoprotic acid–base reactions the calorimetric titration procedure appears to offer significant advantages in terms of speed, while the more traditional ampoule method leads to slightly more precise results.[861]

Subsequent to the pioneering work of Feates and Ives[212] (see also Refs. 399 and 445) there has been considerable interest in ΔC_P° values for ionization reactions. Because calculation of ΔC_P° from K values at several temperatures is done by way of two differentiations with respect to temperature, very accurate K values are required. Investigation of the theoretically interesting temperature variation of ΔC_P° effectively requires still another differentiation, which imposes still more stringent requirements on the accuracy of the K values. It is also possible to obtain ΔC_P° values from ΔH° values determined calorimetrically at several temperatures. This method offers the relative advantage of requiring only a single differentiation. Another calorimetric method is based on determination of partial molal heat capacities in very dilute solutions, which is experimentally difficult but requires no differentiation with respect to temperature. Each of these approaches to ΔC_P° is difficult, and each has so far yielded results that are afflicted with larger uncertainties than are often desired. In spite of these problems, currently increasing interest in ΔC_P° values is likely to be repaid with considerably improved understanding of the interaction of solute species with water.

Volume changes associated with ionization reactions can be evaluated from K values determined at several pressures, as described by Lown *et al.*,[527] who cite earlier investigations carried out by several methods. It is also possible to obtain ΔV° by way of dilatometric measurements of volume changes accompanying neutralization and by apparent molal volume determinations as described and reviewed by King.[446] King[446] has also contributed to and reviewed attempts to gain understanding of solute–solvent interactions from volumetric data.

Izatt and Christensen[402] and Larson and Hepler[493] have compiled extensive tables of pK, ΔG°, ΔH°, ΔS°, and $\Delta C_P{}^\circ$ values for acid ionization reactions in aqueous solution. No similar compilation of ΔV° values is available, but many ΔV° values have been cited.[337,446,462,527,671,795] Jencks and Regenstein[409] have listed pK values for a large number of acids and bases, with emphasis on those that are of biological interest.

3.2. Substituent Effects on Ionization of Organic Acids

We represent the ionization of an acid by

$$HA^z(S) = H^+(S) + A^{z-1}(S) \tag{36}$$

in which (S) indicates that the preceding species is in solution. When our principal interest is in substituent effects, it is useful to compare the thermodynamic functions for ionization of substituted acid HA^z with those for some reference acid HR^z in terms of the proton transfer reaction represented by

$$HA^z(S) + R^{z-1}(S) = A^{z-1}(S) + HR^z(S) \tag{37}$$

Thermodynamic quantities for (37) are obtained as $K_s = K_{HA}/K_{HR}$ and $\delta(\Delta X^\circ) = \Delta X^\circ_{HA} - \Delta X^\circ_{HR}$, in which K_s and the related thermodynamic functions represented generally by ΔX° are the substituent effects of interest. Effects of temperature and solvent are often much smaller for reactions of type (37) than for either reaction (36) or the similar reaction of HR^z.

The most widely used correlation of substituent effects is the Hammett equation, which we write as

$$\log K_s = \varrho\sigma \tag{38}$$

The "substituent constant" σ is intended to depend only on the substituent, while the parameter ϱ depends on the reaction, solvent, and temperature. The wide range of empirical success of the Hammett equation in correlating both equilibrium constants and rate constants provides general support for the idea that certain effects of substituents can properly be regarded as being (almost) independent of solvent and temperature.

Thermodynamic analysis of eqn. (38) in terms of a temperature-independent σ has led[339] to several proportionality relationships between the various thermodynamic functions and σ. For the common case where the accuracy of experimental data limits us to consideration of $\delta(\Delta H^\circ)$ and $\delta(\Delta S^\circ)$ as independent of temperature with $\delta(\Delta C_P^\circ) = 0$, it was found[339] that

$$\varrho = \varrho_\infty[1 - (\beta_i/T)] \tag{39}$$

and

$$\delta(\Delta H^\circ) = \beta_i \, \delta(\Delta S^\circ) \tag{40}$$

in which β_i is an integration constant (not restricted to any particular value) that is identical with the quantity called the isoequilibrium (or isokinetic) temperature by Leffler and Grunwald.[503] When $\delta(\Delta C_P^\circ) \neq 0$ so that both $\delta(\Delta H^\circ)$ and $\delta(\Delta S^\circ)$ vary with temperature, more complicated equations were obtained, as shown elsewhere.[339]

There are a number of proton transfer and other reaction series that do exhibit the various proportionalities such as (40) that have been derived[339] from the Hammett equation. On the other hand, it is becoming well known that the Hammett equation often provides a good correlation of equilibrium constants and free energies even when enthalpies and entropies vary in apparently erratic fashion. We are therefore faced with the problem of understanding why the Hammett equation "works better than it should." Another way of phrasing the problem is to ask why it is possible to ignore both solvent and temperature while explaining substituent effects on equilibrium constants reasonably successfully in terms of potential energies (σ values) of solute species. We approach these problems by way of a particular model for thermodynamics of substituent and solvent effects.

It has been shown previously[65,335,339-340,445,492,493] that it is useful to express thermodynamic functions for reactions of type (37) in terms of "internal" and "environmental" contributions as in the following:

$$\delta(\Delta H^\circ) = \delta(\Delta H_{int}) + \delta(\Delta H_{env}) \tag{41}$$

$$\delta(\Delta S^\circ) = \delta(\Delta S_{int}) + \delta(\Delta S_{env}) \tag{42}$$

King[445] has clearly described these contributions: "Internal effects are those intrinsic to the molecules of the acid and base. Environmental effects are those which result from interaction of the molecules of the acid and base with the solvent." Evidence that $\delta(\Delta S_{int}) \cong 0$ for many symmetric reactions of type (37) has been cited earlier.[335,340] Thus we have $\delta(\Delta S^\circ) \cong \delta(\Delta S_{env})$.

Earlier investigations[65,339] have proceeded on the basis of a proportionality between $\delta(\Delta H_{\text{env}})$ and $\delta(\Delta S_{\text{env}})$:

$$\delta(\Delta H_{\text{env}}) = \beta_e \, \delta(\Delta S_{\text{env}}) + \gamma \, \delta(\Delta H_{\text{int}}) \tag{43}$$

in which β_e is a general "environmental" parameter and γ is a specific parameter that depends on the solvent and the temperature. A variety of evidence has been cited[65,335,339,340,399,492,493] to show that $\beta_e \cong T$. Among the equations that result from this treatment are

$$\delta(\Delta G^\circ) = \delta(\Delta H_{\text{int}})(1 + \gamma) \tag{44}$$

and

$$\varrho\sigma = [C(1 + \gamma)/2.3RT][-\delta(\Delta H_{\text{int}})/C] \tag{45}$$

in which C is an arbitrary proportionality constant. We identify $C(1+\gamma)/2.3RT$ with ϱ and $-\delta(\Delta H_{\text{int}})/C$ with σ. Thus this reaction model leads to a linear free energy relationship with ϱ dependent on solvent and temperature and σ dependent only on the substituent. It is (partial) "compensation" of $\delta(\Delta H_{\text{env}})$ and $\delta(\Delta S_{\text{env}})$ that is responsible for the simple form of eqns. (44) and (45) and the resultant understanding of substituent effects largely in terms of "internal" enthalpy (potential energy). Further, it is this compensation that accounts for the observation that the Hammett equation gives a good approximate correlation of equilibrium constants even when enthalpies and entropies are not in accord with the exact thermodynamic requirements of the Hammett equation.[339]

Differentiation of eqn. (44) with respect to temperature leads[339] to various proportionalities that are identical with those derived from the Hammett equation, and also leads to the isoequilibrium relationship

$$\delta(\Delta H^\circ) = \left(T - \frac{1 + \gamma}{d\gamma/dT}\right) \delta(\Delta S^\circ) \tag{46}$$

Comparison of (40) with (46) shows that

$$\beta_i = T - \frac{1 + \gamma}{d\gamma/dT} \tag{47}$$

Thus the influence of solvent on substituent effects can be expressed in terms of either ϱ or γ. Earlier attempts along this line have been based on treatment of the solvent as a continuous dielectric and have met with some success but have not come to grips with the "structure" of the solvent which is obviously important for aqueous and other hydrogen-bonded systems. We suggest that combination of the treatment outlined here with the "two-state" model for water may be a new fruitful approach.

3.3. Examples of Hydration Effects on Acid–Base Ionization Reactions

Numerous examples of hydration effects on acid–base ionization reactions have been referred to and discussed by Larson and Hepler,[493] King[445] and Bolton and Hepler.[65] Here we can cite only a few further examples and call attention to a few investigations that illustrate useful approaches.

Cox et al.[145] have determined equilibrium constants at several temperatures for ionization of several alkylammonium ions and have calculated values for the various thermodynamic functions for the ionization reactions. We can do no better than to quote their summary of their general approach to interpretation of their data:

"The properties of simple ions and neutral molecules in aqueous solution can be understood broadly in terms of the two following effects: (i) Electrostatic interactions between ionic charges and solvent water leading to orientation of water molecules in the immediate neighborhood of the ion and thus to a decrease in both the entropy and heat capacity of the system; (ii) interactions between inert neutral molecules and the water structure (so-called 'hydrophobic interactions') the exact nature of which has not yet been fully elucidated leading to a decrease in entropy, but an *increase* in the heat capacity of the system."

Because the conclusions reached by Cox et al.[145] cannot be summarized adequately in a short space, we refer the interested reader to their paper and a subsequent paper by Timini and Everett[778] on ionization of some aqueous amino-alcohols. Here we note that these workers have observed [145,778] a general linear relationship between ΔS° and ΔC_P° values for ionization of related acids.

A classic paper that we can refer to only briefly is "The Ionization Functions of Di-isopropylcyanoacetic Acid in Relation to Hydration Equilibria and the Compensation Law" by Ives and Marsden.[399] This paper is an admirable example of careful analysis of experimental results followed by thorough study of the meanings of the various thermodynamic results in terms of molecular structures and interactions of solute species with solvent. Detailed study of this paper is recommended as an early step in any effort to do the same kinds of things for other reactions. Here we have space only for one brief quotation:

"It follows that if any *one* thermodynamic function at a single temperature is to be used in discussing reactions in relation to molecular models (e.g., polar effects of substituents) then, the zero-point energy change, $\Delta H_0{}^\circ$, being inaccessible, ΔG° is better than any other because it is far less sensitive to complications introduced by the solvent in which the reaction is performed. This is in agreement with previous conclusions, but it is evident that there is no thermodynamic function which can truly eliminate all the effects of 'solvent participation,'... ."

We note here that the above conclusion about ΔG° is consistent with our eqn. (44) and that the "complications" of solute–solvent interactions that affect other thermodynamic functions make ΔS° and $\Delta C_P{}^\circ$ particularly useful in studies of hydration effects.

Schwartz and Howard[729] have reported pK_1 and pK_2 values for ionization of 1,2-dihydroxycyclobutenedione ("squaric acid") at several temperatures. The unusual strength of this acid ($pK_1 = 0.6$ at 25°C) may be partly due to unusual interactions of either the acid or its anion with solvent water. Although calorimetric investigation of "almost strong" acids is difficult, such measurements[5,104,106,109,110,312,381] have yielded satisfactory ΔH° and ΔS° values for other acids with small pK values. We suggest that the ΔS° value may prove particularly enlightening in this case.

Fong and Grunwald[220] have determined ionization constants for 2,4,6,2′,4′,6′-hexanitrodiphenylamine in water and in various aqueous acetone mixtures. Equilibrium constants for ionization were found to *increase* with increasing acetone concentration in the solvent, which is directly contrary to the often useful (oversimplified) charge-in-dielectric-continuum model. Fong and Grunwald[220] have explained this interesting result in terms of dispersion forces. We suggest that ΔH° and ΔS° values for ionization of this acid in water and in aqueous acetone will be useful. Although this particular acid is in some senses an extreme example, it is evident that similar interactions may be nonnegligible for many other reactions and deserve more consideration than they have usually received.

Middleton and Lindsey[568] determined acid ionization constants for aqueous hexafluoropropan-2-ol (HFIP) and hexafluoropropan-2,2-diol (HFPD) and commented on the "abnormal acidity" of HFPD in terms of hydrogen bonding. Subsequent measurements[858] have led to $pK = 9.42$ for HFIP and to $pK = 6.65$ for HFPD, both at 25°C, to confirm that HFPD is a distinctly stronger acid in aqueous solution than is HFIP. Calorimetric measurements[858] have led to $\Delta H^\circ = 6.37$ and to $\Delta H^\circ = 6.00 \, \text{kcal mol}^{-1}$ for ionization of aqueous HFIP and HFPD, respectively. Corresponding entropies of ionization are $\Delta S^\circ = -21.7 \, \text{cal deg}^{-1} \, \text{mol}^{-1}$ for HFIP and $\Delta S^\circ = -10.3 \, \text{cal deg}^{-1} \, \text{mol}^{-1}$ for HFPD. These results show that the difference in pK values is mostly due to the difference in entropies of ionization. It has been suggested[858] that the entropy of ionization of HFIP is "normal" and that the considerably less negative entropy of ionization of HFPD is due to an unusually large partial molal entropy for the anion of HFPD, which could result from an unusually small amount of "ordering" of water molecules around this ion. The ordering of solvent molecules near anions of this type can be expressed in terms of the effective electrostatic field about the $-O^-$ group. In the case of the HFPD anion the effective field near the $-O^-$ group may be diminished (compared to the field near the $-O^-$ of the anion of HFIP) by the presence of the geminal $-OH$ group that is oriented by intramolecular charge–dipole interaction. In molecules with appropriate geometry this interaction leads to intramolecular hydrogen bond formation.

Öjelund and Wadsö[608] have determined pK and $\Delta H°$ values for ionization of glyoxylic acid, pyruvic acid, and α-ketobutyric acid. These acids all have larger ionization constants than do formic, acetic, and propionic acids. The $\Delta H°$ of ionization values are positive for the α-keto acids and are negative for the other acids mentioned above. On the other hand, the $\Delta S°$ values for the α-keto acids are much less negative than are the corresponding entropies of ionization of formic acid, etc. Öjelund and Wadsö[608] have interpreted these results in terms of the idea that there is a change of degree of hydration accompanying ionization and have discussed their conclusions in relation to various spectroscopic investigations. This general approach is also appropriate to investigation of various dicarboxylic acids and to gem-diols such as HFPD discussed in the preceding paragraph.

In conclusion, we should like to emphasize that investigations of ionization reactions (pK *and* such other thermodynamic functions as $\Delta H°$, $\Delta S°$, $\Delta C_P°$, and $\Delta V°$) yield results that shed useful light on the relationship between structure and chemical reactivity and also on solute–solvent interactions. Because solute–solvent interactions depend on the detailed properties of the solvent as well as on the solute, such investigations can be used as an effective probe into the "structure" of water and aqueous solutions.

Ionic Transport in Water and Mixed Aqueous Solvents*

Robert L. Kay

Chemistry Department
Carnegie-Mellon University
Pittsburgh, Pennsylvania

1. INTRODUCTION

The mechanism of ionic transport was one of the earliest problems to interest the physical chemist. This early interest resulted not only from the intrinsic importance of the problem but also from a number of important developments before the turn of the last century. The Arrhenius theory[26] established the dissociated nature of electrolytic solutions and precise methods for the measurement of both conductances[456] and transference numbers[378] were developed quite early. The introduction of transference data was the most significant contribution since they permitted an unambiguous split into ionic conductances to be made and thereby allowed comparisons between anions and cations and between all ions at different temperatures and pressures. Until very recently ionic mobilities were the only ionic property that could be obtained independent of any arbitrary split.[†]

After a good procedure was found for the extrapolation of conductance data to infinite dilution verification of Kohlrausch's law of independent limiting ionic mobilities[455] soon followed. This led directly to the formula-

* The preparation of this manuscript was financed by a grant from the National Science Foundation.

† Ionic vibration potentials now permit ionic volumes to be determined.[877] Other methods of a semiempirical nature[132] have also been devised to establish a split of partial molar volumes of salts into their ionic components.

tion of Walden's rule[806] from Stokes' law,[755] namely the limiting conductance–solvent viscosity product depended only on the size of the moving ion. It soon became obvious that few systems followed Walden's rule, water possibly being the system with the most abnormal behavior.

For nearly 50 years little further progress was made in elucidating the mechanism of ionic transport. The reason for this lack of progress was possibly the introduction of the Onsager theory[611,612] and its success in predicting the concentration dependence of conductance, which led to considerable emphasis on theories of ionic association and the constitution of concentrated solutions. Gurney[302] and Frank[228] were among the first in the modern era to reconsider the problem from a different point of view. It was their contention that all theories that ignored the effect of the ionic charge on the solvent in the environment of the ions (the ionic cosphere) and that were based solely on a continuum model for the solvent were doomed to failure, particularly where aqueous solutions were concerned. Subsequently a number of publications from this laboratory substantiated this claim. It is with this aspect of the problem that this chapter will be concerned.

Since this is not a general review, no attempt will be made to include every investigation of conductance in aqueous solutions, and for the most part data are taken from the work in this laboratory or from recent publications of former co-workers. In general, this discussion will be restricted to spherical, univalent ions and the proton is specifically not considered owing to its unique transport mechanism.

2. MEASUREMENT

The measurement of ionic conductances has undergone considerable improvement during the past ten years. Although the conductance measurement has been perhaps one of the easier measurements to carry out with high precision, it was tedious and required considerable skill on the part of the operator. As a result, the number of systems that could be investigated in a reasonable time was limited and this led to a general lack of comprehensive investigations particularly as far as temperature and pressure dependences were concerned. The complete lack of precise temperature coefficient data for ionic mobilities for any solvent other than water until just recently* leaves one to wonder at the several attempts to determine the mechanism of ionic conductance without such basic information.

* G. A. Vidulich, R. L. Kay, and G. P. Cunningham, accepted, *J. Solution Chem.* (1973).

It soon became apparent that two bottlenecks existed in the measurements. First, the skill and time required for precise conductance measurements had to be reduced substantially; and second, better detectors were required for the moving boundary method of measuring transference numbers so that these important data could be measured readily under a variety of conditions. The first requirement was met by designing equipment capable of measuring each quantity by at least a factor of five better than was required (0.02%). In particular, this improvement eliminated operator skill as a limiting factor since the instruments did not have to be pushed to their limit. Fortunately, modern technology has provided improved microbalances, constant-temperature baths, cup dropping devices,[430] and computer-analyzed output.[422] Working in completely closed systems for solvent[202] and salt[435] at all times was of particular importance for aqueous solutions and this aspect added greatly to the internal agreement that could be obtained. A complete description of the apparatus and procedure has recently been reported.[201]

One of the most important recent achievement has been the development of the radio-frequency detector for moving boundaries[649] which permits transference numbers to be measured under conditions never before possible. The principle on which the detector is based is shown in Fig. 1. The moving boundary tube contains five platinum film rings fused onto the *outside* of the glass. Two of the inner rings are connected to a bridge operated at 10.7 MHz to reduce the impedance of this resistance–capacitance circuit. The signal from the center electrode is compared to a reference signal from the bridge generator in a phase-sensitive detector in order to produce a recorder signal that reflects the magnitude of the re-

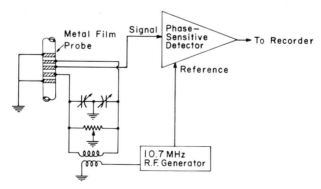

Fig. 1. Schematic diagram of the radio-frequency moving boundary detector.

sistive component of the bridge off-balance. Initially the bridge is balanced by the variable resistance and capacitors. As the boundary passes the second ring the bridge becomes unbalanced owing to the different resistances of the leading and following solutions in a transference experiment. This off-balance reaches a maximum value as the boundary passes the center ring, after which it decreases, leaving the bridge balanced again as it passes the fourth ring. Then two outer grounded rings are added to isolate the probes (set of three rings) from each other and thereby stabilize the baseline. A typical signal is shown in Fig. 2.

Calibration runs on dilute aqueous solutions indicate a standard deviation of 0.02% is possible without extending the detector to its full sensitivity. Since the corresponding rings in each probe can be joined in parallel, only three leads are required for the detector circuit. This is particularly important in high-pressure work, where the leads must be passed through the high-pressure vessel. It is in such measurements that this instrument has been used most extensively.

Another improvement in measuring techniques is in conductance cell design for measurements at high pressures. Although conductance measure-

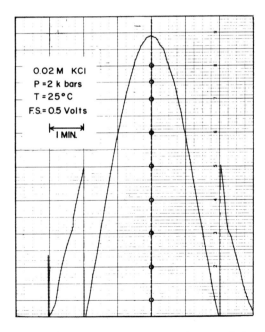

Fig. 2. Typical trace from the RF detector of the
passage of a boundary.

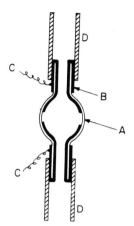

Fig. 3. High-pressure conductance cell; (A) Pyrex glass, (B) platinum film electrodes, (C) leads, (D) Teflon tubing.

ments are perhaps one of the easiest and most precise measurements that can be made at 1 atm, at high pressures a number of problems are encountered. These have been described recently by Gancy and Brummer[252] along with a design that overcomes some of the difficulties. Unfortunately, their cell design suffers from two deficiencies. They contain complicated and expensive metal-to-glass seals that are prone to cracking since it is difficult to locate different materials with both the same temperature and pressure coefficients. Also, they used mercury to transmit the pressure to the cell contents, thereby requiring the removal of all oxygen from the solutions owing to a mercury–oxygen reaction at high pressures.

These complications are eliminated in the very simple high-pressure cell we have designed as shown in Fig. 3. Essentially, it consists of a glass pipet A, about 5 cm long, which is coated inside and out with several layers of a platinum film fused onto the glass as shown at B. Electrical connection is made on the outside at C. Pressure is generated to the cell contents by means of the flexible Teflon tubes D which are closed with standard tapers. Four of these cells can be placed in a single pressure vessel. After platinizing the electrodes in the usual manner, the conductance measured in these cells had an acceptable frequency dependence, the resistance of the electrode films can be reduced to an acceptable value, and the change of dimensions with pressure can be readily calculated. A precision of 0.1% is indicated at pressures up to 3 kbar. Complete details have been reported elsewhere.[823]

3. LIMITING IONIC CONDUCTANCES IN BINARY SOLUTIONS

The limiting ionic conductances for a number of the more common spherical ions have been collected in Table I. Included are recent data for the trialkylsulfonium[198] and the hydroxylated tetraalkylammonium[151,199] ions. In Table II are listed similar data for D_2O solutions at 25°C. These data have been corrected, where necessary, to bring them into conformity with the same primary standard.[415] For this purpose the following equation

TABLE I. Limiting Ionic Conductances in Aqueous Solutions[a]

Ion	10°	25°	45°
Li^+	26.37	38.66	58.02
Na^+	34.93	50.20	73.83
K^+	53.08	73.55	103.61
Rb^+	56.2_8[480]	77.2_0[480]	107.5_8[480]
Cs^+	56.50	77.29	107.56
Me_4N^+	30.93	44.42	65.01
Et_4N^+	21.90	32.22	47.95
$n\text{-}Pr_4N^+$	15.33	23.22	35.78
$n\text{-}Bu_4N^+$	12.56	19.31	30.40
$i\text{-}Am_3BuN^+$	—	18.3[429]	—
$n\text{-}Am_4N^+$	—	17.5[429]	—
$i\text{-}Am_4N^+$	—	18.0[429]	—
$(EtOH)Me_3N^+$	—	38.21[199]	—
$(EtOH)_2Me_2N^+$	—	33.58[199]	—
$(EtOH)_4N^+$	18.3_5[151]	27.0[199]	40.1[199]
Me_3S^+	33.32[198]	47.51[198]	—
Et_3S^+	23.56[198]	34.53[198]	—
Pr_3S^+	17.53[198]	26.01[198]	—
Bu_3S^+	—	22.59[198]	—
F^-	—	55.32	—
Cl^-	54.33	76.39	108.96
Br^-	56.15	78.22	110.69
I^-	55.39	76.98	108.76
ClO_4^-	—	67.2[154]	—

[a] Unless otherwise noted the data were taken from a compilation in Ref. 428.

TABLE II. Limiting Ionic Conductance in D_2O at 25°C[a]

Ion	$\lambda_0^{\pm}$
Na^+	41.62
K^+	61.40
Cs^+	64.44
Me_4N^+	36.61
Et_4N^+	26.44
$n\text{-}Pr_4N^+$	18.84
$n\text{-}Bu_4N^+$	15.62
F^-	44.79
Cl^-	62.83
Br^-	64.67
I^-	63.79

[a] Taken from a compilation in Ref. 428.

has been proposed[516] as a secondary standard for the molar conductance of KCl in water at 25° at concentrations (mol liter^{-1}) up to 0.01 M:

$$\Lambda = 149.94 - 94.65C^{1/2} + 58.74C \log C + 198.46C \tag{1}$$

Where necessary, all conductances were recalculated to bring them into conformity with the Fuoss–Onsager[249] equation in the form

$$\Lambda = \Lambda_0 - (\alpha\Lambda_0 + \beta)C^{1/2} + (E_1\Lambda_0 - E_2)C \log C + JC \tag{2}$$

where

$$\alpha = 0.8204 \times 10^6/(\varepsilon_0 T)^{3/2}; \qquad \beta = 82.501/\eta(\varepsilon_0 T)^{1/2}$$
$$E_1 = 6.7747 \times 10^{12}/(\varepsilon_0 T)^3; \qquad E_2 = 0.9977 \times 10^8/\eta(\varepsilon_0 T)^2$$

and ε_0, η, and T are the solvent dielectric constant, viscosity, and temperature, respectively. The functional form of the parameter J is not important since, for extrapolation purposes, Λ', given by

$$\Lambda' \equiv \Lambda + (\alpha\Lambda_0 + \beta)C^{1/2} - (E_1\Lambda_0 - E_2)C \log C = \Lambda_0 + JC \tag{3}$$

is plotted versus C and the intercept gives Λ_0. Equation (3) gives a sufficiently linear plot for all the aqueous solution data for extrapolation purposes.

Other extensions[418] and revisions[250,251] of eqn. (2) produce essentially the same extrapolated values.

In general, the data in Table I are based on the consistent set of limiting salt conductances and transference numbers reported by Harned and Owen[317] and were calculated at the temperatures indicated by means of their interpolation equation, which is a cubic expression in temperature. A recent measurement[431] of the transference numbers for KCl indicates the error at 10°C attributable to this interpolation equation is no more than 0.1% and a recalculation is not warranted.

The most extensive measurements of salt conductances in aqueous solutions at elevated pressures are those of Brummer and Gancy.[80] These are given in Table III as a ratio of the limiting salt conductances at any pressure to its value at 1 atm for each temperature. The density data from a compilation by Kell (see Volume 1, Chapter 10) were used to convert the reported specific conductance ratios to equivalent conductance ratios. The extrapolation of these data to infinite dilution has been discussed in considerable detail by Gancy and Brummer.[253]

4. MECHANISM OF IONIC CONDUCTANCE

A number of the factors affecting ionic mobilities have been characterized and some progress has been made in evaluating the magnitude of these effects. It is convenient to identify them by writing the limiting conductance–solvent viscosity product, or Walden product[806] as it is generally known, as a sum of terms

$$\lambda_0\eta = (\lambda_0\eta)_{\text{CS}} + (\lambda_0\eta)_{\text{CE}} \tag{4}$$

where, for convenience, the charged-sphere contribution is written as

$$(1/\lambda_0\eta)_{\text{CS}} = (1/\lambda_0\eta)_{\text{IS}} + (1/\lambda_0\eta)_{\text{DR}} \tag{5}$$

Here, the term subscripted IS gives the contribution to the conductance to be expected from inert spheres, while the term subscripted DR gives the contribution to the conductance resulting from charging the spheres and is identified below as a solvent dipole relaxation effect. The last term, subscripted CE, is a term that contains the effects resulting from changes in the solvent in the vicinity of an ion (the ionic cosphere[302]) that result from the presence of either a charged or an inert sphere.

TABLE III. Λ_P/Λ_1 **for Aqueous Solutions**

	P, kbar	$10°$	$25°$	$45°$
LiCl	0.5	1.044	1.021	1.004
	1.0	1.070	1.032	1.002
	1.5	1.080	1.033	0.996
	2.0	1.080	1.032	0.986
NaCl	0.5	1.031	1.012	0.997
	1.0	1.047	1.015	0.989
	1.5	1.048	1.009	0.978
	2.0	1.039	0.997	0.963
KCl	0.5	1.032	1.013	0.998
	1.0	1.047	1.016	0.992
	1.5	1.050	1.011	0.981
	2.0	1.042	1.009	0.966
RbCl	0.5	1.029	1.010	0.995
	1.0	1.041	1.011	0.986
	1.5	1.041	1.004	0.974
	2.0	1.031	0.991	0.958
CsCl	0.5	1.024	1.007	0.992
	1.0	1.033	1.004	0.980
	1.5	1.029	0.995	0.965
	2.0	1.015	0.979	0.946
KF	0.5	1.031	1.013	0.998
	1.0	1.047	1.017	0.992
	1.5	1.052	1.015	0.983
	2.0	1.048	1.008	0.974
KBr	0.5	1.026	1.008	0.994
	1.0	1.036	1.007	0.984
	1.5	1.033	0.998	0.969
	2.0	1.019	0.982	0.951
KI	0.5	1.017	1.000	0.986
	1.0	1.018	0.991	0.969
	1.5	1.007	0.976	0.948
	2.0	0.987	0.955	0.926

In the following sections each of these effects will be considered separately, followed by a discussion of their temperature and pressure dependences. In order to illustrate the magnitude of the various terms in eqn. (4), the Walden products for the alkali metal, the halide, and the tetraalkylammonium ions (R_4N^+) in four solvents will be used. Their magnitudes are shown in Fig. 4, with a more detailed plot for the R_4N^+ ions in Fig. 5. The solvents were chosen carefully. In each case they consist of small, fairly simple molecules for which precise transference and conductance data are available. Also, they cover the complete scale of hydrogen-bonding capabilities: Acetonitrile is aprotic and an unassociated polar liquid; methanol can form two hydrogen bonds per molecule, permitting one-dimensional structures or chains; formamide can form a maximum of three hydrogen bonds, thereby producing a two-dimensional or sheetlike structure; and, of course, water can form four tetrahedrally oriented

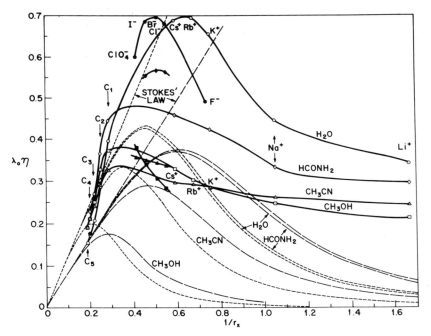

Fig. 4. The Walden product for alkali metal, halide, and symmetrical tetraalkylammonium ions in water, formamide, methanol, and acetonitrile at 25°C as a function of the reciprocal crystallographic radii. The tetraalkylammonium ions are identified by the number of carbon atoms in a single side chain. (- - -) Stokes' law for the slip case [eqn. (7)]; (— —) Stokes' law for the stick case [eqn. (6)]. The dashed curves give the results of the Zwanzig calculation, eqns. (9) and (12).

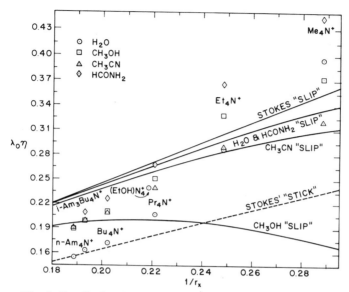

Fig. 5. Details for the tetraalkylammonium ions from Fig. 4.

hydrogen bonds, permitting three-dimensional structures. By including this variety of solvents, the peculiarities of water stand out most vividly. The physical properties of the solvents are given in Table IV. The ionic conductances at infinite dilution were taken from the following sources: methanol, Kay and Evans;[428] acetonitrile, Springer et al.;[427] and formamide, Thomas and Evans.[773]

TABLE IV. Physical Properties of Solvents at 25°C

	ϱ, g cm^{-3}	ε_0	ε_∞	$10^{11}\tau$, sec	$10^2\eta$, P	$12^{30}\phi$ [eqn. (12)]
Water	0.99707	78.40[432]	5.3[645]	0.823[645]	0.8903[766]	1.260
Formamide	1.1296[773]	109.5[410]	3.0[36]	3.9[36]	3.301[410]	1.205
Methanol	0.78658[152]	32.62[152]	5.6[86]	4.77[86]	0.5428[152]	25.34
Acetonitrile	0.7767[152]	35.95[152]	2.0[470]	0.39[470]	0.341[152]	3.416

4.1. Stokes' Law

The simplest substitution for the inert sphere term in eqn. (5) comes from an application of Stokes' law,[755] which can be expressed as*

$$\lambda_0 \eta / Z = Fe/6\pi r_S = 0.820 \times 10^{-8}/r_S \tag{6}$$

where Z is the absolute magnitude of the ionic valence, e the electronic charge, F the Faraday, and r_S the hydrodynamic or Stokes radius of the ion with limiting molar conductance λ_0. This equation predicts the terminal velocity resulting from equating the electrical and viscous forces acting on a charged sphere situated in a viscous medium under the influences of a potential field. It was derived for the movement of inert spheres through a solvent continuum with the boundary condition that the solvent must adhere perfectly to the surface of the sphere (perfect sticking). If, on the other hand, the spheres are assumed to move without any adhesion to the solvent, so that no radial velocity gradient is imparted to the solvent (perfect slipping), eqn. (6) becomes[486]

$$\lambda_0 \eta / Z = Fe/4\pi r_S = 1.230 \times 10^{-8}/r_S \tag{7}$$

Inherent in the model on which eqn. (6) is based is the requirement that the moving entity must be large compared to the solvent molecules. Perrin[629] has made the necessary corrections for nonspherical ions with various geometries but this complication is avoided here by considering only spherical ions.

The requirement of the large size disparity between solvent and solute molecules has not received much attention. If the solvent molecules consist of a long-chain polymer or an extensively hydrogen-bonded chain, the solvent viscosity would be expected to be exceedingly high; but since the ions could move by displacing only segments of the long chains, they could avoid the work required to overcome the macroscopic viscosity. Consequently, in these cases Stokes' law would not be expected to hold. Treiner and Fuoss[783] have illustrated this phenomenon most effectively from conductance measurements in cyanoethyl-sucrose–acetonitrile mixtures. It is interesting to note that Krishnan and Friedman[477] have presented thermodynamic evidence that nonelectrolytes can fit between the chains of hydrogen bonds in alcohols without disrupting the chains. Of course,

---

* For simplicity, the conversion factor 1/300 has been eliminated in writing the quantity Fe. For numerical calculations it must be included.

a solid aqueous gel as a solvent system should be the ultimate in showing derivations from Stokes' law and for this reason this effect will be called the gel effect. In order to minimize this gel effect, the solvent systems investigated here consist of the smallest molecules possessing the required hydrogen-bonding properties.

The well-known fact that eqns. (6) and (7) do not fit the limiting ionic conductances for both aqueous and nonaqueous solutions is shown in Figs. 4 and 5. For aqueous solutions the alkali and halide ions show both positive and negative deviations (Fig. 4), while the tetraalkylammonium ions (R_4N^+) approach Stokes' law only as they become very large (Fig. 5). Allowing for complete slippage constitutes some improvement, although the large R_4N^+ ions show increasing deviations with increasing size.

The most obvious explanation for the negative deviations from Stokes' law is solvation of the ions: The moving entity, on the average, is larger than its crystallographic size, owing to one or more spheres of coordinated solvent molecules. This is a reasonable hypothesis since the smallest, and presumably most solvated, ions show the greatest deviations. However, no quantitative theory of solvation has proven satisfactory, although experimental methods have recently been devised that permit coordination or hydration numbers to be measured for ions of higher valence. Unfortunately, the exchange rate of water molecules with univalent ions is too fast for the various direct methods, such as NMR chemical shifts or isotopic dilution, to be applicable. Consequently, not even the hydration numbers, much less the hydrodynamic size, are known for the ions of real interest to transport problems. Little confidence can be placed on the results of various indirect methods,[515] as illustrated by the fact that they have predicted hydration numbers varying between 16 and 31 for the Al^{3+} ion, whereas the exact methods agree on a value of six. Solvation in general has recently been the subject of a comprehensive review.[515]

Attempts have been made to account for even the positive deviations from Stokes' law by hydration. It was noted that eqn. (6) fits the $i\text{-}Am_4N^+$ ion at 25° in aqueous solution. Consequently, it was assumed that all the tetraalkylammonium ions were unhydrated but the ionic conductances deviated from Stokes' law owing to a size deficiency.[693] From the magnitude of these size deficiencies a calibration curve was constructed,[597] with the result that all ions then showed negative deviations from the predicted curve which could then be explained by hydration. Subsequently it was shown that this calculation was not valid since the Walden products for the R_4N^+ ions in aqueous solutions were found to be temperature-dependent,[428] indicating that even these inert ions are hydrated (see Fig. 7) although we are dealing with an example of hydrophobic hydration, as will be discussed below.

There is a good possibility that some of the deviations from Stokes' law are the results of this size discrepancy effect. Indeed the type of correction postulated by Robinson and Stokes[693] might have some validity for nonaqueous solutions since it has been found that the Walden product for the R_4N^+ ions are relatively temperature-independent in at least methanol (see Fig. 7). However, the construction of those calibration curves is quite arbitrary and, furthermore, all deviations from ideal behavior should not be attributed to any one source when it is known that more than one factor is in operation.

Perhaps the most promising approach to an evaluation of the inert-sphere term in eqn. (5) is to avoid using Stokes' law and instead obtain it directly from experiment by measuring diffusion coefficients of uncharged compounds comparable in size to the ions. After extrapolation to infinite dilution, limiting salt conductances could be obtained from the Nernst–Einstein equation[61]

$$(\lambda_0\eta)_{IS} = D_0\eta(ZFe/kT) \tag{8}$$

Owing to the absence of a charge, the limiting diffusion coefficient D_0 will contain no contribution from the dipole relaxation or cosphere effects. However, this will not be true for aqueous solutions, owing to the possibility of hydrophobic hydration, and in this case this approach will not be directly applicable. A project to obtain these data is now underway.*

4.2. Solvent Dipole Relaxation Effect

For many years it was suspected that the dielectric constant of the solvent should play an important role in the ionic transport process. Finally, in 1920 Born[69] described a dielectric effect that could affect the magnitude of ionic mobilities. He suggested that a moving ion could orient the solvent dipoles in its cosphere and, since such oriented dipoles require a finite time to relax to their random orientations after passage of the ion, a net retardation should be imposed on the moving ion. Unfortunately, this observation of Born received little attention until it was rediscovered independently by Fuoss[248] almost 40 years later and presented in a semi-empirical form. Subsequently the effect was evaluated by Boyd[70,71] and more exactly by Zwanzig,[881] who obtained an equation of the following

* D. F. Evans, private communication, 1971.

form:

$$\lambda_0 \eta / Z = Fe / [A_s \pi r_i + (A_z \phi / r_i^3)] \tag{9}$$

where $A_z = 2/3$,[881] ϕ is given by

$$\phi = (Ze)^2 (\tau/\eta)[(\varepsilon_0 - \varepsilon_\infty)/\varepsilon_0^2] \tag{10}$$

and, since the equation was derived for the case of complete sticking, $A_s = 6$. Here, τ is the dielectric relaxation time and ε_0 and ε_∞ are the low- and infinite-frequency dielectric constants, respectively. An inspection of eqn. (9), which will be designated as the Zwanzig equation, shows that, essentially, another term which is inversely proportional to the cube of the ionic radius has been added to the Stokes radius.

Frank[228] has pointed out that eqn. (9) correctly predicts the existence of the maximum found in the $\lambda_0 \eta$ versus $1/r_x$ curve for aqueous solutions but pointed out that no positive value of r_i when substituted into this equation would reproduce the magnitude of the measured $(\lambda_0 \eta)_{max}$. The maximum calculated value of λ_0 for aqueous solutions is 29.4, whereas several observed conductances of univalent ions are as high as 75. Furthermore, he noted that the predicted $(r_i)_{max}$ is not in agreement with experiment, nor is a correction for "slippage" sufficient to help.

In order to evaluate the ionic radius parameter in the Zwanzig equation, it was rearranged by Atkinson and Mori[29] by taking reciprocals

$$ZFe / \lambda_0 \eta = A_s \pi r_i + (A_z \phi / r_i^3) \tag{11}$$

so that from a plot of the left-hand side versus $A_z \phi$ a value of the ionic radius can be obtained from both the intercept and the slope. Unfortunately, their test of this equation contains a number of plotting errors and the apparent agreement with experiment is erroneous. When plotted correctly it has been found that, although a reasonable straight line is obtained for the tetraalkylammonium and halide ions in a series of homologous alcohols,[200] vastly different values of r_i are obtained from the slope and the intercept. Furthermore, Kay et al.[425] earlier had shown that, when applied to a variety of ions in water, acetonitrile, and methanol, a linear relationship was not obtained, the deviations in the case of water being particularly large. As a further test of the theory, it was shown that no value of r_i in eqn. (9) could even predict the ratio of conductances in methanol and acetonitrile.[425]

Agar[3] and Frank[228] both pointed out that this evaluation of ϕ did not take into account the fact that a moving ion imparted a velocity gradient to the solvent in its cosphere. When allowance was made for this fact as well as for a deficiency in the previous derivation and the effect recalculated[882] for the two extreme cases of complete slipping and complete sticking of the solvent molecules at the surface of the ion, eqn. (9) was found not to change form but ϕ became

$$\phi = (Ze)^2(\tau/\eta)[(\varepsilon_0 - \varepsilon_\infty)/\varepsilon_0(2\varepsilon_0 + 1)] \tag{12}$$

with

$$A_s = 6; \quad A_z = 3/8 \qquad \text{for complete sticking}$$
$$A_s = 4; \quad A_z = 3/4 \qquad \text{for complete slipping}$$

This constitutes a substantial change from the earlier eqn. (10) and has necessitated a complete reevaluation of the theory. The revised $A_z\phi$ term is now only about one-third its previous value for large ε_0. Since ϕ is always positive, this equation predicts negative deviations from Stokes' law that increase as the ionic radius r_i decreases.

The predicted behavior for water, formamide, methanol, and acetonitrile for both the slip and stick cases can be seen in Figs. 4 and 5. As was true of the earlier expression, a conductance maximum is predicted, the value of which can be calculated from

$$(\lambda_0\eta)_{\max} = (FeZ/4)(3/A_s\pi)^{3/4}(A_z\phi)^{1/4} \tag{13}$$

$$(r_i)_{\max} = (3A_z\phi/A_s\pi)^{1/4} \tag{14}$$

It is clearly observable in Fig. 4 that, although the Zwanzig theory correctly predicts the general shape of the conductance curves, the quantitative agreement is poor. It is interesting to note that the observed conductances of the ions increase with the degree of hydrogen bonding existing in the solvents,* that is, the Walden product decreases in the order H_2O > $HCONH_2$ > CH_3OH > CH_3CN, whereas the order predicted by the Zwanzig equation would be $HCONH_2 \approx H_2O$ > CH_3CN > CH_3OH. The best agreement between experiment and theory is for the aprotic solvent CH_3CN, in that the conductance maximum is predicted at the correct radius (slip case only) for at least the cations. The same behavior was also noted by Fernández-Prini and Atkinson,[215] who claimed that for the slip case the Zwanzig theory provided a reasonable fit of the experimental

---

* The crossover at K^+ for methanol and acetonitrile has been attributed to the much stronger basic properties of methanol compared to acetonitrile.[430] At small distances acid–base effects could predominate and enhanced solvation by methanol would account for the lower mobilities in that solvent.

data for at least the Me_4N^+ and Et_4N^+ ions in the polar aprotic solvents acetonitrile, dimethylformamide, acetone, and nitrobenzene, whereas the agreement with experiment was very poor for protic solvents and for metal ions in all solvents. However, their preference for the slip ion is based on very tenuous evidence. In the four aprotic solvents investigated the fit was very similar to that shown for acetonitrile in Fig. 4. As can be seen in Fig. 5, Et_4N^+ and Me_4N^+ do indeed fit the slip case not badly but, as the R_4N^+ ions get larger, Stokes' law for the slip case is not approached by the experimental points but rather these points decrease to a position midway between the slip and stick cases. Consequently, these two cases must be considered at the present time as merely the two extremes expected from Stokes' law.

Although the Zwanzig equation alone is not successful in explaining ionic mobilities of electrolytes, particularly in water, its derivation was a most important step in the formulation of a comprehensive quantitative theory. There is good reason to expect that it accurately describes the effect of the ionic charge on the mobilities with the added assumption that the charge in no way affects the physical properties of the solvent in the ionic cosphere. The fact that the theory seems to fit data for aprotic, unassociated solvents best with increasing deviations as the degree of hydrogen bonding of the solvent increases strongly suggests that changes in the ionic cosphere of associated liquids resulting from the presence of a charged ion plays an important role in the conductance process.

4.3. Cosphere Effects

The next step in the formulation is to consider the effect of both an inert and a charged sphere on the physical properties of the solvent in the cosphere. Three of these cosphere effects have been recognized at the present time as being important to the ionic transport process.

Fernández-Prini and Atkinson[215] speculated that the dielectric constant in the cosphere could be substantially altered from the value for the pure solvent owing to the high charge density at the surface of small ions. By assuming a step function in which there is dielectric saturation in the solvent at all points up to a distance r_0 from the ion, they were able to show that a constant r_0 equal to approximately 2 Å accounted for the mobilities of the larger alkali metal cations in a number of aprotic solvents. On the other hand, the variable values of r_0 obtained for ions in protic solvents could be a manifestation of the other cosphere effects influencing the transport mechanism in such solvents.

   This is an important result in spite of the arbitrary nature of the calculation, since it gives a feasible explanation for the large deviations of the larger metal ions such as Cs^+, Rb^+, and K^+ from the Zwanzig equation in aprotic solvents. In an unassociated solvent such as acetonitrile the effect of the ionic charge on the solvent viscosity is expected to be small. Since these ions are not considered to be hydrated in water, there is little reason to believe they would be extensively solvated in acetonitrile. Consequently, this demonstration of possible dielectric saturation in the cosphere is impressive and adds to the increasing list of facts that indicate that further progress on this problem of the mechanism of ionic transport will require a knowledge of the structure of the ionic cosphere.

   Besides affecting the dielectric constant of the ionic cosphere, it is even more feasible to assume that the ionic charge affects the viscosity of the cosphere. Any change in viscosity would affect the mobilities directly, whereas the dielectric saturation effect is a second-order correction. This is a particularly attractive assumption with which to account for the extremely high mobilities of the larger alkali metal and halide ions in water. It is an experimental fact that these ions decrease the viscosity at least of cold water and they have become known as structure-breakers for want of a better name. As early as 1934 Cox and Wolfenden[146] discussed the possibility of the depolymerization of water by ions, but it was Frank and Evans[230] and later Gurney[302] who considered this problem in depth and clarified the concepts involved. Other direct evidence that these ions decrease the amount of long-range order in the cosphere in aqueous solutions comes from NMR[195,196] and dielectric[271] relaxation measurements which indicate that water molecules in the vicinity of these ions have greater rotational and translational freedom than in bulk water.* Furthermore, this assumption is in agreement with an overwhelming mass of thermodynamic data, much of which is to be found in other chapters of this treatise.

   In the Frank–Wen model[232] for ionic solutions the cosphere is divided into two regions, a region A adjacent to the ion surface, containing water molecules that on the average move with the ion (electrostrictive hydration), and a region B, further removed from the ion, in which the ionic charge has disrupted to some extent the normal tetrahedrally coordinated hydrogen bonding that is believed to exist in pure water. Ions of high charge density should have an appreciable region A with possibly a small region B, whereas the reverse should be true of ions of low charge density.

   The application of this model to ionic transport has been discussed

---

* Editor's note: These topics are discussed in detail in Chapters 7 and 8.

in considerable detail by Kay and Evans.[428] In each pure protic solvent the viscosity is determined to a considerable extent by long-range hydrogen bonding and the degree of such bonding could be influenced considerably by the ionic charge. For this reason it is reasonable to assume that in all protic solvents the large univalent ions move in a cosphere in which the viscosity is lower than the bulk solvent viscosity. Consequently, such ions could be said to have a positive excess conductance.

A third cosphere effect is also considered in the Frank–Wen model. If an ion is sufficiently large and contains an inert surface, such as the larger tetraalkylammonium ions, it is possible that the solvent molecules at the surface of the ion will not be influenced at all by the ionic charge and consequently can be oriented into favorable positions by neighboring water molecules more easily than if the ion was not present. Whether this explanation is correct or not does not alter the fact that the cosphere of the larger tetraalkylammonium ions appears to contain greater order or a greater degree of hydrogen bonding than does bulk water. This is reflected in the fact that as the R_4N^+ ions become larger their Walden products for aqueous solutions become increasingly smaller and diverge from the common nonaqueous line (Fig. 5). It would appear that these ions have a negative excess mobility in aqueous solutions owing to what has been called hydrophobic hydration and for this reason are referred to as hydrophobic structure-makers.

This effect is unique to aqueous solutions and it appears to be associated with the ability of water to form three-dimensional hydrogen-bonded structures. These ions have been found to increase the viscosity of water well beyond that which would be accounted for by their large size,[434] and the NMR[195,196] and dielectric[271] relaxation behavior of such solutions indicates decreased rotational and translational freedom of the water molecules in the cosphere of these ions. Further evidence for the above conclusions is provided by other aspects of their solution properties.[233]

Perhaps the most convincing evidence to substantiate the behavior postulated for these large hydrophobic ions comes from a consideration of transport data for the $(EtOH)_4N^+$ ion which is formed by substituting a hydroxyl group for each terminal methyl group in the Pr_4N^+ ion. The insertion of this functional group into the side chain of the Pr_4N^+ ion should result in little change in size, and if size is the only criterion, both these ions should have the same mobility. If anything, the substituted ion should have a lower mobility owing to the possibility of hydration of the hydroxyl group. As can be seen in Fig. 5, the opposite effect is observed, the $(EtOH)_4N^+$ ion is almost 20% faster than its alkyl analog due, presumably, to the absence of water-structure enforcement around this ion that is present in the hydrocarbon side chains of the Pr_4N^+ ion in aqueous solutions.[199] Confirmation of the result is obtained from the fact that the $(EtOH)_4N^+$ ion increases the viscosity of water by an amount substantially less than does the Pr_4N^+ ion.[199] These results are of considerable importance to the development of a theory of cosphere behavior because they rule out the possibility that the abnormally low mobilities of the large R_4N^+ ions in aqueous solutions are solely the result of their size.

Another demonstration of these structural cosphere effects has been reported by Kay and Evans[427] from conductance data for D_2O solutions. It is reasonable to assume that the 23% higher viscosity of D_2O over that of H_2O at 25°C can be accounted for by a difference in their hydrogen-bonding characteristics. Consequently, ions should have a greater effect on the cosphere properties in D_2O than in H_2O. By defining R as a ratio of Walden products,

$$R = (\lambda_0\eta)_{D_2O}/(\lambda_0\eta)_{H_2O} \tag{15}$$

Kay and Evans were able to show that structure-breakers indeed had a larger excess mobility in D_2O than in H_2O, with the reverse behavior for the hydrophobic structure-makers. Subsequently, a most interesting result was reported by Krishnan and Friedman.[474] Their plot of ionic enthalpies of transfer of ions from H_2O to D_2O versus the reciprocal crystallographic radii is shown in Fig. 6. (See also Chapter 1.) Their ionic values were obtained by assuming equal contributions for $AsPh_4^+$ and BPh_4^- ions. Their data are given by the points, whereas the line represents the values of $R - 1$ obtained from eqn. (15), after normalization of the scale so that the Walden product line passes through the heat of transfer datum point for the Cs^+ ion. It can be seen in Fig. 6 that both the heat of transfer and the conductance data show surprisingly identical behavior. This strongly suggests that the two sets of data have a common underlying physical basis.[474]

4.4. Temperature Coefficient

The cosphere viscosity effects outlined in the previous section involve hydrogen bonding and should be particularly temperature-sensitive since thermal energies are comparable to the hydrogen bond energy. On the other hand, electrostrictive hydration and solvation in general should be relatively insensitive to small temperature changes, owing to the much larger energies involved. For this reason temperature coefficients of ionic

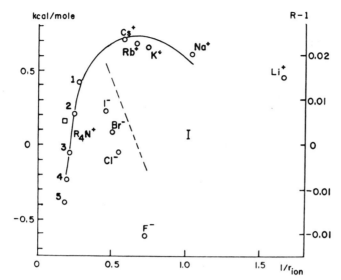

Fig. 6. Heats of transfer from H_2O to D_2O (points and left scale) and solvent-isotope effect on the Walden product (lines and right scale). R is defined by eqn. (15) and the $R - 1$ scale set so the line passes through the point for the Cs^+ ion. The abscissa is the reciprocal crystallographic or estimated ionic radius. (Reproduced from Ref. 474 with the permission of the authors and copyright owners.)

conductances have been very useful probes with which to study structural effects in the ionic cosphere.

In Fig. 7 are plotted the temperature coefficients of the Walden product for the alkali metal, the halide, and the R_4N^+ ions as a function of the crystallographic radius. The small and highly hydrated Li^+ and Na^+ ions have temperature coefficients close to zero. The Cl^-, Br^-, I^-, K^+, and Cs^+ ions have increasingly larger negative temperature coefficients as their size and structure-breaking ability increases. As the temperature increases, the hydrogen-bonding properties of the bulk solvent should approach that of the cosphere solvent and structure-breakers should lose their excess mobility. It is interesting to note that this effect is more pronounced at the lower temperature, where the cospheres are expected to contain more structure. It would appear that the maximum structure-breaking power would be exhibited by an ion of radius approximately 2 Å. The Me_4N^+ ion is larger than this, although it still definitely acts as a structure-breaker.

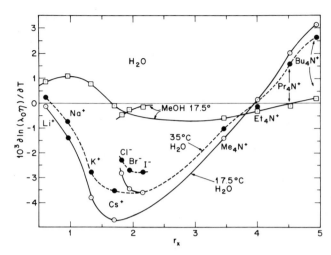

Fig. 7. The temperature coefficient of the Walden products for the alkali metal, halide, and tetraalkylammonium ions at the average temperatures 17.5 and 35°C in aqueous solutions and at 17.5°C in methanol solutions.

The onset of hydrophobic hydration in the Pr_4N^+ and Bu_4N^+ ions is evident from the large positive temperature coefficient. The negative excess conductances of these ions is eliminated with increasing temperature, as the enforced excess water structure about their hydrocarbon chains "melts." Particularly significant is the crossover in the temperature coefficient curve for the two temperatures. At higher temperatures both the structure-breaking and structure-making effects decrease in magnitude, presumably as the amount of three-dimensional hydrogen bonding in the cosphere is significantly reduced. The Et_4N^+ ion appears to be of a size that renders a net effect in the cosphere of zero.

Supporting evidence for the conclusion arrived at from these temperature coefficients for ions in aqueous solutions is obtained from similar data for the $(EtOH)_4N^+$ ion. The Walden products for this ion are 0.240, 0.240, and 0.239 at 10, 25, and 45°C, respectively,[151,199] indicating a temperature coefficient of zero and an absence of any hydrophobic hydration.

The temperature coefficients of the Walden products in methanol solutions are, in contrast to the aqueous solution results, quite small.*
It is interesting to note that this is the only comprehensive study of tem-

---

* G. A. Vidulich, R. L. Kay, and G. P. Cunningham, unpublished data.

perature coefficients based on accurate transference numbers that has been reported for any solvent other than water. Since temperature dependence is such a basic property in the study of any mechanism, it is not surprising that so little progress was made over so many years in our understanding of the mechanism of ionic transport.

It should be pointed out that the solvent dipole relaxation effect will not account for the temperature coefficients shown in Fig. 7. Substitution of the proper data in eqn. (12) indicates that ϕ for water increases slightly with temperature, although the exact magnitude is in doubt owing to the uncertainty in the dielectric relaxation times. Consequently, the Zwanzig theory [eqn. (9)] predicts a small negative temperature coefficient that increases with decreasing ionic size. As can be seen in Fig. 7, the change of the temperature coefficient with ionic size is just opposite to this predicted change for the alkali metal and halide ions, and the temperature coefficients of the larger R_4N^+ ions are positive rather than negative as predicted by the theory.

The sensitivity of these aqueous ionic cosphere structural or association effects to small temperature changes is equally reflected in the temperature dependences of the viscosity B coefficients, where B is defined by[416]

$$(\eta - \eta_0)/\eta_0 = AC^{1/2} + BC \qquad (16)$$

In aqueous solutions structure-breakers have negative or abnormally low B values that tend to increase as the temperature increases, owing to the decrease in the degree of hydrogen bonding in the cosphere.[419] By a comparable argument hydrophobic structure-makers should have negative temperature coefficients for B, a fact which has been verified by experiment.[434] A plot of the change of the Walden product with temperature versus the change of the viscosity B coefficient with temperature for various ions in aqueous solutions is given in Fig. 8. The viscosity B coefficients for the R_4N^+ ions in methanol have been shown to be independent of temperature in the interval 10–45°C and to be equal in magnitude to the predictions of eqn. (16) with A obtained from a Debye ionic atmosphere calculation and B estimated solely from the size of the ions.[434] Although viscosity B coefficients are extremely sensitive to structural effects in the ionic cosphere, it should be pointed out that they suffer from two limitations. They cannot be split into ionic contributions without some arbitrary assumption and they are not infinite-dilution values and consequently are subject to interionic interactions that can be difficult to evaluate in the measurable concentration range, particularly in nonaqueous solvents.

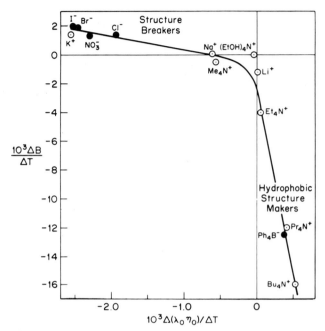

Fig. 8. A comparison of the temperature dependence of viscosity B coefficient and limiting ionic conductance–viscosity products for aqueous solutions. (Reproduced from Kay.[423] with the permission of the copyright owners.)

4.5. Pressure Coefficient

In Section 4.1 it was suggested from the Walden product data that ions in associated liquids appear to have excess mobilities in proportion to the complexity of the hydrogen-bonding potential of the solvent. In Section 4.3, however, it is pointed out that only in water can the structural cosphere effects be detected by the temperature coefficients of ionic mobilities. Presumably this is due to the fact that these effects depend on the particular three-dimensional structure that can exist only in water and such structures are particularly sensitive to small temperature changes. It is reasonable to assume that this formation of long-range, three-dimensional order in water is associated with a volume increase and consequently it should be extremely pressure-dependent. Such explanations have been suggested to account for a unique property of cold water, namely the decrease in viscosity with increased pressure,[53] as shown in Fig. 9.

Ionic transport data at elevated pressure are now available which permit this hypothesis to be tested. Precise transference numbers have been reported[619] for aqueous KCl and KBr at 10 and 25°C, from moving boundary measurements using the radio-frequency detector described in Section 2. A precision of better than 0.05% is claimed at pressures up to 3 kbar. Typical results for KCl at 25°C are shown in Fig. 10 where T_{obs}^+, the measured transference number at 0.02 M, is shown at pressure up to 3 kbar. The difference between T_{obs}^+ and T_0^+ is due to the compressibility of the solution and the moving boundary tube, while the further correction to give the limiting value of T_0^+ was obtained by a Kay–Dye extrapolation[426] at each pressure.

Since both limiting conductances[80] (Table III) and transference data[649] are available for KCl and KBr at 10 and 25°C, a stringent test of both sets of data can be obtained by comparing the λ_0^+ for the K$^+$ ion obtained from both salts. The average difference at four pressures up to 2 kbar was 0.06% and 0.15% at 25 and 10°C, respectively.

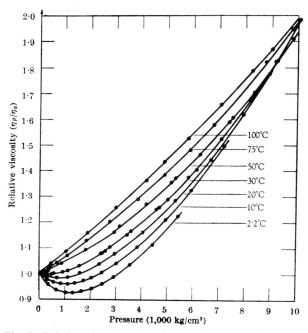

Fig. 9. Relative viscosity of water as a function of pressure at several temperatures. (Reproduced from Ref. 53 with the permission of the authors and copyright owners.)

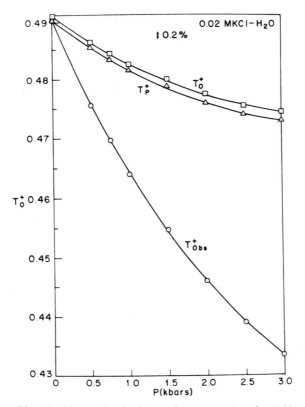

Fig. 10. Observed cationic transference numbers for KCl
in aqueous solutions as a function of pressure. T_P^+ and
T_0^+ are the result of correction for compressibilities and
extrapolation to infinite dilution.

An interesting fact is illustrated in Fig. 11, in which a typical set of
data for KI in H_2O at 25°C is plotted as a function of the pressure. The
difference between the specific and equivalent conductance ratios is the
result of the correction for the compressibility of the solvent, while a further
large correction is required for the viscosity changes. It is interesting to
note that the Walden product ratio is very close to unity at all pressures
and consequently a consideration of salt conductances alone would leave
one with the impression that Stokes' law gives an excellent estimate of the
pressure dependence of ionic conductances. However, such a conclusion
is not confirmed when the ionic conductances for aqueous solutions are
examined in Fig. 12. It can be seen that a whole spectrum of deviations

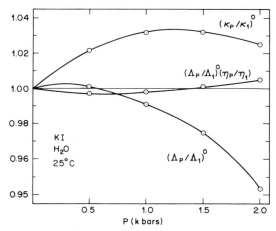

Fig. 11. Limiting specific conductance, equivalent conductance, and Walden product ratios for KI as a function of pressure.

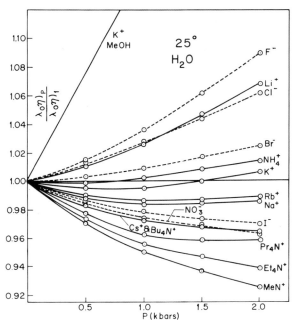

Fig. 12. The effect of pressure on the limiting Walden products for the alkali metal, halide, and tetraalkylammonium ions in water at 25°C. Preliminary results for the K⁺ ion in methanol at 25°C are also included.

exist covering the range from $+8\%$ to -8%. Included in this figure are data for the tetraalkylammonium ions obtained in recent measurements[823] using the conductance cells described in Section 2.

Although the pressure dependence of the Walden product for aqueous solutions as shown in Fig. 12 is more complicated than anticipated, some recognizable features do stand out. There appears to be a negative component in the pressure coefficient for the common structure-breaking ions, such as K^+, Rb^+, Cs^+, and I^-, that could be the result of a decrease in the amount of hydrogen bonding in water, due to the increased pressure. Thus, at elevated pressures structure-breakers appear to lose their excess mobility to some extent. Also, although the Na^+ ion is out of place, situated as it is between the Rb^+ and Cs^+ ions, it is certainly feasible to attribute this to increased hydration with increased pressure, since the process of electrostriction takes place with a large volume decrease and any ion with an unfilled coordination sphere should readily hydrate under increased pressure.

Another puzzling feature of the data in Fig. 12 is the fact that the large R_4N^+ ions have negative pressure coefficients whereas the smaller ions, such as Li^+ and F^-, have large, positive coefficients. It is most difficult to explain the very low Walden products for the Me_4N^+ ion at high pressures unless it is assumed that this ion undergoes hydrophobic hydration as the pressure increases. Such an assumption requires these hydrophobic effects to take place with a volume decrease and such a result was not anticipated. The fact that the larger R_4N^+ ions have less negative pressure coefficients than the Me_4N^+ ion could be a reflection of the fact that these ions are already considerably hydrated at normal pressures.

A result that was even less anticipated is the large, positive pressure coefficient for the small, highly hydrated ions. The preliminary results for the K^+ ion in methanol, also shown in Fig. 12, suggested one explanation. In methanol solutions the Walden product increases by over 10% while the viscosity of the solvent increases by almost 50% at a pressure of 1 kbar. This strongly suggests that another effect, which could be called pressure-induced structure-breaking, is in operation. The forces exerted by an ionic charge on its cosphere are considerably larger than the forces exerted to produce the pressure changes involved here. Consequently, it is doubtful whether the induced pressure would affect the cosphere properties to anywhere near the same extent as it affects the properties of the bulk solvent. As a result all small ions with large surface charge densities should be expected to act as structure-breakers in methanol at high pressures. In order to test this hypothesis, both conductance and transference measurements

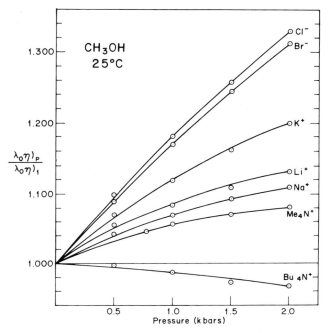

Fig. 13. The effect of pressure on the limiting Walden products for several ions in methanol at 25°C.

were carried out in anhydrous methanol (Fig. 13).* The order of the ions is approximately that expected on a size basis; but of particular interest are the results for the Bu_4N^+ ion. Owing to its large size, this ion should produce no cosphere effects in methanol and, consequently, should not be able to exhibit pressure-induced structure-breaking. It should be noted that the Walden product for this ion, in contrast to the smaller ions, is indeed almost independent of pressure.

Attractive as this pressure-induced structure-breaking hypothesis may be, it has at least one serious weakness. At a pressure of 2 kbar the viscosity of water has increased by only 3%, yet the Walden products for the Li^+, F^-, and Cl^- ions show an increase of over double that amount. Since the cosphere pressure effect cannot be greater than the increase in the bulk viscosity produced by pressure, some other factor would seem to be in operation here that contributes a positive component to the pressure coefficient for these ions. It is difficult to associate this effect with the

* B. Watson and R. L. Kay, unpublished data.

disappearance of the three-dimensional structure that occurs with increased pressure, since such an assumption conflicts with the zero temperature coefficient of the Walden products for the Li⁺ and F⁻ ions.

One major difficulty with these interpretations is that we do not know the pressure derivatives of the terms in eqn. (5). Diffusion coefficients of uncharged spheres have not been measured and the necessary dielectric relaxation data at high pressures are not available for a calculation of the dipole relaxation effect.

Some interesting information can be obtained from an inspection of the temperature dependence of Walden products as a function of pressure for aqueous solutions, as shown in Fig. 14. At elevated pressures all the ions can be classified into the same three groups that are obtained from their temperature coefficients at normal pressures. The electrostrictive structure-makers and the Et_4N^+ ion have close to a zero temperature coefficient at all pressures. It is interesting that the Na^+ ion has the negative temperature coefficient characteristic of structure-breakers but, as the pressure increases, its value approaches the zero-temperature-coefficient line characteristic of highly hydrated ions. This confirms our earlier conclusion arrived at from a consideration of the isothermal pressure coefficient for this ion.

The normal structure-breaking ions have negative temperature coefficients at atmospheric pressure, their magnitudes reflecting their net effect on their cospheres. As the pressure increases, the temperature dependence becomes the same for all the ions but still in the direction of structure-

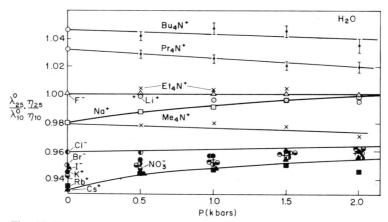

Fig. 14. The temperature dependence of the Walden product between 10 and 25°C as a function of pressure for several ions in aqueous solutions.

breaking, indicating that our postulate of pressure-induced structure-breaking may well be an important factor for these ions at elevated pressures. The fact that the Me_4N^+ ion has not reached the same value as the other ions may be an indication that this ion does indeed exhibit an increased hydrophobic effect at elevated pressures.

In a third group can be placed the larger R_4N^+ ions that have the positive temperature coefficient characteristic of hydrophobic structure-makers. The fact that their positive temperature coefficients diminish only slightly with pressure suggests that these ions maintain their hydrophobic properties even at elevated pressures, where the normal three-dimensional structure of water has essentially disappeared. This is rather an unexpected result, but it does confirm the conclusion reached from our isothermal pressure coefficient data.

Further evidence is available to support the postulated pressure-induced structure-breaking hypothesis. This effect is so large in methanol solutions at high pressures that even at normal pressures the ions must be exerting some influence on the viscosity in the cosphere. Recently, the limiting diffusion coefficient of the uncharged spherical analog, Me_4Sn, has been determined in methanol at 25°C.* A value of $(\lambda_0\eta)_{IS}$ for inert spheres of the same size as Me_4Sn can be obtained from these data and the use of eqn. (8) and Stokes' law can be avoided. By using the Zwanzig equation in the form given in eqn. (11), a value for the Walden product for charged spheres can be calculated from

$$(1/\lambda_0\eta)_{CS} = (1/\lambda_0\eta)_{IS} + 300(A_z\phi/Fer_i^3) \qquad (17)$$

where ϕ is given by eqn. (12). After substitution in eqn. (4), the cosphere contribution to the conductance of the Me_4N^+ ion can be calculated and is given in Table V for both the stick and slip cases. A small extrapolation of the diffusion data was required to estimate $(\lambda_0\eta)_{IS}$ for an ion the size of Me_4N^+ from the diffusion data for Me_4Sn. Since the $(\lambda_0\eta)_{CE}$ term is positive in both cases, it confirms the assumption that even in methanol the Me_4N^+ ion has a positive excess conductance and appears to be a structure-breaker. Presumably, the small negative temperature coefficient (Fig. 7) for the ion in methanol reflects this fact. For the stick case the cosphere effect accounts for about 23% of the total conductance of the Me_4N^+ ion and it is interesting to note that the values of the Walden prod-

---

* We are indebted to Dr. Evans and Dr. Lammantine for making their data available prior to their publication.

TABLE V. Calculation of the Cosphere Effect for the Me_4N^+ Ion in Methanol at 25°C

	$(\lambda_0\eta)_{IS}$	$(\lambda_0\eta)_{CS}$	$\lambda_0\eta(Me_4N^+)$	$(\lambda_0\eta)_{CE}$
Stick	0.500	0.288	0.373	0.085
Slip	0.500	0.168	0.373	0.205

uct predicted for inert spheres from diffusion data are 100% and 43% higher than the values predicted by Stokes' law for the stick and slip cases, respectively. However, in the slip case the cosphere effect would account for 55% of the total conductance and this appears too high for a cosphere effect that shows only a small temperature dependence. Consequently, the above must be considered positive evidence that the stick calculation in the Zwanzig equation is to be preferred.

5. LIMITING IONIC CONDUCTANCE IN AQUEOUS SOLVENT MIXTURES

The nature of the increased long-range order encountered in aqueous mixtures of the alcohols received considerable attention in a review by Franks and Ives,[234] who pointed out the existence of extrema in many properties in the water-rich region of such systems. The large maximum in the ultrasonic absorption of 10 mole % of tert-butyl in an aqueous mixture[57] is one case, although perhaps the most striking is the 600 cal deg^{-1} mol^{-1} reported by Arnett and McKelvey[24] for the change in the heat capacity of sodium tetraphenylborate when the solvent composition is changed from 0.5 to 8% tert-butanol in water. The detection of an ethanol clathrate[646] at low temperatures and at a composition corresponding to these extrema suggested that the long-range order might have some type of clathrate geometry, but no development has resulted along these lines as results of further investigations have become available.

Three solvent systems containing mixtures of tert-butanol, ethanol, and dioxane with water are considered here because these are the only aqueous systems for which precise transference data[78,238] are available. The aqueous dioxane[424] and ethanol[125] mixtures both exhibit large viscosity maxima near 80 mole % water amounting to, respectively, 2.2 and 2.7

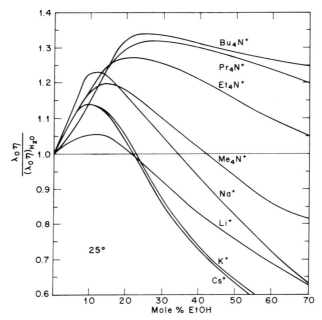

Fig. 15. The limiting Walden products for several ions in ethanol-water mixtures at 25°C as a function of mole percent ethanol.

times the viscosity of pure water at 25°C. The *tert*-butanol–water mixtures[78] have a broad maximum around 65 mole % water with a viscosity about five times that of pure water. The conductance measurements for the ethanol–water mixtures cover the whole composition range, whereas they were restricted to compositions containing at least 70 mole % water in the case of the other two mixtures in order to avoid complications resulting from ionic association as the dielectric constant became low.

A representative picture of the behavior of ionic mobilities in the aqueous solvent mixtures is given in Fig. 15, where limiting cationic Walden products at 25°C in ethanol–water mixtures[327,742]* are shown for a large portion of the composition range. One feature stands out and is characteristic of all ionic conductance data for aqueous mixtures. All curves show a maximum in the water-rich region after which the conductance decreases to the much lower values of the Walden products in pure ethanol. This fact was often missed in the past, because the measurements were carried

* Also T. Broadwater and R. L. Kay, unpublished data.

out at too wide a composition interval. The conductance of anions presents essentially the same type of picture.

Comparable data to those shown in Fig. 15 were collected or measured for the *tert*-butanol–[78] and dioxane–water[424] mixtures and showed essentially the same behavior except that the maxima for *tert*-butanol solutions were substantially greater and for dioxane solutions substantially smaller than those for ethanol–water. The results in Fig. 16 for the K+ ion show the typical behavior found for both structure-breaking and electrostrictively hydrated ions. On the other hand, as shown in Fig. 17, ions such as Bu_4N^+ that are subject to hydrophobic hydration show almost the same behavior in all the solvent mixtures.

From these observations it would appear that the behavior of the Bu_4N^+ ion is independent of the nature of the organic component of the mixtures, and the ion merely undergoes a dehydration in which the hydrophobic effects exhibited in pure water continuously disappear as the concentration of the organic component increases. Arnett[19] reached the same conclusion from his heat of solution measurements on aqueous alcohol solutions. It would appear that the Bu_4N^+ ions cannot compete with the organic components for the available water molecules.

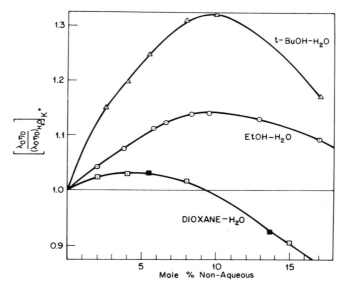

Fig. 16. The limiting Walden products for the K+ ion in *tert*-butanol–, ethanol–, and 1,4-dioxane–water mixtures at 25°C in the water-rich region.

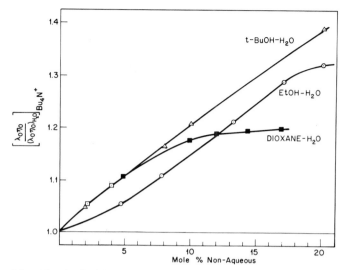

Fig. 17. The limiting Walden products for the Bu_4N^+ ion in *tert*-butanol–, ethanol–, and 1,4-dioxane–water mixtures at 25°C in the water-rich region.

The results for the structure-breaking ions are not as easily explained. The maximum mobility effect was studied in some detail in *tert*-butanol mixtures, where it is most pronounced. The results are shown in Fig. 18 for cations and Fig. 19 for anions. At first glance it would appear from the excess conductance evident in the water-rich region that these ions are better structure-breakers in these mixtures owing to the increased structure that is known to exist in these solutions. However, the ionic size dependence is in the wrong order. All other criteria set the order of decreasing structure-breaking as $Cs^+ > K^+ > Me_4N^+ > Na^+ > Li^+ >$ and $I^- > Br^- > Cl^-$, whereas on the basis of the excess mobilities in these mixtures the order would become $Na^+ > K^+ > Cs^+ > Me_4N^+ > Li^+ >$ and $Cl^- > Br^- > I^-$. Thus the order is reversed for all the anions and the cations Na^+, K^+, and Cs^+.

The facts that the smaller ions show the largest maxima in the Walden product in these mixtures and that the magnitudes of the maxima vary with the degree of hydrogen bonding believed to exist in the solutions suggest that the maxima may be the result of a gel effect which permits the smaller ions to move between the hydrogen-bonded chains without having to overcome the rather high solvent viscosity. However, the experimental result that the very large BrO_3^- ion shows substantially the greatest maxima in aqueous dioxane[124] mixtures would seem to rule out this possibility.

The Zwanzig equation, with ϕ calculated from eqn. (10), does not predict the conductance maxima found in the ethanol[425] and dioxane[424] aqueous mixtures, the only systems for which the necessary dielectric data are available. The revised expression for ϕ given in eqn. (12) does not alter this result. Consequently, the conductance maxima cannot be explained by solvent dipole relaxation, although more extensive and precise dielectric relaxation data are badly needed in this area. Engel and Hertz[196] have reported the interesting result from NMR relaxation studies that the addition of KI to water, containing a few per cent of ethanol, causes the water molecules to become more mobile. Consequently, since the conductance studies indicate that these ions also become more mobile than the bulk viscosity would predict, it would appear that a structure-breaking role for the ions is still

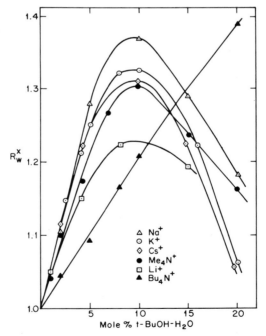

Fig. 18. The limiting Walden products for cations in the water-rich region of *tert*-butanol–water mixtures. $R_w^x = (\lambda_0 \eta)_{\text{mixture}} / (\lambda_0 \eta)_{\text{H}_2\text{O}}$. (Reproduced from Ref. 78 with the permission of the authors and copyright owners.)

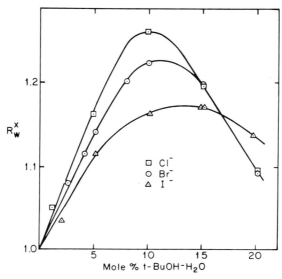

Fig. 19. The limiting Walden products for anions in the water-rich region of *tert*-butanol–water mixtures. R_w^x is given in the caption to Fig. 18. (Reproduced from Ref. 78 with the permission of the authors and copyright owners.)

the preferred explanation in spite of the "wrong" observed order. It is interesting to note that the pressure effects reviewed in the preceding section produced an order similar to the one found in the solvent mixtures. Additional evidence for the structure-breaking concept comes from the observation that a decrease in the temperature increases the magnitudes of the maxima in the Walden products for ethanol–water mixtures.* However, no definite conclusion can be reached, at least until diffusion data on uncharged molecules in these mixtures become available along with more comprehensive and precise dielectric relaxation data. There is a good possibility that $(\lambda_0\eta)_{IS}$ will turn out to be substantially greater than the values predicted by Stokes' law, particularly for small ions, so that the cosphere effects will then show the normal size dependence characteristic of structure-breakers.

* Also T. Broadwater and R. L. Kay, unpublished data.

CHAPTER 5

Infrared Spectroscopy of Aqueous Electrolyte Solutions

Ronald Ernest Verrall

Department of Chemistry and Chemical Engineering
University of Saskatchewan
Saskatoon, Canada

1. INTRODUCTION

In the past few decades a study of the role of solvent in the neighborhood of solute particles has been the preoccupation of a large number of experimentalists and theorists engaged in liquid-phase solution chemistry.

Solutions are very important in many areas of science and therefore are requisite of this extended interest. However, notwithstanding the concerted effort by researchers, the study of liquid systems has been formally difficult and the results prone to equivocal interpretation.

If one establishes a hierarchy of importance among solvents, water must certainly be appropriated a prime position because of its major role in natural processes. Many lines of enquiry have focused on the phenomena associated with solvation in water and the interpretation in terms of structural changes in the solvent about solute particles. Simultaneously and of necessity the problem of the structure of water and its modification with temperature has been vigorously attacked.

While it is generally agreed that the study of aqueous solutions reveals important contributions to the properties brought about by changes in hydrogen bonding in the solvent in the proximity of dissolved solute particles, the contentious issue arising out of these studies has been a choice of a suitable chemical model which adequately describes the various manifestations of these structural effects.

The study of aqueous solution properties may be approached through the use of various experimental procedures designed to provide evidence for structural changes. Transport, thermodynamic, scattering, and spectroscopic experiments provide evidence for these structural effects and it would be intuitively reasoned that the latter two ought to be more decisive in resolving structural problems. As pointed out by Friedman et al.,[474] this has not been largely the case. In fact, whereas thermodynamic and transport results tend to corroborate each other qualitatively and offer a more acceptable basis for rationalization of ion–solvent interactions, the spectroscopic evidence has not been as decisive a tool in resolving structural problems.

This result at first sight appears unreasonable, in that spectroscopic studies at frequencies in excess of 10^{12} Hz should show the presence of different structural environments about a given species in solution. The answer to this dilemma lies in the fact that, although the hydrogen-bonded structures are of longer lifetime (10^{-7}–10^{-11} sec) than the impinging radiation, the spectroscopic[818] and scattering[706] experiments give results that show structural changes which are actually averages over the contributions of several structures. The interpretation of the various contributing structures in terms of chemical models is the current point of debate.

The aim of this chapter is to give a concise account of the experimental methods under discussion and to draw attention to the information that can be derived from infrared results. Finally, a reasonable and objective review of the data in the light of current structural models for aqueous solutions will be presented. It is the author's sincere hope that a failure to satisfy these goals to the mutual satisfaction of all whose work will be reviewed may be compensated for by demonstrating the importance, difficulty, and challenge that infrared studies of aqueous electrolyte solutions present.

2. INFORMATION ON AQUEOUS IONIC SOLUTIONS OBTAINABLE FROM INFRARED ANALYSIS

2.1. General

Infrared spectroscopic studies of aqueous ionic solutions can be a useful tool in the characterization of structural changes in the solvent. The effects of solute–solvent interactions can modify vibrational spectra of the solvent in a number of ways. The intensities can be changed by orders of magnitude; the frequencies of normal vibrational modes of a molecule may be shifted to higher or to lower values; the bandwidths at one-half maximum absorbance, $\Delta\nu_{1/2}$, being measures of the uncertainties in energy, may be broadened considerably; and combination bands involving simultaneous transitions in a molecule may be altered. One must realize that in order to obtain observable changes, relative to the pure solvent (or some isotopic mixture*), large salt concentrations are utilized so that the interactions between structural effects around neighboring ionic particles become significant and tend to complicate the microscopic picture about them. This difficulty is overcome to some extent in the case of thermodynamic and transport measurements in that dilute solution studies (in the ideal limit, infinite dilution properties) can lead to an easier rationalization of solvent–ion interactions because of the damping out of complicating factors such as ion–ion interactions. Comparison of observed data with those calculated on the basis of simple theories known to prevail under low-concentration conditions can then be made. Notwithstanding the aforementioned difficulty, it will be shown that some degree of success has been achieved in the rationalization of aqueous salt solution spectra in terms of specific solute–water interactions.

The infrared spectral region covers the electromagnetic radiation range from about 0.75 μm† to 1 mm and is subdivided into the near-infrared (0.75–3 μm), the fundamental infrared (2–25 μm), and the far-infrared (25–1000 μm) regions. The main reasons for this subdivision are that there are variations in the instrumentation employed in each subdivision and, more important, the information derived from each region can be different. Not only does the information obtained from each of the spectral subdivisions provide specific evidence as to the nature of the vibrational states under consideration, but the analyses of spectra in any two of the subdivisions may provide complementary and/or additional information concerning the effects of ionic solutes on the vibrational states of water. Therefore it seems appropriate that this division be retained in the present discussion with the exception that when the far infrared is discussed these limits will be relaxed and the discussion will include studies carried out from 1000 to 10 μm.

2.2. The Effects of Ions on the Structure of Water

Before considering the infrared studies of aqueous ionic solutions it is helpful to briefly state the controversy that exists regarding the details of

---

* A complete discussion of the relevance of this point is made in Section 3.6.
† The units μm and cm^{-1} are both used throughout the text: $(1/\mu m)\times 10^4 = cm^{-1}$.

the intermolecular structure in liquid water. The status of research on the structure of water has been thoroughly summarized in several monographs[190,421] and reviews[227,592,834] and is dealt with in detail in Volume 1 of this treatise. The current models of water structure may be broadly divided into two categories: the essentially "uniformist" average model,[505,639] in which hydrogen bonds are bent but not broken as temperature increases, and the mixture models, which postulate an equilibrium between two or more types of water molecules corresponding roughly to a transition from a hydrogen-bonded species to a nonhydrogen-bonded one as temperature is increased. It will be seen in later discussions that these models have been used extensively to interpret changes in the infrared spectra of aqueous ionic solutions relative to the pure solvent.

One of the basic factors that makes investigations of homogeneous liquid electrolyte solutions so difficult is that in addition to the strong electrostatic ion–dipole interaction that causes simple dielectric polarization, solvation, and electrostriction of nearest-neighbor molecules, there are water molecules which are barely or not at all influenced directly by ions. The infrared spectra can thus be interpreted in terms of a statistically weighted sum of absorption bands for the various interacting solvent molecules with their different neighboring environments.

Further to this point, Frank and co-workers[230,232] have postulated a model for aqueous solutions in which there are envisaged three concentric regions about an ion: (1) an innermost region of polarized, immobilized,* and electrostricted water molecules, (2) an outer region containing water having the normal liquid structure, and (3) an intermediate structure-broken region in which the normal tetrahedral structure-orienting influence of the bulk water is in competition with the radially orienting electric field of the ion. This concept leads to a classification of ions based on their capabilities of either promoting structure ("structure makers") by reinforcing either the inner or outer regions, or breaking structure ("structure breakers") by making the intermediate region about the ion most important. This brief introduction should help set the tenor for the discussion of infrared absorption results. Whereas infrared absorption spectra of ionic solutions might be expected to assist in resolving the controversial issue on the structure of water, it will be seen that their interpretation has not led to complete acceptance of any one model.

---

* One should not be misled by this word as inferring permanency of attachment of water to the ion. These molecules exchange rapidly with the bulk solvent molecules and only in a few exceptional cases[389,390] are the lifetimes of molecules adjacent to the ion of long duration.

2.3. Far Infrared

Most of the early infrared investigations covered frequencies below 10 μm and the observed absorption bands involved transitions between intramolecular vibrational states. Any intermolecular structure in the liquids caused perturbations of the frequencies and intensities of the intramolecular bands and these provided a secondary means of obtaining information regarding intermolecular interactions in the liquid phase.

Above 10 μm one can observe the direct transition of entire water molecules restrained by the relatively weaker intermolecular forces. A clear distinction between intramolecular and intermolecular vibrations is not always possible. In hydrogen-bonded systems the intermolecular interaction can be considered to a first approximation as a loose complex. In principle, where two nonlinear molecules are linked together in a complex there will result a loss of three translational and three rotational degrees of freedom. In the liquid phase the six new vibrational modes arising from the loose complex may be described as librations or restricted rotations, hydrogen-bond stretchings or restricted translations, and hydrogen-bond bendings. The use of isotopic shift information[813] aids in distinguishing between the librational and stretching intermolecular modes. The librational frequencies of water shift to lower frequencies by the approximate ratio of $\sqrt{2}$ or less in D_2O, and the hydrogen-bond bending and stretching vibrations show nearly identical frequencies in water and heavy water. In theory, the latter would be expected to show a shift equal to $(20/18)^{1/2}$. Studies of aqueous electrolyte solutions in this spectral region involve considerable experimental difficulty. The bands are broad and weak, the temperature is an important variable, and complementary Raman measurements are often essential. The effect of dissolved ions on the frequency positions and intensities of the bands therefore reflect changes in the equilibrium configuration of the intermolecular complex and movement of charge associated with perturbation of the equilibrium state.

2.4. Fundamental Infrared

The effect of electrolytes on the fundamental region of the infrared spectrum of water has been extensively studied. The greater interest in this region of aqueous electrolyte solutions may be due to a better intrinsic understanding of the force field and energy function for the lowest vibrational energy change of the water molecule itself. Changes in the spectroscopic parameters used to characterize the infrared absorption bands (Section 2.1) are observed in a number of cases and there appears to be a correlation between modification of band parameters for the pure solvent and the surface charge density of the ionic species in solution.

Theoretical predictions of the influence of solute species on absorption band parameters have not been too successful. The procedures are mathematically difficult and of more import is the fact that they have the fault of being based on a subjective choice of microscopic model on which to carry out the calculation. Various theoretical treatments[39,84,85,448,653,654,774] have been proposed to account for solute-induced changes in the fundamental bands of solvent molecules in nonionic solution media. Although taking us closer to a rational explanation of changes in band parameters (compared to pure gaseous solvent), these treatments have fallen short of a complete interpretation. They have tended to account quantitatively for specific kinds of intermolecular forces such as dipole–dipole, while omitting special cases of specific interactions such as hydrogen bonding. Partly for this reason and partly because of the added complexity of ionic systems, those familiar with the basic theory of molecular vibration often have regarded the semiquantitative treatments of ionic solutions with a justifiably critical appraisal. Some degree of success has been achieved in predicting the frequency shifts[587,716,717] of the fundamental vibrations of the free water molecule from calculations of the vibrational frequencies of water in aquocomplexes. These have been very approximate treatments but the agreement between experimental results and theoretical predictions indicates that the ion–water interaction is strong enough in certain instances so that the pair can be treated as a vibrational unit.

One of the most delicate probes of bonding conditions that can be obtained experimentally is the integrated absorption intensity B. It reflects the strength of the intermolecular interactions and is represented by the following expression:

$$B = (1/cd) \int_{\nu_1}^{\nu_2} (I_0/I)_\nu \, d\nu \tag{1}$$

where I_0 and I are respectively the intensities of the incident and transmitted light of frequency ν, c is the concentration, and d is the path length of the sample. Unfortunately, the integrated intensity is one of the most difficult parameters to measure accurately in spectroscopic work and in a number of instances the relative change in peak absorption intensities of bands have been used as a less acceptable criterion in discussing intermolecular interactions. Since it can be shown from quantum mechanical theory that the intensity B of a fundamental vibration is related to the change in dipole moment with respect to normal coordinate ($d\mu/dr$) and, because of the likely change in polar character and electron mobility within the water molecule due to the influence of the electric field, it is readily apparent that the integrated intensity should be somewhat sensitive to the nature and concentration of the ionic solute.

The bandwidths of the vibrational fundamental of the solvent are also sensitive to the type of environment in which the molecules find themselves. The large differences in bandwidths between the crystal hydrates (a few cm^{-1}) and liquid solutions (a few hundred cm^{-1}) suggest that structural order must be a significant contributor to this difference. In the case of ionic solutions it is likely that the relatively stronger electrostatic ion–dipole interactions can accentuate structural differences and hence contribute to further modification of the broad bands in the pure liquid phase. Therefore, variation of bandwidth of the fundamental vibrations of ionic solutions, compared to the pure liquid, can be interpreted qualitatively in terms of structural changes.

Closely associated with bandwidths is the intensity distribution across the absorption band. This is essentially a measure of the distribution of bond strengths of the intramolecular vibration under observation and any inflections in this distribution would suggest preferred bond strengths. Such observations are considered by the proponents of the mixture model for pure liquid water as indicating the presence of specifically favored structures in the solution.

The most easily identifiable change in the infrared absorption of solvent bands in ionic solutions is the shift in the band frequency. These shifts are also a measure of the strength of the ion–water interaction as it affects the valence electron density of the intramolecular vibrational mode of the solvent molecule.

It has been shown above that there are several absorption band parameters which can provide some insight into changes in intermolecular interactions upon going from pure liquid water to aqueous ionic solutions. It is important and must be insisted upon, however, that the conclusions about molecular interaction and structure drawn from such observations be mutually consistent. Furthermore, they should be consistent with other observed data derived from thermodynamic and transport phenomena. However, as was pointed out earlier, to obtain observable changes in infrared work, somewhat concentrated conditions are required and it is very difficult to extrapolate these results to the lower concentration regions where other nonspectroscopic data can normally be obtained.

The study of the fundamental vibration region of aqueous ionic solutions has not been without difficulties, some of which are of an experimental nature and will be discussed further in Section 3, and some of which are inherent in the nature of the infrared absorption of water itself. For the latter the description of atomic motion leads to the observation that bands can be described that are combinations of fundamentals and are called combination bands. Fundamentals are given the notation ν_n, where $n = 1, 2, \ldots$ and refers to the designated fundamental frequency. Combination bands can be written as a sum band $(\nu_1 + \nu_2)$ or a difference band $(\nu_1 - \nu_2)$. There also exists overtones designated as, e.g., $2\nu_n$ to indicate that they appear at approximately twice the fundamental frequency ν_n.

The first overtone of the fundamental bending vibration of the water molecule, $2\nu_2$, has led to some difficulties in interpreting changes in the band shape of the fundamental stretching mode of pure H_2O. These additional difficulties stem from the fact that the frequencies of the two O—H stretching vibrations, ν_1 and ν_3, and that of the first overtone of the bending vibration, $2\nu_2$, are close to each other. The overlapping ν_1 and $2\nu_2$ modes may be in Fermi resonance with each other. Fermi resonance is a vibrational perturbation and occurs when two vibrational levels in a molecule may have nearly the same energy. If these two vibrations are of the same symmetry species of vibration, a perturbation of the vibrations can occur and the result is the shifting of one of the levels to a higher energy while the other falls to a lower energy. There effectively results a "borrowing" of intensity by $2\nu_2$ from ν_1. The same arguments can be made concerning work in pure D_2O. Obviously one has to choose experimental conditions to counteract or reduce such problems in order to facilitate interpretation of results and this is done by judicious choice of solvent conditions as described in Section 3.6.

2.5. Near Infrared

In the last decade there has been a renewal of interest in near-infrared studies of aqueous electrolyte solutions despite the fact that combination and overtone bands generally have medium to weak intensities and this makes them more difficult to study. As pointed out by Frank,[229] the lack of understanding of the laws governing the addition of intensities in overtone and combination bands has made rationalization of intensity changes in terms of ion–solvent interactions less convincing than in the case of the fundamental bands. The large number of possible combination and overtone bands that are infrared-active, coupled with the uncertainty in anharmonicity factors affecting changes in the absorption frequencies of these bands, make some assignments in this region difficult. Extensive studies by Luck[528-531, 534,535] on the near-infrared spectrum of pure water have probably led to the recent renewed interest in the effect of electrolytes on the combination and overtone bands in this region (see Volume 2, Chapter 4). Recently McCabe and Fisher[538,539] have resolved and interpreted near-infrared difference spectra of aqueous alkali halide solutions in terms of: (1) water excluded by the hydrated solute, (2) water of hydration, and (3) a final component consisting of the absorption by the solute itself. By means of such considerations,[539] these workers were able to attempt a quantitative evaluation of a number of parameters related to the hydration of ionic species in aqueous solution. This constitutes a substantial extension of normal procedures of extracting information from infrared absorption bands and will be given closer scrutiny in Section 4.3.

3. EXPERIMENTAL METHODS

3.1. General

The practical difficulties of examining the infrared spectra of aqueous solutions have deterred many spectroscopists from attempting to obtain information by this physical method. However, during the course of the past several decades techniques and instrumentation have been developed which in turn encouraged the pursuit of such studies in order to provide some information on structural characteristics of aqueous solutions.

The validity of spectra depends upon many factors, of which selection of proper instrument operating variables is one of the most important. No attempt will be made to describe the interrelation between operating variables and the selection of proper instrumental operating and recording variables since these have been adequately described in a number of texts, of which one of the latest and most lucid presentations is that of Alpert et al.[8] Suffice it to say that improper utilization of instrument variables may well lead to artifacts from which false inferences can be drawn. Hopefully, workers in this area are cognizant of such instrumental peculiarities and take the steps to reduce potential anomalies to a minimum. To say with certainty that it has always been done is obviously impossible since some workers have seldom made the effort to describe carefully the precise operating conditions of the instrument under which spectra have been obtained. Finally there is the inherent problem that no two instruments, even if they are similar models supplied by the same manufacturer, have the same optical characteristics. This raises the point of the desirability of obtaining spectra on different instruments to assure the validity of unusual results.

In the following parts of this section attention is drawn to certain practical difficulties that arise when working with water and the experimental methods used to reduce these as much as possible.

3.2. Atmospheric Absorptions

Atmospheric absorption from CO_2 and H_2O may present serious problems because, in contrast to the other atmospheric components, they absorb infrared radiation. In single-beam instruments all atmospheric absorptions will appear as bands the intensities of which will be a function of their concentrations. These absorptions can partially or totally mask out any sample absorption if the sample happens to absorb in the same region as these gases. The most troublesome regions are 2.67 and 5.5–7.5 μm for H_2O vapor and 4.25 and 14.98 μm for CO_2.

Double-beam instruments are designed to minimize some of these problems. The lengths of sample and reference beams are equal, so that for uniform atmosphere concentration the detector will not see any difference in the energy of the two beams. However, for relatively strong absorbers, even though bands do not appear, there is a considerable reduction in energy and this reduces the proper operating conditions of the instrument. Consequently, it is a wise procedure to eliminate such gases either by appropriate adsorption techniques or purging with an inert gas.

3.3. Cells and Cell Path Lengths

The intense absorption by water throughout much of the infrared region seriously hinders the registration of good infrared spectra without appropriate measures to reduce sample optical path lengths. In addition, many of the usual window materials are unusable because of their solubility,* most notably NaCl, an inexpensive and readily repairable window material. To overcome the strong absorption characteristics of water, capillary thicknesses are used. In the near-infrared region, where the

---

* The Irtran series available from Eastman Kodak is available for infrared windows and are essentially unaffected by water. These are pressed polycrystalline materials and have great resistance to thermal shock.

absorption bands are less intense, it is possible to use relatively long path length glass cells but elsewhere it is impractical to exceed 0.075 mm, with usual thicknesses ranging between 0.001 and 0.050 mm.

The most widely used window material in the fundamental and near-infrared regions is CaF_2, which can be used to 8 μm. BaF_2 is also a useful window material to about 12 μm; however, it is rather brittle and care must be taken in handling the windows. LiF can be used to 6 μm in aqueous solutions that are not basic in nature, since the windows tend to become slightly opaque under the latter conditions due to enhanced hydrolysis of lithium ion in a basic aqueous medium.

There are several materials that are available for windows in the far infrared. Crystal quartz can be used between 50 and 1000 μm, although some slight wedging of windows is necessary to reduce interference effects due to its high reflectivity. Polyethylene has some attractive features such as high transparency over a wide wavelength range and low cost; however, certain mechanical disadvantages[217] and uncertainty in absorption characteristics as a function of temperature[759] reduce its usefulness for quantitative analysis.

Silicon has good potential[217] despite high reflection losses. For wavelengths beyond 17 μm silver chloride can be used, although the high refractive index of this material causes greater surface reflection losses than for barium fluoride.

The use of short path lengths in infrared studies contributes largely to the difficulty in obtaining quantitative information from absorption bands. Techniques have been developed for preparing thin cells of known thickness below 0.002 mm and these provide some acceptable basis for quantitative studies.

Spacers fabricated from chemically inert materials are often used to establish set path lengths. Materials such as lead and Teflon are commonly used and the spacer thicknesses can be accurately measured outside the cell. If studies are carried out as a function of temperature, the relatively high coefficient of expansion of Teflon may lead to path-length changes with temperature. An initial thickness for a spacer obtained outside the cell at ambient temperature may be in error when applied to the same spacer at some other temperature in the cell. Furthermore, the inability of Teflon to resist compressive forces may also allow for a decrease in spacer thickness when the cell is assembled or even variation of path length within the assembled cell due to uneven forces exerted over the surface area of the spacer. These sources of error appear to be very small. However, when dealing with a strongly absorbing medium and extremely short path lengths the small changes mentioned above can seriously restrict the reliability of any quantitative information.

To obtain spacers with thicknesses of the order of 0.001 mm creates practical difficulties in itself. They are not normally available from commercial suppliers, which leads to improvisation. Gold and silver leaf can be used as spacer material, but at these thicknesses they are very fragile and somewhat difficult to use. Teflon spacers of slightly greater thickness can be inserted in the cell and compressed. However, accurate thicknesses are temporarily lost. The following technique developed by Thompson[775] and used by the author on several occasions works very well. Spacers are prepared by spraying one of the windows with a polytetrafluoroethylene aerosol. The window is first covered with a template, preferably made of soft material in order not to damage the window. The form should be suitable for allowing maximum incident beam transmission and injection of sample into the cavity. Through trial and error one can become adept in making uniform gasket thicknesses.

Using the above procedure it is obvious that the path length of the assembled cell must be determined by the interference fringe pattern method. For instruments calibrated in linear wavenumbers we have

$$d = n/2(v_2 - v_1) \tag{2}$$

where d is the path length of the cell and n is the number of maxima (or minima) between the wavenumbers v_1 and v_2. For linear wavelength instruments we have

$$d = n\lambda_1\lambda_2/2(\lambda_2 - \lambda_1) \tag{3}$$

where n is the number of maxima (or minima) between the wavelengths λ_1 and λ_2. It is desirable, for reasons mentioned previously, that the cell path length be determined at every temperature when infrared studies are conducted as a function of temperature.

An alternative method of calibrating the cell is to fill it with some standard material of known absorbance and calculate the path length from the resulting intensity data. The latter procedure is useful when dealing with long path lengths or where the windows happen to be non-parallel.

3.4. Temperature Variation of Sample

Temperature is a very important variable in the study of aqueous systems. The structural characteristics of water, because they depend upon relatively weak intermolecular forces, are susceptible to considerable change over a narrow temperature range. The temperature variation of measured absorption band parameters may give information about changes in electrolyte solutions which are dependent upon the weaker intermolecular forces as opposed to the temperature-independent ionic charge effects.

Several problems arise, however, when employing this experimental technique. First, when thermostated cells below ambient temperature are used an inactive infrared gas, free of water vapor, should be flushed through the cell compartment to prevent water from the air condensing on the cell windows. Second, one has to consider the problem arising from the sample radiating more energy to the surroundings than it receives. In the case of heated cells placed in the optical path, thermal radiation from the heated cell must be allowed for. Conversely, when a sample is cooled below ambient temperature the detector may effectively radiate more energy to the cold sample than it receives. In all infrared instruments the optical path is carefully shielded from the sources of heat such as the hot electronic components; therefore, heated or cooled sample cell holders can be potentially troublesome at extreme temperature conditions. At the short-wavelength end of the infrared spectrum, within the temperature range 0–100°C, the error arising from such a source is less than 2%, i.e., at a given wavelength the radiation peak of the heated sample has an intensity less than 2% of that due to an instrument source whose temperature is roughly four times greater. At longer wavelengths this situation is less acceptable since the lower-temperature source shows a slower decrease of intensity with wavelength than does the instrument source, and the percentage contribution of the former may be two to three times greater than that listed above for shorter wavelengths. Single-beam measurements can be more readily corrected than double-beam measurements for extraneous radiation from a sample.

3.5. Reflection Losses

Another problem which should be considered when good spectral data are required is that of reflection losses occurring at the window–air and window–aqueous solution interfaces. The former can be measured with a single window in air, while the latter may be calculated from the mean refractive indices for the window material and the solution. Wyss and Falk[871] have shown that this correction can amount to about 5% when the peak absorbances are of the order of 0.5.

3.6. Isotopically Dilute HDO as a Solvent

The reasons were noted in Section 2.4 why pure H_2O or D_2O creates difficulty in so far as interpretation of fundamental vibrational absorption bands is concerned. To alleviate this problem, it has become the established practice to use isotopically dilute solutions of HDO in D_2O or H_2O rather than pure H_2O or D_2O as solvent. In the HDO molecule the three vibrational modes appear in separate regions of the spectrum and are not coupled together, thereby providing a simpler shape of the OH and OD valence stretching bands for interpretation. In addition, through knowledge of the equilibrium constant for the reaction $H_2O + D_2O \rightleftharpoons 2HDO$, the concentration of HDO in either H_2O or D_2O can be controlled, enabling the use of cells with longer path length.

Two values, 3.96[784] and 3.8,[451] have been reported for the equilibrium constant of the above reaction. These values have been obtained from statistical mechanical calculations of the gas-phase equilibrium, but there appears to be independent evidence that they are applicable to the condensed-phase equilibrium. It has been pointed out by Swenson[765] that there is reason to believe the equilibrium constant is not very different in aqueous solutions since nuclear magnetic resonance studies[379] of this equilibrium in acetone solutions have shown the concentration equilibrium constant to be 4.0 ± 0.4. Furthermore, it can be shown by combining vapor pressure data[451] of the appropriate phase equilibria with the gas-phase constant that the liquid-phase and gas-phase equilibrium constants are about the same at $30°C$. The temperature dependence of the equilibrium constant for the gas-phase equilibrium varies slightly from 3.94 to 4.00[784] and 3.76 to 3.89[451] in the temperature range $273–400°K$. As a consequence, within the uncertainties of the above values of the constants, the use of a single value of K for all temperatures would appear to be appropriate and up to now this has been done.

Although the use of HDO as solvent has its advantages, it also has some disadvantages in terms of background absorption. It is apparent upon quantitative examination of the equilibrium exchange reaction that the equilibrium concentration of H_2O or D_2O can contribute to the respective intensities of the OH or OD valence stretching vibrations of the HDO molecules in solution. When recording the spectrum without any reference cell judicious choice of starting concentrations and cell path length are required to obtain optimum results. For example, if one were considering the fundamental OD stretching vibration of HDO derived from a solution of D_2O in excess H_2O the band to be observed has its maximum at 3.98 μm and because of the excess H_2O, there are additional bands about 4.65 μm (fairly weak) and at 2.92 μm (very intense). The result of having a very low D_2O concentration with a long compensating path length to make the band observable is a very large curved background which must be taken into account. At the other extreme of the scale, i.e., a high D_2O concentration, the solution contains a considerable number of D_2O molecules which absorb radiation exactly at the same frequency as the HDO molecules. Verrall and Senior[731] have reported an optimum concentration range of 5–10 mole % D_2O in H_2O with a cell thickness range of 1–2 μm which reduces the correction due to the above effect to less than 2% of the observed band intensity. Similar difficulties can arise in the complementary case of H_2O in excess D_2O.

Some spectra have been recorded in the double-beam mode using a reference cell filled with pure H_2O or D_2O of path length equal to that of the sample cell. A problem arising in this instance is that, although somewhat longer path lengths can be utilized, it is still difficult to obtain equal path lengths for the sample and reference cells at such small path lengths. Furthermore, if equal path lengths are used, this introduces an error in the resulting difference spectra since the concentration of the major component in the sample cell will be slightly less than in the reference cell. In fact, one should reduce the reference path length to compensate for this concentration imbalance. The method of recording spectra without a reference cell, at the same time correcting for background absorption, is probably the most useful in the fundamental and far-infrared regions when quantitative intensity data are required.

Finally in preparing the solutions it is advisable to make them up by weight and preferably in a dry box in order to reduce exchange with atmosphere moisture. If one wishes to express the concentrations in terms of moles per liter, then density data on H_2O–D_2O solutions are required.

4. CRITICAL REVIEW OF AVAILABLE INFRARED DATA

4.1. Far Infrared

There has been limited experimental work conducted on the effect of ions on the far-infrared spectrum of water. Some of the most extensive far-infrared studies of the pure solvent phase and ionic solutions have been carried out by Draegert and co-workers.[179,180] Prior to that there had been some limited work by Swain and Bader[764] and Falk and Giguère.[210] The far-infrared work of Draegert and co-workers[179] on pure H_2O and D_2O

revealed general absorption throughout the region above 10 μm with at least two well-defined absorption bands, one, a broad, intense band near 14.7 μm (20 μm in D_2O), the other a much weaker band appearing as a shoulder centered near 58.8 μm (60.6 μm in D_2O) on the low-frequency side of the major band. For reasons mentioned in Section 2.3 the major band has been attributed to hindered rotation of water molecules and the minor band to hindered translation of water molecules. Focusing attention on the four-coordinated structure of water in terms of a simple C_{2v} symmetry designation, it would appear[180] that two of the librational modes of the central water molecule should be infrared-active.* From the infrared work to date there is no clear evidence for the presence of more than one component in this major librational absorption band of water, although strong evidence for the existence of more than one component from inelastic neutron scattering,[494,705] hyper-Raman scattering,[772] and Raman[816] studies would lead to such speculation.

Due to the broad nature of the librational band for water and the general absorption throughout the region, it is difficult to locate new points of maximum absorption in solutions and hence detect any fine structure. The conclusions drawn from the far-infrared work to date on ionic solutions therefore have been of a qualitative nature, but it is interesting to note that the perturbations of the pure solvent spectral band parameters of these intermolecular modes show some systematic correlation with the nature and concentration of the ions.

A comprehensive study of the effect of varying concentrations of ionic solutes on the far-infrared spectrum of D_2O[180]† at ambient temperature has led to the following observations: (1) There is a rather striking resemblance of the observed spectra of solutions, even at the highest concentrations, to the spectrum of pure heavy water, (2) in certain cases the changes in temperature and the addition of salts to pure D_2O cause similar spectral changes, and (3) the spectra of solutions have certain features which cannot be duplicated by changing the temperature of pure water.

The conclusions arrived at above were based on an arbitrary self-consistent method of obtaining some pseudo band parameters in order to overcome the practical problems of intense absorption and inability to

---

* Vibrational assignments for the five-molecule hydrogen-bonded structure of water according to C_{2v} symmetry indicate that the $\nu_5 a_2$ assignment (nonsymmetric deformation of the A_2 species) is infrared-forbidden and corresponds to the 22.2-μm band observed only in the Raman spectrum of H_2O.[813]

† Liquid D_2O was employed as solvent because polyethylene was used as window material in most cases and it is transparent above 15.4 μm.

measure precise sample thickness. These parameters included (1) ν_L and ν_T, the frequencies of maximum absorption of the librational and hindered translation bands, respectively, (2) "width index" $\nu_1 - \nu_2$, the bandwidth of the librational band at a normalized transmittance of 0.2, (3) "asymmetry index" $\nu_L - \bar{\nu}(0.2)$,* a measure of the difference in frequency between ν_L at 0.1 normalized transmittance and the mean of $\nu_1 - \nu_2$, $\bar{\nu}$, and (4) a_T/a_L, the relative values of the absorption coefficients of the shoulder band (hindered translation) and major band taken at the estimated frequencies of maximum absorption of the shoulder and major band, respectively.

Figures 1–3 summarize some of the results of the work. Figure 1 indicates that the gross features of the 4 M electrolyte solutions are somewhat similar to those of the D_2O solvent. Unfortunately, without integrated absorption intensities the decisiveness of this comparison is weakened. For example, it would be of interest to know whether the total intensity within the librational frequency region of pure D_2O increases or decreases with different salts and varying concentrations. Such changes would give some insight into the relative strengths of the ion–water hydrate bands in comparison to the pure water bands.

Analysis of Fig. 2 suggests that the cations in the alkali halides effect band parameter changes of the order produced by anions. This was not clearly evident in previous work[764] and has been less apparent in the case of Raman studies[813,814] of salts containing a variety of cations. However, this discrepancy is perhaps not too surprising in view of the fundamental difference between these spectroscopic phenomena, namely polarization versus dipolar change. It is likely that the anion–water mode of interaction is susceptible to polarization changes which would be more effectively induced by larger anions, whereas the cation–water interaction would be more dependent upon dipolar change in the vibrational mode of interaction and this would be readily expected of small cationic species. Results indicate such a rationalization is far too simple and that factors like size, charge on the ion, and orientation of water molecules about ions create a configuration of minimum energy for the local structure.

At this point it is worthwhile briefly to draw attention to the problem of accounting for the differences between the interactions of water molecules with cations as opposed to anions. Within the uncertainty of application of the Born model for ion–water interactions there are thermodynamic

---

* For a symmetric band this index is zero, whereas the implication of a positive index is that the spectral absorption from the higher frequency ν_1 to $\bar{\nu}(0.2)$ is greater than from $\bar{\nu}(0.2)$ to ν_2, the lower frequency.[180]

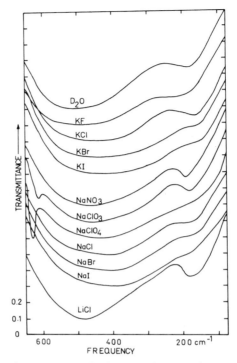

Fig. 1. Normalized spectral transmittance curves for D_2O and $4\,M$ solutions of strong electrolytes. The interval between marks on the transmittance axis is 0.1; the upper curves have been displaced by multiples of this interval. The sharp bands near $600\,cm^{-1}$ in $NaClO_3$ and $NaClO_4$ solutions are due to normal vibrations of the anion groups. (Reproduced from Ref. 180 with permission of the authors and *J. Chem. Phys.*)

indications[83,498,600] that the anion–water interaction is somewhat weaker than the cation–water interaction for a given charge/size ratio.

It was first suggested by Bernal and Fowler,[46] and generally continues to be accepted as a viable model,* that the cation–water interaction results in the coplanar arrangement of cation, oxygen, and hydrogen atoms. This

---

* Alternatives to this proposal have been put forward by Verwey[800] based on considerations of the interaction with neighboring water molecules as well as by Vaslow[791] using Buckingham's[83] ion–quadrupole interaction data. Both workers suggest that small cations lie at a substantial angle to the dipole axis.

prevents a water molecule from rotating except about the dipole axis,[421] a mode which does not contribute to orientation polarization, and might account for the relatively smaller effect of cations on the hindered rotational band of water as observed from Raman studies.[814] Noyes[600] has suggested that such an orientation places less restrictive demands upon the configuration water molecules must assume about cations of varying size, in order to fit into the local water structure. It would appear the latter argument places more stress on entropic contributions to minimization of free energy, whereas other arguments[791,800] are primarily based on enthalpic contribu-

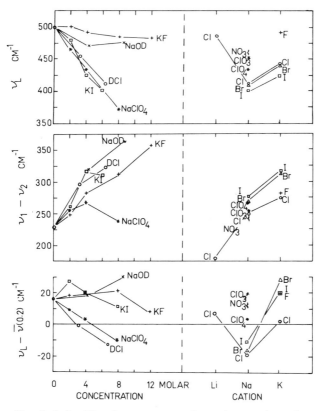

Fig. 2. Left: The frequency ν_L of maximum absorption, the width index $\nu_1 - \nu_2$, and asymmetry index $\nu_L - \bar{\nu}(0.2)$ of the hindered rotation band as a function of concentration. Right: The same parameters for 4 M solutions of compounds of lithium, sodium, and potassium. (Reproduced from Ref. 180 with permission of the authors and *J. Chem. Phys.*)

tions. Presumably, because a water molecule bound to a cation in a co-planar manner can bind to only two other water molecules, it is a less stringent configurational requirement to meet than having to satisfy a three- or four-coordination arrangement, although at the expense of forming two fewer interactions.

In the case of water molecules about anions it is generally agreed that water molecules tend to orient themselves with one of their O—H bonds normal* to the anion surface and this would leave them freer to rotate than in the case of water molecules about cations. This would result in a contribution to orientation polarization and account for the greater anionic effects on Raman molar librational intensities.[814]

Returning to the previous discussion, it can be seen from Fig. 2 that the frequencies ν_L of all electrolyte solutions are lower than ν_L for pure D_2O. The same effect was obtained with AgCl windows for aqueous solutions of NaOH and HCl relative to pure H_2O. For 4 M solutions containing a common cation the relative order of decreasing ν_L for potassium solutions is D_2O, KF, KCl, KBr, and KI, whereas for the sodium series it is D_2O, NaOD, $NaNO_3$, $NaClO_3$, $NaClO_4$, NaCl, NaBr, and NaI. For solutions with a common anion the order of decrease in ν_L is LiCl, DCl, KCl, and NaCl. Increasing salt concentration also decreases ν_L except for the cases of KF and NaOD. $NaClO_4$ shows an unusually large effect of reducing the librational frequency at its highest concentration, a decrease of some 130 cm^{-1} relative to D_2O. The absolute values of ν_L and the shift of the ν_L band toward lower frequencies for solutions of HCl in H_2O in this work agree with previous work.[210]

The results of the effect of salts on the hindered translation mode of D_2O are less systematic. It can be seen from Fig. 3 that ν_T for some solutions is lower than ν_T for D_2O; others are higher. Solutions containing a common potassium cation at 4 M show a decrease in ν_T in the order KF, D_2O, KCl, KBr, and KI, whereas for sodium solutions the decrease is $NaNO_3$, NaOD, NaI, $NaClO_4$, NaCl, D_2O, $NaClO_3$, and NaBr. For the chloride solutions ν_T decreases in the order DCl, NaCl, D_2O, LiCl, and KCl. The apparent difference in the order of the effect of sodium salts on the shift of ν_T as compared to ν_L is insignificant, as the shifts are smaller and the estimated uncertainties of the same magnitude as for ν_L.

Figure 2 indicates a broadening of the librational band in 4 M ionic solutions, a notable exception being LiCl. Increasing solute concentration

---

* An alternative view[83] suggests that the water dipole should be at an angle of 0° to the electric line of force for minimum dipolar interaction with the field.

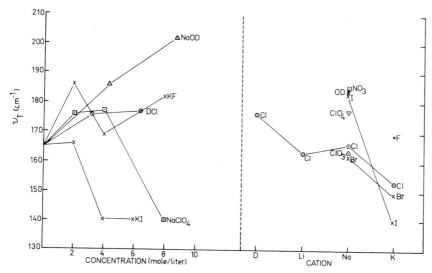

Fig. 3. Left: The frequency of ν_T of maximum absorption of the hindered translation band as a function of concentration. Right: ν_T for 4 M solutions of compounds of deuterium, lithium, sodium, and potassium. (From data in Ref. 180 with permission of the authors and *J. Chem. Phys.*)

has a similar effect; however, KI and $NaClO_4$ appear to narrow the band again above 4 M. A broadening of the librational band and a shift of ν_L to lower frequencies has also been shown[179] to occur for D_2O with increase in temperature. The apparent similarities in the effects of change in temperature and addition of some solutes upon the librational band of water enables the ions to be classified as "structure breakers" or "structure makers" (Section 2.2), similar to other spectral studies.[92,102,255,375,727,813, 814,837,873] The far-infrared work of Draegert and Williams[180] indicates that Li^+ and F^- should be regarded as structure makers whereas Na^+ and K^+ are structure breakers. For the sodium series of salts Fig. 2 indicates that the effectiveness of anions breaking structure, as partially measured by the shift in ν_L to lower frequencies, increases in the order NO_3^-, ClO_3^-, ClO_4^-, Cl^-, Br^-, I^-.

The above classification is in general agreement, excepting Na^+, with the predictions of Samoilov[712] based on the ease of translational movement of water molecules as a whole in the first coordination layer around an ion. In addition, the far-infrared results concur with Swain and Bader's[764] classification based on an ion's ability to change the structure difference between light and heavy water.

A further interesting feature of their work relates to the shapes of the v_L band for $NaClO_4$ and KI solutions. Figure 2 indicates that for these solutions the asymmetry index decreases rapidly with increasing concentration, whereas the width index increases and then decreases. One explanation for such an observation[180] is that there may be unresolved low-frequency components that are increasing in intensity with accompanying decrease in the high-frequency component associated with normal water structure. Although this is a logical supposition, it relies on information derived from other spectroscopic studies which to date have yielded somewhat better resolution than infrared work.

An attempt was made to compare the effects of various solutes with the effect of temperature, using the concept of a "structure temperature," which defines the temperature at which pure water would have the same properties as those of a given solution.[180] Analyses of several concentrated salt solutions led to structure temperatures greater than 100°C. It would appear that it is inappropriate to consider such a scheme when discussing concentrated salt solutions. This is not unexpected[180] since at high concentrations one envisages rather direct interaction of increasing numbers of water molecules with ions and there is no longer the effect of ions simply disturbing the intermolecular solvent structure which is measured effectively by temperature change.

There is general agreement among the various studies of aqueous acid and base solutions. Earlier far-infrared studies[210,764,878] of these solutions were not as extensive in scope, however. Swain and Bader[764] observed that the hydronium and hydroxide ions left the librational band of liquid water relatively undisturbed except for some broadening. Falk and Giguère[210] observed that aqueous HCl, and Zengin[878] observed that HF solutions, shifted v_L to lower frequencies. These observations have been partially confirmed by the work of Draegert and Williams (Figs. 2 and 3). The librational bandwidth for acidic and basic solutions is large and increases with solute concentration. The frequency v_L decreases for increased acid concentrations, as observed previously;[210,878] however, the base changes slowly under the same conditions. The asymmetry index increases slowly for the base but decreases rapidly as acid concentrations are increased. Increasing acid concentration showed little effect upon v_T, whereas increased base concentration caused v_T to increase. The invariance of v_T with increasing acid concentration may be real or it could be an artifact due to broadening and rapidly decreasing asymmetry index of the librational band. These effects could cause the v_L band to overlap the v_T band making background correction somewhat more difficult in locating v_T.

The far-infrared region of aqueous electrolyte solutions provides some interesting features that relate to the effect of ions on the solvent intermolecular interactions, but improved techniques and instrumentation are required for better quantitative analyses. In a private communication, Professor Williams has informed the writer that such an attempt has been partially carried out in his laboratories. Reflectance spectra of aqueous alkali halide solutions from 5000 to 500 cm^{-1} have confirmed the observed shifts for ν_L.[180] What is more important, however, is that this technique makes possible the evaluation of absolute values of absorption coefficients. When completed this work should provide a more quantitative probe in assessing intermolecular structure changes in aqueous electrolyte solutions.

4.2. Fundamental Infrared

Some recent studies of the fundamental region have led to a more quantitative analyses of the water absorption bands than was possible in the far infrared. Practical problems are still prevalent and this is indicated by a number of studies that have simply concerned themselves with the effect of ions on broadening and frequency displacement of the maximum absorption of bands in this region.

No attempt will be made to review all the data pertaining to work carried out prior to 1945. Much of the work to that time was done in pure H_2O or D_2O and the interpretive problems inherent under these conditions have been discussed. For an extensive bibliography on the infrared spectra of aqueous solutions prior to 1945 one should consult Barnes et al.[35]

It is well established for the infrared valence stretching vibration of the OH bond that increased strength of hydrogen bonding is accompanied by a shift in band maximum to lower frequency and an increase in the integrated intensity and half-bandwidth ($\Delta\nu_{1/2}$).[634] An increase in frequency of the bending mode is also observed. Interpretation of the data therefore centers upon how the concentration and type of salts affect the above parameters and hence the strength of the hydrogen bonds formed between the water molecules.

Most of the work to date shows that the band parameter changes are less sensitive to cationic than to anionic species.[203,204,320,807,837] Generally, the bands are broad and single-peaked, resembling the bands of HDO in pure water.[203,204,320,807,871] However, aqueous solutions of perchlorate,[76,183,320,441] tetrafluoroborate,[76,184] and trichloroacetate[184] salts show splitting in the valence OH or OD stretching vibrations. The same phenomenon has been observed in Raman[442,821] studies of perchlorate salts.

TABLE I. The Shift in Frequency $\Delta\nu_1$ of the $\nu_{\text{O-D}}$(HDO in H_2O salt solution) Band from $\nu_{\text{O D}}$(HDO in pure H_2O) for Different Concentrations of Dissolved Salts[a]

Solution	Concentration, mol liter^{-1}	$\Delta\nu_1$,[b] cm^{-1}	Ref.
LiCl	4.7	+7	441
	8.3	+10	441
	13.1	+16	441
	Saturated	−5	807
LiBr	3.7	+18	441
	6.7	+22	441
	9.3	+27	441
LiI	3.5	+21	441
	7.3	+37	441
NaF	2.0	0	96
NaCl	2.5	+20	96
	5.5	+14	871
	Saturated	+6	807
NaI	3.7	+41	441
	6.4	+46	441
NaOH	1.0	0	96
	5.0	0	96
KF	Saturated	−50	807
KCl	2.5	+5	96
KI	Saturated	+39	807
CsCl	Saturated	+8	807
MgCl$_2$	Saturated	−18	807
CaCl$_2$	2.5	+20	96
	Saturated	+13	807

[a] Temperature is 25°C except for results from Ref. 96, which are for 31°C.
[b] Positive sign signifies displacement toward higher frequencies.

Some workers[96,184,203,204,320,776,807,837] have studied primarily the effect of salts on the frequencies of maximum absorption, whereas others[76,183,441,871] have investigated, in addition, the effect of ions on the integrated intensity.

Table I contains some results of the frequency shift of the ν_1 vibration (OD valence stretch) of the HDO molecule in H_2O salt solutions from the frequency in pure H_2O. Table II contains similar results for the ν_3 vibration (OH valence stretch) of the HDO molecules. Figure 4 shows results of the same parameter as in Table II, obtained by Hartman[320] for some electrolyte solutions at 25°C. In the few instances where similar salts were studied there is general agreement in the sign of $\Delta\nu_3$, although the value for the 2.5 M NaCl solution[96] appears to differ markedly from Hartman's value.

TABLE II. The Shift in Frequency $\Delta\nu_3$ of the $\nu_{O\text{-}H}$(HDO in D_2O salt solution) Band from $\nu_{O\text{-}H}$(HDO in pure D_2O) for Different Concentrations of Dissolved Salts[a]

Solution	Concentration, mol liter^{-1}	$\Delta\nu_1$,[b] cm^{-1}	Ref.
LiCl	Saturated	+5	807
NaF	0.5	+10	96
	1.0	+20	96
	2.0	+20	96
NaCl	2.5	+40	96
	5.0	+20	96
	Saturated	+20	807
	5.5	+20	871
KF	Saturated	−110	807
KCl	2.5	+40	96
KI	Saturated	+55	807
CaCl$_2$	Saturated	+3	807

[a] Temperature is 25°C except for results from Ref. 96, which are for 31°C.
[b] Positive sign signifies displacement toward higher frequencies.

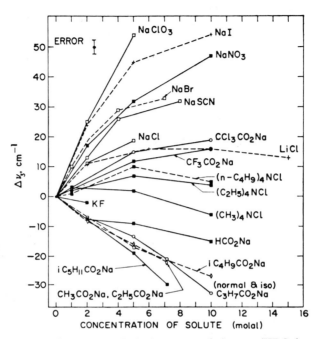

Fig. 4. $\Delta\nu_3$ is the shift in frequency of the ν_{O-H}(HDO in D_2O salt solution) band from the ν_{O-H}(HDO in pure D_2O) band plotted versus the molal concentration of dissolved salt. (Reprinted from Ref. 320. Copyright 1966 by the American Chemical Society. Reprinted by permission of the copyright owner and the author.)

Williams and Millet[837] carried out an extensive study of the effects of various ions on the frequencies of a number of infrared bands of H_2O in the region 1–7 μm, studying 16 alkali and alkaline earth halides. They systematically determined the effect of various cations on a given band by taking a sequence of solutions of compounds at the same molar concentration containing these cations in combination with a common anion, and reversed the procedure to determine the effect of the anions. A comparison of some of their values for the frequency shift of the 3-μm OH vibration band relative to H_2O with those of $\Delta\nu_3$ in Table II and with results from Ceccaldi et al.[96] on precisely the same vibration show some major differences. For example, the values for the frequency change from Millet and Williams are: LiCl, $-32\,\text{cm}^{-1}$ $(0.1\,M)$ and $+22\,\text{cm}^{-1}$ $(2.5\,M)$; NaF, $-128\,\text{cm}^{-1}$ $(0.1\,M)$; NaCl, $-85\,\text{cm}^{-1}$ $(0.1\,M)$ and $+16\,\text{cm}^{-1}$ $(2.5\,M)$; KCl, $-74\,\text{cm}^{-1}$ $(0.1\,M)$ and $+22\,\text{cm}^{-1}$ $(2.5\,M)$; KF, $-80\,\text{cm}^{-1}$ $(0.1\,M)$;

KI, -50 cm^{-1} $(0.1 \, M)$; $CaCl_2$, -110 cm^{-1} $(1.25 \, M)$. The values for the shifts from Ceccaldi *et al.* are: NaF, 0 cm^{-1} $(0.1-1.0 \, M)$; NaCl, 0 cm^{-1} $(0.1 \, M)$, $+15 \text{ cm}^{-1}$ $(1.0 \, M)$, $+25 \text{ cm}^{-1}$ $(2.5 \, M)$, and $+25 \text{ cm}^{-1}$ $(5.0 \, M)$; KCl, 0 cm^{-1} $(1.0 \, M)$, $+20 \text{ cm}^{-1}$ $(2.5 \, M)$, and $+20 \text{ cm}^{-1}$ (saturated); $CaCl_2$, 0 cm^{-1} $(1.25 \, M)$. It is not correct to anticipate the effect of various salts on the ν_{OH} band of H_2O and HDO to be the same, but a direct comparison of the results of Ceccaldi *et al.* and Williams and Millet is in order. The marked differences still prevail and are particularly noticeable at the lower concentrations. If the low-concentration values[837] are correct, then the effect of small concentrations of salts on the 3-μm band is greater than has been acknowledged. Considering the difficulties that arise when working with the 3-μm band of H_2O and the magnitude of the changes in Table II, it appears likely that the low-concentration results[837] are in error. Furthermore, of the eight sequences studied[837] for the 3-μm band, six were in disagreement with the Bernal and Fowler[46] theory of aqueous solutions, which qualitatively states that the effects of ions on the water are proportional to the polarizing powers of the ions.

Further studies[203,204] of the displacement of the 3412-cm^{-1} peak of the pure water absorption band in the 3-μm region have been carried out as a function of salt concentration for the potassium halide series KF, KCl, KBr, and KI. Displacements were toward higher frequencies for all salts. The relative order followed the sequence $F^- < Cl^- < Br^- < I^-$. In addition, graphs of frequency displacement versus solute/solvent mole ratio showed a tendency for all salts to achieve a constant maximum value above a given mole ratio. Solutions containing equimolar amounts of KI and KBr up to a solute/solvent mole ratio of 0.12 were also examined. For these ternary systems, up to 0.04 mole ratio, the displacement is equivalent to that obtained from addition of the individual displacements for the binary solutions. Above 0.04 mole ratio the frequency displacement of the mixture becomes increasingly smaller than the value obtained on the basis of the binary solution additivity scheme. The binary solution results obtained in this work also disagree with those of Millet and Williams.[837]

From Fig. 4 one can extract the effect of a number of salts at varying concentrations on the average hydrogen bond strength in water. The sequence[320] in order of increasing average hydrogen bond strength at $4.0 \, m$ is: $NaClO_3 < NaI < NaBr \leq NaNO_3 \cong NaSCN < NaCl < CCl_3CO_2Na \cong LiCl < CF_3CO_2Na \leq n\text{-}Bu_4NCl < Et_4NCl \ll RCO_2Na$, where R represents alkyl groups up to C_6. Although $NaClO_4$ is not listed, an average value for the two bands in this region indicates it has the greatest weakening effect on the average hydrogen bond strength. An examination of the data in Tables I and II for approximately similar molar concentrations of salts shows that irrespective of the monovalent cation present, the following series is observed for the order of increasing hydrogen bond strength: $I^- < Br^- < Cl^- < F^-$. This is in agreement with the halide subseries taken from Hartman[320] and Fabbri and Roffia.[203,204]

The effect of cations is less systematic. In some series with common anions the changes in the position of the frequency maximum are small, e.g., the alkali chloride series in Table I. The largest shifts toward strengthening hydrogen bonding in water occur for the cations having a higher surface charge density, such as Li^+ and Mg^{2+}.[807] It has been stated[627] that the larger the surface charge density of a group, the greater its hydrogen-bond-accepting strength. Qualitatively, the results indicate surface charge density of ionic species is one of the factors that affects internal solvent vibrations. It would appear that of the multiplicity of probable arrangements for neighboring ions or molecules about any given water molecule,[807] a greater contribution to change in the intramolecular valence stretching vibration is made by water–anion interaction. The various configurations in which a cation or hydrogen of another water molecule is interacting with the oxygen atom in a particular water molecule apparently have only secondary effects upon the stretching vibration of an OH bond of the same molecule simultaneously interacting with an anion.

It is difficult to speculate upon the microscopic picture of these relatively concentrated ionic solutions in terms of band frequency changes. Bandwidths at one-half maximum absorption can be a further aid; it has, however, not been extensively measured in much of the work. Half-bandwidths[807] for saturated solutions show that the ν_1 vibration of HDO in ionic solutions of LiCl, $MgCl_2$, and KF broaden the band, whereas KI, NaCl, CsCl, and $CaCl_2$ narrow the band relative to the pure solvent. Similarly for the ν_3 band of HDO; LiCl, NaCl, $CaCl_2$, and KF broaden, whereas KI narrows, the band. The wide variation in concentrations of these salts at saturated conditions prevents a thorough comparison under unique concentration conditions. Suffice it to say that significant changes arise with KF and KI. In the case of KF the bandwidths of ν_1 and ν_3 increase by $+165$ and $+100$ cm^{-1}, respectively, whereas for KI they decrease -20 and -30 cm^{-1}, respectively, relative to the pure solvent. Taking the $\Delta\nu_1$ and $\Delta\nu_3$ band shift data for these salts from Tables I and II along with the above values for bandwidth change, one is able to appreciate the structure-breaking and structure-making characteristics of KI and KF solutions, respectively, or in terms of their effect on the average hydrogen bond strength in water, the strengthening effect of KF and the weakening effect of KI.

A systematic study of fundamental band parameters as a function of salt concentration for a series of salts has been lacking. Clarke,* in some unpublished work, has observed for KI solutions a rapid change in half-bandwidth and frequency of maximum absorption of the ν_1 band of HDO around 4 M KI. This seems to occur at about the same concentration at which the viscosity passes through a minimum and the self-diffusion coefficient of water reaches a maximum. This is in accordance with a general "tightening" of the structure and Clarke suggests that the solvated ions are in contact in this region, and from this condition rearrangement is in the direction of a "loose" salt hydrate structure. This argument seems plausible, in view of the fact that the OD stretch of HDO in NaI $\cdot$ 2H$_2$O doped with D$_2$O occurs at 2587 cm^{-1},[724,725] which is a reasonable extrapolation of his data for NaI to that mole ratio.

Clarke has observed increased bandwidths for KI solutions below 4 M, indicating a greater range of OD band lengths in solution as compared to the pure liquid or perhaps the presence of two overlapping bands, one representing hydrated water, the other "free" water. If the frequency separations are appreciably less than the bandwidths, a single envelope would be observed, the band parameters of which would depend on the band parameters of the individual components. It appears that careful studies of band parameters over a wide concentration range are useful in revealing changes in the local structure of solvent about ions. Effects begin to appear which intuitively would be anticipated as the solvent/ion mole ratio is reduced in solution.

The broad bandwidth of the ν_1 and ν_3 vibrations of HDO in KF solutions is of special interest. Clarke suggests a bridging-type structure exists similar in nature to the structure postulated by Mohr et al.[583] from their studies of the OH stretching region of Bu$_4$NF–water solutions in CCl$_4$ for various concentrations of salt and water. The favored structure is a bridged cyclic water dimer[789] associated with a fluoride ion. Mohr et al. argue that the absence of free OH in their spectra, which they observed for other tetraalkylammonium halide salts, and the broad shifted band, possibly resulting from the overlapping of several hydrogen-bonded OH groups, imply such a structure.

Various studies of the HDO[96,411,807] and H$_2$O[837] ν_2 bending vibration have been carried out for a number of alkali and alkaline earth halide solutions of varying concentration. The results of the shift of this band are shown in Fig. 5 and, with the exception of NaF and KF, the shift is to lower

---

* D. E. Clarke, private communication.

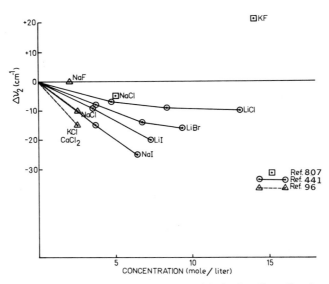

Fig. 5. $\Delta\nu_2$ is the shift in frequency of the bending vibration of HDO (HDO in H_2O salt solution) from the bending vibration (HDO in pure H_2O) plotted versus the molar concentration of dissolved salt.

frequencies. The results[837] of the shift of the ν_2 band in electrolyte solutions show similar trends within a given series of salts, although the absolute values differ for low-concentration conditions. Figure 5 clearly indicates the opposing effects of the F⁻ and I⁻ ions on solvent hydrogen bond strength. The relative trends in other ionic species are in general agreement with the classification made on the basis of the fundamental stretching vibration results.

Studies[96,171,638,836,837] of the variation in the position of the absorption maximum at 4.7 μm in H_2O due to dissolved electrolytes have been carried out. This band is considered to be an association band, the assignment of which is still a point of debate. It generally is considered to be $(\nu_2 + \nu_L)$, an association of the bending vibration with a far-infrared hindered rotational band. On the other hand, Williams,[835] based on his most recent work,[179] suggests that the assignment $\nu_2 + \nu_L - \nu_T$ (ν_T is infrared-active hindered translation frequency of the molecule as a whole) is more plausible, in view of the rather good agreement between observed and predicted values. The choice of assignment is not crucial to this discussion. In view of the fact that most salts tend to shift the positions of both ν_L (Section 4.1) and ν_2 to lower frequencies, one would predict a substantial shift of the associational band in the same direction. The results[96,837] clearly indicate this, with the exception of the fluorides of the alkali metal series. These salts shift ν_L and ν_2 to higher frequencies and hence the position of the associational band is shifted to higher frequencies also.

No effect of sodium deoxyribonucleate on the position of the 4.7-μm band of water has been found.[171] This is contrary to what had been expected, considering Jacobson's[404-407] hypothesis that DNA, when dissolved in water, should promote extensive intermolecular ordering in the solvent, thereby shifting the 4.7-μm band to higher frequencies as well as increasing its intensity. Although solutes with alkyl residues are considered to be water structure promoters from a variety of transport and thermodynamic studies, the evidence for such behavior from infrared studies of the fundamental region of water is minimal. Hartman's[320] study of the effect of a series of sodium carboxylate salts on the position of the ν_3 band of HDO indicates only secondary effects associated with the hydrocarbon chains, whereas the results of his study of the homologous R_4NCl series of salts shows that increasing chain length increases the frequency maximum of the ν_3 band. It appears that the intramolecular bands are affected less by ions which contain large alkyl groups.

The effect of ions on the intensity of fundamental infrared bands of water has not been studied extensively because of technical difficulties discussed in Section 3. In contrast there have been extensive Raman studies[92,442,727,809,810,813,816,821,833] where frequencies and molar integral intensities have been measured and evaluated. Kecki et al.[183] have evaluated the change in molar integral intensity of the OD vibration in H_2O–HDO–D_2O solutions of several electrolytes. Wyss and Falk[871] have studied the OD and OH vibrations of H_2O–HDO–D_2O–NaCl solutions as a function of temperature and HDO concentration. Examination of experimental methods in these studies indicates careful procedures were followed throughout. In particular, the work of Wyss and Falk exemplifies added precautions to account for concentration variation with temperature change as well as corrections for reflection losses at the cell window interfaces.

Kecki[441] observed a decrease in the molar integral intensity of the ν_1 band of HDO with increasing concentrations of LiBr, LiI, NaI, and $NaClO_4$ and an increase for increasing LiCl concentrations. For the lithium halide series the effect of halide ion upon the intensity followed the order $I^- < Br^- < HDO < Cl^-$. This disagrees with the Raman results[442,809,810] of the potassium halide series, for which the following order was obtained: $F^- < HDO < Cl^- < Br^- < I^-$. The method[441] whereby the absorbances were obtained appears to be correct, so that on balance the source of the difference in order between the Li$^+$ and K$^+$ series is not immediately clear. Infrared and Raman[807,821] studies indicate nondirected ionic interactions between small cations and the oxygen atoms of water have little effect* on the OD and OH stretching vibrations transmitted through the oxygen atoms. This would preclude the direct effect of Li$^+$ and K$^+$ ions accounting for this difference. However, both the intensity and frequency changes for the infrared study of the ν_1 band of HDO in salt solutions are indicative of increasing structural breakdown as the halide ion size increases.

Wyss and Falk[871] observed no new features in the HDO spectrum of NaCl solutions over a range of temperatures. The effect of NaCl on the ν_1 and ν_3 stretching bands of HDO is to decrease the band area, to decreaes the bandwidth, and to shift the frequency of maximum absorption upward. These results are in almost total agreement with those of Waldron[807] and Hartman[320] except that Waldron observed an increase in bandwidth for the ν_3 vibrational mode of HDO in a saturated aqueous NaCl solution. The effect of increased temperature upon these band parameters was to reduce the upward shift of the peak frequency caused by NaCl addition such that above 90–100°C the peak frequency is shifted downward. For the other band parameters the temperature effect on the NaCl solution was similar to that of HDO in water. The shape of the band changes slightly when NaCl is added to water, causing a decrease in intensity of the ν_3 band, for example, below 3350 cm^{-1} and above 3600 cm^{-1}, with a corresponding increased peak intensity in the intermediate region. This behavior is suggested by the authors as being consistent with the effect of temperature on the band parameters of these solutions.

The relatively broad bands without overt shoulders are proposed by the authors as being clearly more indicative of a continuum model for aqueous solutions. The comparable broadness of the HDO band in salt solution and in H$_2$O or D$_2$O solution suggest the OH$\cdots$Cl$^-$ and OH$\cdots$O groups have a similar distribution of frequencies of interaction due to a flexible structure of water around ions. As was noted before, apparently the cation has no direct effect on stretching band profiles.

The mixture model of nonhydrogen-bonded and hydrogen-bonded OD and OH stretching components can account equally well for the observed changes in band profile. If it is assumed that strong hydrogen bonds between nearly linear Cl$^-\cdots$HO groups give rise to a strong, narrow

---

* Evidence[366,395] indicates that the cations which are most likely to produce significant effects on the OD or OH stretching vibrations are those that engage in partially covalent cation–oxygen interactions.

hydration band of intermediate frequency, and at the same time the hydration of the solute species can provide a competition for OH groups, then, because of the equilibrium between hydrogen-bonded and nonhydrogen-bonded OH groups (either to H_2O or HDO), the infrared intensities of the nonhydrated OH components should decrease at higher and lower frequencies.

One of the most interesting developments in recent infrared studies of salt solutions has been the study of the fundamental stretching bands of HDO in the presence of polyatomic anions such as perchlorate[76,183,441] and tetrafluoroborate.[76,184] Some earlier studies[320,776] had indicated that the fundamental bands were split in the presence of the perchlorate anion and Fig. 6, taken from Brink and Falk,[76] clearly indicates this. It shows the molar absorptivities of the OD stretching bands of HDO in aqueous solutions of $NaClO_4$ and $Mg(ClO_4)_2$ of varying concentration at 28°C. The spectra for the OD stretching vibration agree with those reported by Kecki et al.[183,441]

The following observations can be made with regard to these solutions:

1. A shoulder on the high-frequency side of the OD bands appears at low salt concentrations and increases to a well-defined peak at higher concentrations.

2. The high-frequency component (referred to as band A by Brink and Falk) increases in intensity, while that of the low-frequency component (referred to as band B by Brink and Falk) decreases as the perchlorate concentration increases.

3. The high-frequency component is narrower (50–60 cm^{-1} bandwidth at one-half peak absorptivity) and less intense than band B (bandwidth of 160–180 cm^{-1}), the bands being separated by about 120 cm^{-1}.

4. The position of band A is independent of perchlorate concentration but shows a slight dependence upon the cationic species. Its bandwidth appears to be independent of the cation and salt concentration,[76] although other work[183] shows a slight dependence on cation.

5. The position of component B varies with ClO_4^- concentration, increasing concentrations of $LiClO_4$, $NaClO_4$, and $Ba(ClO_4)_2$ shifting it to higher frequencies, whereas increasing $Mg(ClO_4)_2$ concentrations slightly lowers the peak frequency.

6. There appears to be a general increase in bandwidth of band B[183] with increasing concentration, the increase being greater for the divalent cation perchlorates.

7. The curves exhibit an isosbestic region for all cations.

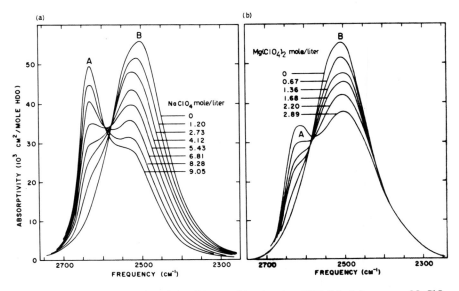

Fig. 6. Molar absorptivities of the OD stretching bands of HDO in (a) aqueous NaClO$_4$ and (b) aqueous Mg(ClO$_4$)$_2$ at 28°C.[76] (Reproduced by permission of the authors and the National Research Council of Canada from *Can. J. Chem.*)

Kecki *et al.*[183] have evaluated the molar absorptivities of the high-frequency band at constant ratios of perchlorate to water concentrations and have shown that the intensities of band A increase in the order Na$^+$ < Ba^{2+} < Li$^+$ < Mg^{2+}. This order is also observed in Fig. 6. A similar order is obtained when a comparison of the displacements of band A to lower frequencies is made for different cations; Na$^+$ < Ba^{2+} < Li$^+$ < Mg^{2+} < Ag$^+$ (Ref. 76 for saturated solutions) and Na$^+$ $\cong$ Ba^{2+} < Li$^+$ < Mg^{2+} (Ref. 183). Kecki *et al.*[183] showed from their results that this order agreed with the charge/radius ratio order evaluated for these cations. It is obvious that Ag$^+$ will not fit into this pattern, in fact, making use of Pauling radii, taken from Cotton and Wilkinson,[135] Ag$^+$ should fall below Na$^+$ for such a correlation. Brink and Falk explain the cation effect on the position of band A in terms of solvent-separated cation $\cdots$ water $\cdots$ anion pairs, as have been postulated to occur in aqueous solution.[121] It is questionable whether this is still a plausible explanation for saturated AgClO$_4$ solutions, where the concentration could attain 26 *m*,[747] leading to a mole ratio for ClO$_4^-$/water of 0.48. Such a high mole ratio could produce some contact cation $\cdots$ anion pairs which may contribute to the anomalous position of Ag$^+$ in a correlation of frequency shifts of band A with cation charge/

radius ratios. Walrafen[821] reports that Li^+, Na^+, and K^+ do not greatly affect the OD and OH stretching vibrations, as observed by Raman studies.

The controversy over the interpretation of the absorption spectra of water in terms of a continuum model or a mixture model for water persists in the interpretation of data from aqueous perchlorate solutions. Brink and Falk[76] interpret the perchlorate results as showing band A to be characteristic of OH groups interacting weakly, through hydrogen bonds, to the ClO_4^- ions and band B corresponds to all remaining types of OH groups.

The authors cite experimental evidence for this assignment of band A. They include the observation[75] of almost identical frequencies for the stretching frequencies of HDO in crystalline $NaClO_4 \cdot H_2O$, $LiClO_4 \cdot 3H_2O$, and $Ba(ClO_4)_2 \cdot 3H_2O$. In addition, using the Badger–Bauer[32] relation and the displacement of band A in aqueous and methanolic solutions relative to the free OH stretch in the vapor, they obtained a value of about 2 kcal mol^{-1} for the $OH \cdots ClO_4^-$ bonds, in comparison to a value of 2.1 kcal mol^{-1} from their work[75] in the hydrate crystals. Also, the relatively narrow bandwidth and invariance of the position of band A with temperature change are characteristic of weakly interacting groups.

An important feature of the perchlorate results is the position of band A. It is almost the same as that for the OD stretching vibration of HDO in liquid H_2O at 400°C and a density of 1.0 g cm^{-3}.[222,519] In addition, it agrees with the frequency of maximum absorption for the nonhydrogen-bonded water molecules inferred by Senior and Verrall[731] from objective analysis of accurate infrared absorption measurements and by Walrafen[817] from Raman studies of the OD stretching band in HDO. Such evidence appears to corroborate the conclusions drawn by Kecki et al.[441] from infrared work and Walrafen[821] from Raman studies that the high-frequency component is due to extensive breakdown of hydrogen-bonded water structure by the ClO_4^- ion, an effect similar to that of increasing temperature.

Recent studies[76,184,822]* on a number of other polyatomic anions have revealed interesting facts. Infrared and Raman studies of ternary H_2O–D_2O solutions of PF_6^-,[822] SbF_6^-,[822] BF_4^-,[76,184,822] and CCl_3COO^-[184] indicate overt splittings in the OD and OH Raman and infrared stretching contours. On the other hand, similar studies of ternary solutions of SO_4^-,[76,822] NO_3^-,[76,822] NO_2^-,[76] ClO_3^-,[76,822] IO_4^-,[76,822] ReO_4^-,[76,822]* MnO_4^-,[822] and SCN^-[822] do not show any overt splitting of the OD or OH contours. The distinguishing feature between these two groups is that the band-splitting anions give rise to very strong acids.

---

* Also R. E. Verrall, unpublished results.

For each of the band-splitting anions a characteristic property of the central atom or group of atoms, e.g., CCl_3 in CCl_3COO^-, can be interpreted as leading to a diminished electron density on the peripheral atoms.[184] Presumably, these features are responsible for the weak hydrogen-bond[76] or non-hydrogen-bond-like[821,822] interactions postulated between water and the anions.

Infrared studies[822] of 2.6 and 5.3 M solutions of $NaPF_6$ reveal pronounced absorption maxima near 2540 ± 10 cm^{-1} and 2670 ± 5 cm^{-1} in the OD stretching region and near 3440 ± 20 and 3640 ± 20 cm^{-1} in the OH stretching region. In addition, infrared shoulders were observed near 3250 ± 20 and 3600 ± 20 cm^{-1}, the latter being very weak. Similarly, for $NaSbF_6$ solutions frequency maxima were observed in the OD stretching region at 2540 ± 20 and 2650 ± 10 cm^{-1} and in the OH stretching region at 3620 ± 10 cm^{-1}, 3440 ± 20 cm^{-1} with a shoulder at 3260 ± 20 cm^{-1}. These results and those obtained for BF_4^{-}[76,184] and ClO_4^{-}[76,184,821] clearly show that the position of the high-frequency component of the OD or OH stretching contours depends upon the anion present. These components for $NaSbF_6$, $NaClO_4$, and $NaBF_4$ occur nearer to those obtained from the binary HDO in H_2O system. Walrafen[822] has observed that the high-frequency components are broader for $NaSbF_6$, $NaClO_4$, and $NaBF_4$ than for $NaPF_6$. Further, some Raman peak heights differ for $NaSbF_6$, $NaClO_4$, and $NaBF_4$ from those of the $NaPF_6$ solutions in the OH stretching region. However, this apparent variation does not appear in the infrared spectra.

Although the studies of the anions that give rise to strong or moderately strong acids do not show overt splittings of the OD and OH Raman or infrared stretching contours, they manifest the basic two-component substructure by exhibiting inflection points, by marked asymmetry, or by unusual contour flattening.[822] For example, Walrafen's[822] study of a 9.5 M solution of $NaMnO_4$ in 10 mole % D_2O in H_2O shows the OD and OH stretching contour is markedly skewed to high frequencies with the OD absorption maximum centered near 2580 ± 10 cm^{-1}. This is 70 cm^{-1} higher in frequency than the absorption maximum for 10 mole % D_2O in H_2O and the latter peak is more symmetric. Similarly, for ReO_4^-, Verrall* has observed for a 3.84 M solution of $NaReO_4$ in 6 mole % D_2O in H_2O that the band is shifted to higher frequencies relative to the binary solvent solution by 25 ± 5 cm^{-1}, and the band in less symmetric.

Analog computer decomposition of the Raman[822] OD stretching

---

* R. E. Verrall, unpublished results.

contours of ReO_4^-, MnO_4^-, IO_4^-, NO_3^-, ClO_3^-, and SCN^- using a two-Gaussian approximation indicates that the high-frequency components are located in the range 2600–2635 cm^{-1}. These are considerably lower in frequency than the positions of the intense high-frequency components obtained from the band-splitting anions. This suggests that these anions interact with water differently from the band-splitting anions.

Based on analog computer decomposition of OD and OH Raman and infrared stretching contours, Walrafen[822] concluded that the Gaussian component substructure for the PF_6^-, SbF_6^-, ClO_4^-, and BF_4^- anions may be described as a superposition of two spectra. One is composed of components thought to arise from weak electrostatic interactions between the anions and HDO or H_2O molecules. These are sharp and occur in the non-hydrogen-bonded frequency regions. The other spectrum arises from altered water structure and contains components occurring in the nonhydrogen-bonded as well as hydrogen-bonded frequency regions of OD and OH. This argument is based primarily on OH band resolution. It appears, with the exception of PF_6^-, that the electrostatic anion–HDO interactions are unresolved from the broad OD stretching component of the altered water structure in the nonhydrogen-bonded region. Walrafen's[822] curve resolution method also reveals bands which he associates with interactions between the first and second hydration layers about the anions.

The anions that do not split the OD or OH stretching contour are envisaged by Walrafen[822] as engaging in distorted nonlinear $X \cdots H$—O and $X \cdots D$—O interactions, where X refers to O or N(SCN^-) of the anionic species. On the other hand, Brink and Falk[76] suggest the lack of discrete structure in the case of these anions is caused by hydrogen bonds, for example, between $SO_4^{2-} \cdots H$—O, being sufficiently strong so that their frequencies merge with the broad band due to other OH groups.

The presence of a pseudoisosbestic point in Fig. 6, also obtained in other infrared[183,441] and Raman work,[821] is relevant. It has been shown by Senior and Verrall[731] that the existence of an isosbestic point, when accompanied by other indications of band complexity, is a strong argument for the existence of a two-state equilibrium. Temperature isosbestic frequencies have been obtained for HDO in H_2O from infrared absorption measurements,[731] 2575 cm^{-1}, and from mercury-excited Raman spectra,[817] 2570 $\pm$ 5 cm^{-1}. In the infrared studies of ternary perchlorate solutions at constant temperature the following values of concentration isosbestic frequencies were obtained: Kecki et al.[183] 2570–2590 cm^{-1}; Brink and Falk,[76] 2580–2585 cm^{-1}. Raman studies[821] reveal isosbestic frequency values from the ternary perchlorate solutions range from 2570 to 2580 cm^{-1} for the OD vibration. The good agreement between the temperature and concentration isosbestic frequency values obtained from Raman and infrared studies suggests that the effects of increasing temperature and perchlorate concentrations are virtually equivalent. Furthermore, the significant opposite variations of the low- and high-frequency component integrated intensities are all offered[821] as evidence that the perchlorate anion exerts a shift in the equilibrium between hydrogen-bonded water to nonhydrogen-bonded water.

The effect of temperature on the parameters of bands A and B, obtained from infrared studies[76] of a 7 M solution of $NaClO_4$, shows that band A remains nearly stationary whereas band B undergoes a temperature shift similar to that of the corresponding HDO band in water. Similar results were obtained from Raman studies.[821]

Brink and Falk argue that this is demonstrative of what would be expected if band A were representative of OD groups involved in weak hydrogen bonding with perchlorate anions. However, Walrafen's[822] recent deductions concerning the superposition of two spectra can account for this intensity behavior. At the relatively high salt concentration utilized structural alteration takes place in the solvent, but as the temperature increases electrostatic anion–water interactions replace nonhydrogen-bonded components from unaltered water at nearly the same frequencies, thereby maintaining the integrated intensity constant.

Independent evidence[183] of the invariance of the high-frequency component in ternary $HDO–H_2O–ClO_4^-$ solutions has been noted under different experimental conditions. It was shown that the concentration of water molecules "bound" to the ClO_4^- ions does not increase with concentration above $C_{ClO_4^-}/C_{water} > 0.1$ (i.e., $C_{ClO_4^-} > 4\ M$). A competing process consisting of cation–anion contact pairs was proposed. One exception to the above observation was $Mg(ClO_4)_2$ and its behavior was rationalized in terms of the great stability of the hydration layer around the Mg^{2+} ion. Subsequent to this the same authors[184] have suggested the high-frequency component in these bands is due to water molecules isolated between the cation and anion in the ion pairs, although their mechanistic description of this process is not too clear. The absence of splitting in the case of other anions like ReO_4^- is attributed to a "hydrogenlike bond" between the anion and water molecule in the cation hydration sphere. Presumably these are similar to Walrafen's[822] distorted hydrogen bond interactions.

There is little doubt that several species of water molecules located in different environments are present in aqueous solutions of salts containing anions of the extremely strong acids. The categorization of these species based on proposed types of interaction is the point of contention. The Raman spectra appear to present more evidence of complexity in such systems, although several examples[822] were noted where the infrared and Raman OH stretching contours showed comparable overt signs of band structure.

Infrared studies[822] of the effects of other salts on the OH and OD stretching contours in 10 mole % solutions of D_2O in H_2O reveal frequencies corresponding to shoulders and to absorption maxima as follows: $NaB(C_6H_5)_4$, 2630 ± 10 and 2510 ± 5 cm^{-1}; $Cl_3CCOONa$, 2660 ± 10 and 2510 ± 5 cm^{-1}; and $(NH_2)_2CO$, 2600 ± 10 and 2500 ± 10 cm^{-1}. The results for $Cl_3CCOONa$ are in good agreement with those obtained by Dryjanski and Kecki.[184] A spectrum of solid urea in a KBr disk showed insignificant absorption near 2600 cm^{-1}, indicating the solution component is unique in the liquid phase. This evidence suggests that uncharged molecules can also produce shoulders on the OD stretching contours. The spectra reveal that a complex situation exists for these solutes and that Walrafen's[822] interpretation seems plausible. The positions of the high-frequency components makes the rationalization[76] of these components in terms of weak hydrogen bonds between the anions and water somewhat doubtful. Arguments[768] have been presented, however, which suggest that weak hydrogen bonds and strong intermolecular interactions of a physical nature are indistinguishable.

4.3. Near Infrared

Early studies of the effect of ions on the near-infrared spectrum of water were primarily concerned with qualitative changes induced in the spectra by such ions. There was uncertainty governing the assignment of a number of bands* in this region which largely have been alleviated by careful studies of isotopic mixtures. Falk and Ford[209] more recently have shown that a complex situation holds for higher overtone and combination bands when dealing with pure H_2O or D_2O, due to a rapidly increasing number of contributing transitional vibrations. In fact, the near-infrared spectrum of the pure water solvents (H_2O and D_2O) show closely positioned bands, which undoubtedly must overlap and thereby make interpretation

---

* Bayly et al.[41] give a useful summary of numerous early experimental investigations, to 1963, of the near-infrared spectrum of liquid water and have discussed the various attempts at making frequency assignments for the observed bands.

of the effect of ions on any one of these bands difficult and prone to equivocation.

Continued progress in experimental and theoretical aspects of the structure of water have led recently to more quantitative studies of the near-infrared overtone and combination bands of water–electrolyte solutions in terms of absolute intensities. Although advances have been made toward a better understanding of the laws which govern the addition of intensities in the overtone bands, the relative values of which are usually an order of magnitude less than for the fundamental bands, there is undoubtedly a great deal more to be known before there is unanimous acceptance of quantitative studies in this region of the infrared.

Early studies of the near-infrared absorption spectra of water have shown that the spectra are influenced characteristically by temperature changes and by dissolved salts. As a consequence, much of this work centered about correlating changes due to the addition of ionic solutes with those arising out of temperature changes. Some of the earliest work conducted on this aspect of infrared studies was that of Suhrmann and Breyer[762,763] and Ganz.[254,256] The former studied the effect of variable salt concentrations as well as individual cationic and anionic species, by using a series of salts of approximately constant composition having a common anion or cation, respectively. The systematic changes in the position of maximum frequency and magnitude of the extinction coefficient for the pure H_2O absorption bands at 1.96 μm (5160 cm^{-1}), 1.45 μm (6890 cm^{-1}), 1.20 μm (8330 cm^{-1}), and 0.97 μm (10,220 cm^{-1}) led them to interpret these changes in the water spectra as being caused by a breakdown of a highly polymerized, hydrogen-bonded form of water and a concomitant hydration of the ions by the dissociated molecules. Ganz carried out similar studies of the effect of some strong acid and sodium hydroxide solutions of varying concentration on the 0.76-μm (13,170-cm^{-1}) far-infrared band of water. Briefly, his results for increasing acid and base concentrations showed the following effect on the 0.76-μm band. HCl affected the molar extinction coefficient very little but shifted the position of maximum frequency to longer wavelengths; H_2SO_4 lowered the extinction coefficient and moved the band to lower wavelengths; HNO_3 increased the molar extinction coefficient and shifted the band to lower wavelengths; and NaOH increased the molar extinction coefficient with the formation of a sharp band at 0.74 μm (13,480 cm^{-1}). Increasing the temperature of a 2.2 M NaNO$_3$ solution showed the same effect as increasing the HNO_3 concentration, thus indicating the nitrate anion was primarily responsible for the observed changes and that increased NO$_3^-$ concentration caused the same effect as increased solution temperature. A sodium series of salts of approximately constant composition showed that the relative order of their capacity to increase the molar extinction coefficient of the 0.76-μm band, and to shift it to lower wavelengths, was $Cl^- < NO_3^- < ClO_3^- < ClO_4^-$. Taking the assignment of the 0.76-μm band as being $3\nu_1 + \nu_3$,[873] the apparent relative structure-reducing nature of these anions correlates with that inferred from studies of the fundamental mode of vibration.

Williams and Millett[837] conducted a study of the effect of various nonhydrolyzing salts on the position of maximum frequency of the 1.96- and 1.45-μm near-infrared bands of water. They correlated these changes with the effects of temperature variation on these bands. Their systematic studies of a sequence of solutions of compounds, at the same molar concentration, containing a cation in combination with a common anion and vice versa showed the following. For cations the relative order of capability to produce a change in the position of the band frequency maximum similar to increasing temperature was $Li^+ < Na^+ < K^+ < Rb^+$ and $Mg^{2+} < Ca^{2+} < Sr^{2+} < Ba^{2+}$. The anions showed the order $F^- < Cl^- < Br^- < I^-$. These results were shown to agree generally with predictions made by Bernal and Fowler[46] based on the polarizing power of ions.

Szepesy et al.[769] have also studied the effects on the absorption bands of water when several electrolytes are added. They observed an increase in the absorbance of the water bands at 0.97 and 1.20 μm with increasing concentration of LiCl and CsCl, the increase for CsCl being greater than for LiCl. These results for the 0.97- and 1.20-μm bands are in agreement with those of Suhrmann and Breyer[762,763] obtained from the effect of LiCl and CsCl on the 1.97-μm band. It appears that the authors rationalized their results, implicitly, in terms of solvent-separated ion pairs in the case of LiCl and the structure-reducing ability of the cesium ion.

Yamatera et al.[873] studied the effect of electrolytes on the 1.2-μm (8300-cm^{-1}) band (assigned to $\nu_1 + \nu_2 + \nu_3$) in H_2O and the 1.6-μm (6200-cm^{-1}) band in D_2O and correlated changes in band shape and position with the observed temperature effect on the pure solvent bands. In their studies of water at room temperature they observed a very broad and deformed band at 1.2 μm (8310 cm^{-1}) with a distinct shoulder around 1.163 μm (8600 cm^{-1}). As the temperature was increased this shoulder became more apparent. Their interpretation of the results of the water spectrum changes with temperature was based on a model which included three distinct groups of water molecules. These were, respectively, molecules where none, one, and both of the hydrogen atoms are interacting through hydrogen bonding. The loss in broadness and the corresponding increased intensity of the high-frequency shoulder with increased temperature was noted as being suggestive of fewer and fewer molecules bound to nearest neighbors by hydrogen bonds.

The effect of dissolved ions on the position and intensity of the above band appeared to be similar to that observed through temperature variation. As a consequence, the relative intensity changes of the absorption in ionic solutions at 8640 cm^{-1} and 8310 cm^{-1} was interpreted in terms of the degree of "structure-breaking" or "structure-making" power of the dissolved species. It was observed that for inorganic halide salts of a common cation and concentration the greatest structure-breaking effect was associated with the largest anions and similar experimental conditions involving cations showed the effect was less pronounced. No appreciable difference was observed between the potassium and cesium ions.

Of particular import were the results of their studies of aqueous perchlorate and tetraalkylammonium salts. Their spectra of aqueous sodium perchlorate solutions at various concentrations clearly indicate a pronounced enforcing of the high-frequency (high-temperature) component as the ClO_4^- concentration is increased. This suggests a high degree of structure-breaking ability by the perchlorate anion. This is in agreement with other interpretations of the effect of this ion, based on results obtained elsewhere in the infrared.

Results of their studies with the tetraalkylammonium salts show that a 2 M solution of $(CH_3)_4NCl$ acts as a "structure-breaker," while a 2 M $(C_4H_9)_4NBr$ solution seems to effect a promotion of structure in the solution. However, the results of the latter studies are clouded by the fact that, owing to the organic nature of these cations, absorption bands of the C—H overtones are present in this wavelength region. Bunzl[89] has shown that, in particular, the 1.186-μm band of $(C_4H_9)_4NBr$ solute will always complicate the interpretation of the 1.2-μm water band. However, it will be seen later that Bunzl's work on R_4NBr salts using the 0.97-μm band of water, which avoids the above-mentioned difficulties, confirms the qualitative conclusions of Yamatera and co-workers.[873]

In conclusion, the order of increased structure-breaking power deduced from spectral studies[873] is $Cl^- < Br^- < I^- < ClO_4^-$ and $Li^+ < Na^+ < K^+ < Cs^+ < (CH_3)_4N^+$. This is in good agreement with relative orders established from studies in the fundamental and far-infrared regions.

Buijs and Choppin[102] carried out a quantitative study of the effect of temperature and concentration of various electrolytes on the absorption bands in the 1.1–1.3-μm region of water. Their study was based on an interpretation of the band parameters of the near-infrared absorption bands of pure water in terms of a mixture model.[87] Ascribing absorption peaks at 8620, 8330, and 8000 cm^{-1} in the near-infrared spectrum of pure H_2O to free molecules, molecules forming one hydrogen bond, and molecules forming two hydrogen bonds, the authors studied the variation in the degree of hydrogen bonding as ions are added. Quantitative calculations of the mole fraction content of each water species in ionic solutions were carried out using the intensity variations of these bands as prime data. The proposal drew sharp criticism[382] from proponents of the continuum model, primarily on the basis that unequivocal assignment of these three bands to the same combination mode, $\nu_1 + \nu_2 + \nu_3$, may be incorrect in view of the fact $2\nu_1 + \nu_2$, $2\nu_3 + \nu_2$, and $\nu_3 + 3\nu_2$ may also lie in this region. Buijs and Choppin refuted these arguments,[88] indicating that the existence of different species of water molecules offered the most plausible explanation for the observed phenomena. What is perhaps a more important criticism of the work concerns the presence of a subtle mathematical pitfall arising from the analytical procedures used to arrive at a solution for the concentrations of various water species. Luck[533] has shown that the particular assumptions used,[87,102] although limiting the number of possible solutions, do not lead to an unequivocal solution.

One of the first attempts to take advantage of the sharpening in band
Although it appears that absolute values for the relative concentrations of the various water species in solution are in doubt, their variation as a function of ionic concentration can give some qualitative estimate of changes in solvent structure induced by the ions. On this basis the results of Buijs and Choppin indicate the following cation effects: in terms of decreasing ability to produce order, $La^{3+} > Mg^{2+} > H^+ > Ca^{2+}$, and to destroy order, $K^+ > Na^+ > Li^+ > Cs^+ = Ag^+$; for the anions in terms of decreasing ability to produce order, $OH^- > F^-$, and to destroy order, $ClO_4^- > I^- > Br^- > NO_3^- > Cl^- > SCN^-$. It is to be noted that the order-producing effect of H^+ is quite different from the electrostrictive ordering caused by the polyvalent cations. The relative order for the univalent cations is surprising and opposed to the order deduced from studies discussed previously. The anions appear to give better correlation with trends observed in the other infrared regions. As in the case of a number of previous interpretations, the perchlorate anion shows strong structure-breaking characteristics. The unusual order for the univalent cations does cast some doubts on the method utilized to arrive at these conclusions. From what has been proposed by others on the basis of independent measurements of spectroscopic and other properties, Li^+ and Cs^+ should show different behavior.

One of the first attempts to take advantage of the sharpening in band structure of near-infrared bands of water observed with dilute solutions of H_2O in D_2O (or vice versa) was made by Waldron.[807] His studies of the $2\nu_1$ and $\nu_1 + \nu_3$ bands of HDO in D_2O saturated with metal halides showed little change in the position of the $2\nu_1$ band. However, this band may be susceptible to perturbations by several close-lying overtone and combination bands of pure D_2O or H_2O and thereby offset any affect due to specific hydrogen bonding. The $\nu_1 + \nu_3$ band at 1.68 μm (5970 cm^{-1}) showed marked changes. Whereas KF shifted the position of maximum frequency of the band to lower values, KI and NaCl increased the frequency, the relative order being KF < HDO < NaCl < KI. Similarly the variation in band half-width was observed to follow the relative order KI < NaCl < HDO < KF, KI and NaCl narrowing the band relative to the pure solvent. The above trends concur with observations of the variation in the band parameters of the fundamental vibrations discussed in Section 4.2.

Worley and Klotz[862] examined the effect of temperature and various salts on the near-infrared spectrum, 1.3–1.8 μm, of 5 mole % solutions of H_2O in D_2O. Since they were using a double-beam instrument with a reference cell, particular care was exercised in solution preparation to assure the sample and reference solutions had the same salt concentration. Their temperature studies of HDO dissolved in D_2O revealed four bands at the following wavelengths: 1.416 μm (7060 cm^{-1}), 1.525 μm (6570 cm^{-1}), 1.556 μm (6430 cm^{-1}), and 1.666 μm (6000 cm^{-1}). The sharpest band at 1.416 μm was assigned to the overtone of the nonhydrogen-bonded OH stretching vibration $(2\nu_3)$,[873] while the remaining bands were attributed to overtone bands of different hydrogen-bonded species. Observed intensity variation of the bands with increased temperature and added salts, as well as the presence of an isosbestic point at 1.468 μm, were advanced as strong evidence for a chemical equilibrium between hydrogen-bonded and non-hydrogen-bonded species in the HDO–D_2O system. Applying several assumptions, they arrived at a van't Hoff $\Delta H°$ value of -2.37 kcal mol^{-1} in the temperature range 7–61°C. Similar analyses of several salt solutions revealed salts to have a very small effect on the magnitude of $\Delta H°$. With the exception of $NaClO_4$, which gave a value of -1.64 kcal mol^{-1}, the $\Delta H°$ values for the other salt solutions cluster about -2.4 kcal mol^{-1}.

Bunzl[89] has pointed out the possibility of solute absorption in this region by hydrocarbon-containing salts. This would result in potential interference with the water absorption bands and may have a minor bearing on results involving these salts. Of the 19 salts studied by Worley and Klotz, the sodium salts of carboxylic acids and the quaternary ammonium salts showed increasing structure-making characteristics, generally as the hydrocarbon content increased. The substitution of halogen for hydrogen atoms in the carboxylate anion clearly increases the structure-breaking character of the salts. Highly polarizable anions such as Br⁻, I⁻, and SCN⁻ showed structure-breaking influence.

Very little shifting of band positions in this region was observed by the authors. They concluded that any slight changes were possibly associated with changes in band infringement in the equilibrium under study. The continued presence of isosbestic points in the spectra of various salt solutions, including $NaClO_4$, and a simple change in intensities associated with the equilibrium species support this observation.

Bunzl's[89] work constituted the first systematic attempt to use infrared studies in order to observe whether the proposed structure-making characteristics shown from a variety of other measurements on R_4NX solutions could be confirmed spectroscopically. Using the 0.97-μm near-infrared band of water to avoid the aforementioned interference problems arising from the solute, the temperature dependence of the wavelength of maximum absorption of the band was studied in aqueous solutions of tetraalkylammonium bromides at various concentrations.

The change in the wavelength of maximum absorption in the various solutions was used to characterize band changes induced by the salts. In addition, the solutions were described in terms of their "structural temperature,[46] the temperature at which pure water has the same degree of association. At a given temperature and concentration the quantity "Δt," equivalent to the difference between the solution temperature and structural temperature, serves as a useful probe of the change in the association of water molecules in the solution compared with pure water. Δt was positive for all of the R_4NBr salts studied (where R ≡ methyl, ethyl, n-propyl, and n-butyl), indicating the solution temperature was always greater than the structural temperature, therefore apparently corresponding to an increase in water structure in these solutions.

Bunzl was careful to point out that the multiplicity of possible interactions occurring in ionic solutions and their relative effects in terms of the Frank and Wen[232] model for electrolyte solutions indicate that two categories of induced OH stretching frequency shifts predominate. The "structure-induced shift" relates to changes in water–water interactions induced by different arrangement of water about ions relative to the pure solvent, and "charge-induced shifts" relate to water–ion electrostatic interactions. Since the functions relating the magnitude of these interactions to the shift in OH stretching frequency are not likely to be similar, it is not correct to associate the total frequency shift of a vibrational OH band to a change in structural temperature when both interactions may prevail. The R_4NBr series appears to offer a better basis for rationalizing purely structural effects.

Since the slope of Δt as a function of solution temperature, $\partial(\Delta t)/\partial t$, is an extremely sensitive criterion for the detection of structural effects, Bunzl utilized this probe with the following results. Negative values of the slope were obtained for $(C_3H_7)_4NBr$ and $(C_4H_9)_4NBr$, suggesting a structure enhancement around the cations, with the smaller slopes for $(C_3H_7)_4NBr$ solutions showing it to be less effective in this capacity.

The $(C_2H_5)_4NBr$ solutions show no appreciable dependence of Δt on solution temperature, possibly suggesting cancellation of structure-making and -breaking effects of the salt, in accordance with interpretations based on measurement of other solution properties.[434,517] The positive $\partial(\Delta t)/\partial t$ values for $(CH_3)_4NBr$ solutions classify this salt as a structure-breaker, contrary to a prediction based on the sign of Δt alone. However, as Bunzl has suggested, the "charge-induced shift" is likely to prevail in determining the magnitude of Δt, thereby masking the purely structural effects. A similar conclusion has been reached on the basis of partial molar volume studies.[126]

Previous reference[538,539] was made in Section 2.5 to the use of near-infrared spectra to extract quantitative estimates of hydration parameters of alkali halide salts in water. It was first noted[538] that a solution of protein in water measured differentially against water produces a negative near-infrared difference spectrum. This phenomenon is associated with the excluded volume of the solute in the sample cell, in effect causing a dilution of the solvent. In regions of the spectrum where the solvent extinction is much greater than that of the solute, the solvent absorption will be greater in the reference cell than in the sample cell, thus giving rise to a negative difference spectrum. All solutes should give a negative difference spectrum since all of them occupy volume. Certain anomalies arose when comparing difference spectra of ionic and nonionic solutes, in particular the inorganic alkali halides.[539] This was interpreted in terms of a relationship between the shape of the difference spectra and the type of solute–solvent interaction. The difference spectra were resolved into two components, one representing the hydration of the solute and the other the volume of the hydrated solute. The particular values of the hydration numbers calculated for the alkali halide salt series NaCl, KCl, RbCl, and CsCl agree reasonably well with values derived from compressibility measurements. Less satisfactory agreement was obtained with the anionic series NaCl, NaBr, and NaI and KCl, KBr, and KI. Although the values are of the right magnitude, the relative order of the magnitude of hydration numbers for Cl^-, Br^-, and I^- derived from the near-infrared work is opposed to that obtained from compressibility data. However, since the structure-breaking abilities of these anions are likely to prevail at low concentrations, it seems plausible that this could account for the differences, as was indeed stated by the authors.

The above discrepancy illustrates an important point concerning the interpretation of the near-infrared difference spectra, according to McCabe *et al.*[539] It is tacitly assumed that any change in the absorption properties of water is caused directly or indirectly by the presence of the solute and furthermore that all changes in the properties of water due to the presence of the solute are considered as hydration effects. The implication of hydration numbers has been, and continues to be, a contentious issue in solution studies. The basis of their estimation is made precisely on the same assumption used in this spectral interpretation, namely all changes in solution, due to dissolved species, are rationalized in terms of one particular interaction. This constitutes an oversimplification of aqueous systems.

The resolved hydration spectra reveal several important points. As the concentration of NaCl increases in the range $0.5–5.0\ M$ the absorbance and peak frequency both increase. The hydration spectra for a series of cations with a common chloride anion at a constant concentration of $3\ M$ have absorption maxima in the frequency range $6950–7000\ cm^{-1}$. The integrated absorption B varies inversely with the size of the cation, $B_{Na^+} > B_{K^+} > B_{Rb^+} > B_{Cs^+}$. The hydration spectra for a series of anions with sodium or potassium cations show absorption maxima in the same frequency range; however, for both series the integrated absorption varies directly with the size of the anion, $B_{I^-} > B_{Br^-} > B_{Cl^-}$. The narrow range of frequencies for the absorption maxima of the hydration spectra of various ions indicates a similarity of interaction between water and these salts.

The difference between hydration spectra for $3\ M$ NaI and NaCl reveals a symmetric curve that is readily represented over a broad region of intermediate frequencies by a Gaussian distribution function with the following absorption parameters: (1) frequency of absorption maximum, $6997 \pm 3\ cm^{-1}$, (2) absorbance at the above frequency, 0.1708, and (3) half-bandwidth, $212 \pm 10\ cm^{-1}$. There appears to be considerable residual absorbance in the high-frequency wing and a smaller amount in the low-frequency wing after fitting the Gaussian distribution function. Assuming that the contribution by sodium ion hydrate is similar in both cases, the authors state that the resolved curve represents the difference between the absorption by the iodide and chloride hydrates and consequently the absorption by a single "species" of water molecule, i.e., the anion hydrate. Although the data obtained suggest that there is only one species of water molecule contributing to this absorption peak, the authors correctly point out that this does not constitute formal proof of this relationship. Two points to consider are whether the sodium cation behaves in a manner independent of the anion and, if not, whether the residual absorbance in the high-frequency wing measures such a dependence.

Based on the above result McCabe et al.[539] resolved the hydration spectrum for 3 M NaI into two-component curves, i.e., one component for the anion hydrate and a second component obtained by subtracting the anion hydrate component from the hydrate band of the salt, to yield the cation (sodium) hydrate absorption. The band parameters for the latter are (1) absorption maximum, 6800 cm^{-1}, (2) half-bandwidth, 200 cm^{-1}, and (3) absorbance at 6800 cm^{-1}, 0.206. Strong evidence for this latter component is seen in hydration spectra of 3 M CsCl, 0.5–5.0 M NaCl, and 3 M KCl, where distinct shoulders occur at 6800 cm^{-1}.

Although the near-infrared spectroscopic method for measuring hydration properties of electrolytes in aqueous solutions appears to be an appealing approach to determining gross characteristics of water–solute interactions, it unfortunately includes all types of interactions in one experimental probe. Clearly, further studies on a wider variety of solutes may reveal some characteristic differences that can be used to differentiate between changes in the absorption spectra due to true hydration and to indirect changes.

It is a questionable point as to whether all anion–hydrate absorption components should appear at the same frequency and have the same bandwidths. Unfortunately, the authors did not pursue their hydrate band resolution procedure for salts like NaCl, KCl, and CsCl, where overt shoulders appear in the hydrate bands, in order to determine possible variations in the cation-hydrate absorption component parameters.

Preliminary studies do indicate, qualitatively, different influences of anions and cations with regard to their interactions with water, as demonstrated by the position and absorbance of the respective hydrate components.

Further studies[540] by the same authors using temperature difference spectra, i.e., keeping the reference cell temperature constant and varying the sample cell to higher temperatures with appropriate density corrections to assure constancy of water concentration, have revealed strong support for the mixture model concept for water. Spectral changes of water in the 1.45-μm region were rationalized in terms of structural changes in the liquid.

This was based on strong supporting evidence from comparative changes in the absorption spectrum of a saturated solution of water in $CHCl_3$ as a function of temperature. The temperature difference spectra indicate water to be composed of an equilibrium mixture of two or more components.

A high-frequency component is identified as singly hydrogen-bonded water based on its correlation with singly hydrogen-bonded water in acetone. It shows a Gaussian distribution of absorption with values for the frequency of absorption maximum and half-bandwidth of 7081 ± 10 cm^{-1} and 216 ± 10 cm^{-1}, respectively. In addition, these values are independent of temperature over the range 20–80°C.

The low-frequency component centered at 1.493 μm (6700 cm^{-1}) is not Gaussian, although there are indications that it can be resolved into Gaussian components. The spectrum is such that it shows strong similarity to that of ice[528] and hence is identified as "icelike." There is apparently a third component since, according to McCabe et al.,[540] the total absorption band in the 1.45-μm region cannot be fitted* by a summation of the low- and high-frequency components mentioned above. This third component, which does not show up in the temperature difference spectra, has an absorption centered about 1.46 μm and contributes about 15% to the total absorption, the contribution apparently being invariant with temperature in the range 20–80°C.

The above digression into the far-infrared studies of pure solvent are pertinent as they provide a basis for rationalizing the results obtained by McCabe et al.[540] for water in aqueous perchlorate solution.

Their results of the effect of $NaClO_4$ on the 1.45-μm band of water indicate a shift of the band toward shorter wavelengths, and a shoulder on the longer-wavelength side of the band, as the salt concentration is increased. Difference spectra of aqueous sodium perchlorate solutions versus water at the same temperature show a sharp peak at 1.415 μm (7067 ± 5 cm^{-1}). The intensity, but not the position, depends on perchlorate concentration. The position of this band is close to the one observed for the high-frequency component in the temperature difference spectra previously discussed. Furthermore, the behavior of this high-frequency peak with increased perchlorate concentration is very similar to that observed[76,183,184,441] in the fundamental region.

McCabe et al.[540] suggest that the difference in frequency between this component, observed in perchlorate solutions, and the position of the

---

* A similar conclusion has been arrived at by Luck[532] from studies of different overtone bands of H_2O, HDO, and D_2O.

anion-hydrate component[539] (1.429 μm) clearly indicates that the water species corresponding to 1.415 μm in aqueous perchlorate solution is not hydrogen-bonded to the anion. This supports certain previous interpretations[76,183,184,441] from fundamental region studies.

Although the interpretation given by McCabe *et al.*[540] of their aqueous perchlorate data generally agrees with that of Walrafen,[821] the identity of the high-frequency absorbing species differs in each case. Whereas Walrafen[821] contends that this component is nonhydrogen-bonded water, McCabe *et al.*[540] believe it to be a singly hydrogen-bonded water molecule. The latter base this conclusion on evidence derived from infrared studies of dilute solutions of water in CCl_4,[320,751] and in acetone, and various cyclic ethers.[274] In CCl_4, water is considered to be nonhydrogen-bonded (monomeric), whereas in acetone and ether it is regarded as a singly hydrogen-bonded species. The near-infrared absorption peaks for water in CCl_4 at 7165 cm^{-1},[752] and 7148 cm^{-1}[540] and for water at its critical temperature, 7148 cm^{-1},[528] are all interpreted as being indicative of conditions under which the water molecules are nonhydrogen-bonded. Therefore, since the observed high-frequency peak in their various experiments[540] consistently occurs at 1.415 μm (7080 cm^{-1}), they conclude that the frequency difference supports the assignment of the high-frequency peak to a singly hydrogen-bonded species.

A similar correspondence between water-in-acetone, the high-frequency peaks in water–$NaClO_4$, and temperature difference spectra is maintained in the $\nu_1 + \nu_2 + \nu_3$ band.[540] A major peak occurs at 1.157 μm and its position is considerably removed from the monomer peak[751] of water in CCl_4 at 1.147 μm and in liquid water at the critical temperature,[528] 1.148 μm.

At the moment it appears that at least two spectroscopically distinguishable species in the liquid structure of aqueous perchlorate solutions are inferred from Raman[821] and various infrared studies.[183,441,529,540,862] However, the assignment of the high-frequency absorption component differs depending upon whether one looks at the fundamental infrared vibration band or the combination near-infrared bands. In each case a comparison of the position of this component with the absorption of water near or at the critical temperature suggests the assignment should be to a nonhydrogen-bonded or singly hydrogen-bonded water species based on fundamental and near-infrared absorption spectra, respectively.

Intensity changes of the 1.45-μm (6900-cm^{-1}) band have been used[615] to estimate hydration effects of amino acids. The aliphatic amino acids, irrespective of the length of the chain of carbon atoms, appear to be hydrated by one water molecule with the hydration simultaneously involving the NH_3^+ and COO^- groups. However, the probable interference of hydrocarbon C—H overtones in this region[89] makes these results prone to uncertainty.

4.4. Effect of Water on Infrared Spectra of Inorganic Ions

Infrared studies of complex or polyatomic ions in aqueous solution have revealed that the spectra of the isolated ion can be changed considerably due to the interaction potential of neighboring solvent molecules or ions. More extensive use of Raman spectroscopy has occurred because it provides the added benefit of being able to evaluate depolarization ratios, which can assist in the correct choice of symmetry point groups. The infrared results complement the Raman work either by confirming bands that are both Raman- and infrared-active or producing information on bands that are infrared-active only.

Many inorganic compounds in water show infrared absorption bands below 1550 cm^{-1}, where their spectra can be easily examined. Furthermore, in the absence of solid-state crystal fields, which tend to broaden the absorption bands, the solution bands appear over a limited frequency.[287]

Work carried out on aqueous solutions of various nitrates indicates that all nitrate ions in aqueous media produce spectra which are inconsistent with the selection rules for an unperturbed nitrate ion of D_{3h} symmetry. The nitrate ion is considered to be subjected to a distorting potential which results in C_{2v} symmetry and a loss of degeneracy from the E' modes of the free anion.[77,363,365] The removal of the degeneracy has been attributed to an anisotropic ionic atmosphere[561] or to solvent-separated ion pairs.[502]

Irish and co-workers[393,394,396] have shown that Raman and infrared results of dilute aqueous alkali metal nitrates reveal a splitting of the ν_3E' band into a doublet split by 56 cm^{-1}. The ν_4E' mode remains a singlet, except in the case of $>7.0 M$ $LiNO_3$ and $NaNO_3$, where this mode is split into 720- and 740-cm^{-1} components. Experimental evidence from dilute solution studies appears to indicate that the symmetry of the nitrate ion is perturbed by the field due to surrounding water molecules.[393,396] At moderate concentrations solvent-separated ion pairs predominate, with contact ion pairs becoming important in very concentrated solutions.[393] Based on this experimental approach, classification of a number of metal nitrates into contact or solvent-separated ion pairs is possible.[362]

Infrared studies of chloroform and methanolic solutions of tetraphenylarsonium nitrate[158] have also shown a perturbed nitrate spectrum in dilute solutions. Evidence for hydrogen-bonded interactions in these solutions suggests that this form of interaction is likely in the case of aqueous solutions[362] as well.

An alternative interpretation to the symmetry point group treatment has been suggested by Hester[362] based on work[133] on crystalline $Hg(NO_3)_2$ monohydrate. He uses a quasilattice treatment in which correlation field coupling effects between neighboring anions in solution play a significant role. It would appear that the model envisaged by Irish and Davis[393] for concentrated solutions is quasilattice in nature but that these authors find no evidence for such effects.

Further evidence of specific interactions in aqueous solutions affecting the fundamental frequencies of polyatomic anions has been observed for potassium phosphate solutions.[97] Infrared absorption results show a strong band at 1012 cm^{-1} due to the ν_3 out-of-phase P—O stretching vibration.[97,635] In addition, a very weak shoulder is apparent at 938 cm^{-1}. This has been attributed to the theoretically infrared-inactive ν_1 vibration becoming active, due to a creation of asymmetry in the anion point group through hydration.[97] The strength of the interaction between the triply charged anion and the protons of neighboring water molecules is evidenced by the low O—H stretching frequency of the interacting water molecules.

Studies of the infrared spectra of highly enriched ^{18}O-labeled potassium phosphate in aqueous solution show an anomalous increase in the P—O stretching frequencies of these ions. Results indicate a strong absorption at 1050 cm^{-1} corresponding to the ν_3 stretching vibration, with a weak band at 948 cm^{-1}. Additional shoulders appear at 1000, 1083, and 1110 cm^{-1} and are due probably to isotopic modifications.[635] Theoretically, inserting a heavier mass in the isotopic oscillator should result in a frequency decrease, as was observed for the same exchange in the case of crystalline barium phosphate.[630] Obviously, in solution specific anion–water interactions exist, and in the case of the ^{18}O-labeled ions the interaction between the labeled anion and the water molecules is relatively weaker than for normal phosphate. This is corroborated by the fact the labeled phosphate solution shows no trace of an O—H vibration attributable to a strong interaction.[630] It would appear that strongly directed hydrogen bond interaction occurs between the phosphate anion and protons of neighboring water molecules in the case of normal phosphate. The behavior of the labeled anion appears to indicate distorted nonlinear anion$\cdots$H—O interactions, although it is not immediately clear why this should be so.

These few examples indicate the importance of infrared work on structures of polyatomic ions in aqueous solutions, in addition to information derived from the solvent bands themselves. Together, results of these experiments could give a better overall view of the interactions occurring in solution.

5. CONCLUSION

It is apparent that our understanding of aqueous electrolyte solutions is intrinsically tied to our knowledge of the details of the pure solvent. Since water is a complex liquid whose structure is considered in circumstantial terms on the basis of hypothetical models, the projection of these vagaries in the interpretation of infrared spectroscopic data of aqueous solutions is hardly surprising. Theoretical calculations have been of little assistance in supplying useful answers about the molecular nature of water, although recent studies[162,163,177,310,311,457] show some promising leads for the future (see Volume 1).

Many of the studies discussed in Section 4 involved correlation of infrared absorption band changes caused by ionic solutes with those arising from temperature variation. Assuming that water possesses "structure" of some form, due to specific hydrogen bond interactions, it is possible to make rather sweeping inferences about ionic effects in terms of "structure-making" and "structure-breaking" characteristics. The fact cannot be disputed that the multiplicity and complexity of interactions in concentrated solutions such as are required for the observation of changes in solvent infrared bands preclude such a simplified approach. The use of a single parameter, such as the "hydrogen bond strength," to describe changes in water structure due to solutes is argumentative. However, as repugnant as this approach may seem to some scientists, it has been the only useful means of interpreting solution phenomena in a number of instances. More cause for concern has been the use of single experimental observables to arrive at these general conclusions. The mutual consistency of conclusions drawn from changes in several band parameters tend to make them more credible.

Infrared studies of polyatomic anions have been the focal point of recent studies. Aqueous solutions of some of these anionic species show overt signs of band structure in the fundamental absorption region of the solvent. However, their interpretation has been equivocal. On balance, it appears that the majority of interpretations favor the mixture model concept of water.

There has been disagreement among workers as to the precise nature of the components based on such a model. The high-frequency component in near-infrared results for NaClO$_4$ solutions is interpreted as an absorption due to singly hydrogen-bonded water molecules, whereas interpretation of results from the fundamental absorption region are rationalized in terms of nonhydrogen-bonded water molecules. The less well-understood laws governing the intensity behavior and positional assignments of solvent bands in the near infrared may contribute to this.

The practical difficulties inherent in the study of infrared spectra of aqueous solutions are considerable. With continued advances in techniques and instrumentation, however, it is likely that the far-infrared region will reveal more quantitative aspects about intermolecular motions of water molecules. In this regard interferometric and reflectance studies appear to be the most promising.

The near-infrared region of the solvent has been examined extensively to obtain information about structural changes in solution. Recent studies have been used to determine quantitative estimates of hydration parameters for the alkali halide salts. The agreement with data obtained on the basis of thermodynamic measurements has been reasonable in the case of cations, but more studies are required on a variety of ionic solutes before a complete assessment of the method can be made.

Finally, it can be argued that the conclusions arrived at through infrared studies rely to some extent on information drawn from other fields. Current knowledge of water and its solutions dictates that researchers must draw upon the resources of several fields if a comprehensive model is to be attained which will be acceptable to everyone.

CHAPTER 6

Raman Spectroscopy of Aqueous Electrolyte Solutions

T. H. Lilley

Department of Chemistry
The University
Sheffield, England

1. DISCUSSION

The usefulness of Raman spectroscopy in physicochemical studies arises from two facts. First, a consideration of the occurrence of Raman lines and their depolarization ratios, in conjunction with infrared spectral studies, allows the assignment of a molecular species to a symmetry group. A possible complication in the investigation of Raman-active species in solution arises in that the selection rules for transitions may be less rigorously obeyed than they would be in situations where the molecular species is more easily defined. Second, in solution the integrated intensity of a given Raman line is, to a very good approximation, linear in the molar concentration of the species which gives rise to the line.[98,665] Most of the studies on aqueous ionic systems have exploited the latter observation. Several discussions exist on the proportionality between concentration and line intensity[6,666,672,735,874] and although the nature of the corrections necessary to bring about precise proportionality are not completely understood, appropriate adjustment of the experimental optical arrangement does, to a considerable extent, obviate the necessity for such corrections. Thus Raman spectroscopic measurements give an indication of the nature of the molecular species and in solution give a measure of its concentration.

In any given solution many possible interactions between the solute species and the solvent are possible and there is a distribution of many molecular configurations and subspecies. It is, however, customary and convenient to consider a somewhat more limited number of species. Figure 1 shows diagramatically the various states which ionic solutes can take when present in solution.[232] Figure 1(a) indicates the condition when no short-range ionic interactions are occurring and the cospheres[302] of the ionic species are fully developed. The solvent around an ion is considered to be made up of three regions. The first of these (region A) consists of solvent adjacent to the ion and may be called the primary hydration sphere of the ion. Region B consists of water which is under the opposing orienting influences of the ion and of the bulk solvent (denoted region C). The nature and extent of regions A and B would be characteristic of the ionic species and the temperature and pressure of the solution. One imagines that strong electrolytes at not too high concentrations would exist in solution in this way. As the concentration of solute is increased cosphere overlap necessarily becomes more important and configurations as in Fig. 1(b) attain greater prominence (see Chapter 1). There is overlap of the disordered regions of

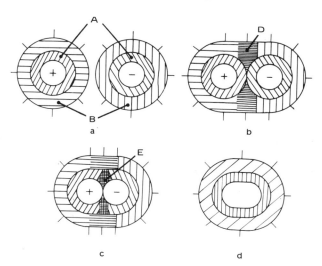

Fig. 1. Diagrammatic representation of ion–solvent interactions in solution. Bulk solvent is denoted region C (not marked). (a) Primary hydration spheres for cation and anion (region A) and outer cospheres for cation and anion (region B). (b) Outer cosphere overlap (region D) (solvent-separated ion pair). (c) Primary cosphere overlap (region E) (contact ion pair). (d) Uncharged species.

the cospheres of the two ionic species with a consequent perturbation in the properties of the solvent molecules in the region of overlap (D in Fig. 1b). Further increase in the concentration may lead to situations like that depicted in Fig. 1(c), becoming more important where primary-hydration-sphere overlap occurs (region E), as well as overlap of the disordered regions (D). Figures 1(b) and 1(c) represent ionic configurations which can be classified as solvent-separated ion pairs and contact ion pairs, respectively. The concentrations at which such configurations are of importance depends upon the nature of the ionic species. In some instances further interaction between the ionic species can occur (Fig. 1d) to give an uncharged molecular species. Such a species would be a nonelectrolyte and rather large modifications in the cosphere would be expected to result. In Fig. 1 only interactions between oppositely charged ions have been considered but, of course, interactions of like-charged species will also gain in importance as the concentration is increased, as will contributions from triplet and higher aggregates of ionic species. The ideas presented in Fig. 1 are easily extended to cope with such situations.

Studies of aqueous ionic solutions by Raman spectroscopy can be subdivided into two broad categories: (1) investigations of the Raman-active transitions of solute species in solution; (2) investigations regarding the changes induced by ionic solutes on the Raman bands of the solvent. This review will be similarly subdivided. It should be mentioned that in some instances overlap occurs between the two subdivisions. These will be considered as they arise.

2. RAMAN BANDS ARISING FROM SOLUTES

This type of investigation has obvious limitations, since only those solutes can be investigated which have Raman-allowed transitions. This means that only those solutions containing polyatomic ionic species are amenable to study and the theoretically important and simplest ionic solutions, namely those containing monatomic ions, are precluded from such investigations.

The first quantitative applications of Raman spectroscopy to aqueous ionic systems were attempts[98,675] to measure the degree of ionization of nitric acid as a function of the formai concentration of the acid. In principle the procedure is simple. As the concentration of nitric acid is increased new bands appear and those which were present in dilute solution diminish in intensity or disappear. Obviously one is observing the gradual association of nitrate and hydrogen ions to give nitric acid molecules. Consideration of the concentration variation of band intensities in conjunction with activity coefficient data allows the evaluation of the equilibrium constant for the process considered. The application of this technique in the determination of the ionization constants of acids in solution has been fully described (see, e.g., Refs. 6, 112, 144, 261, 333, and 676). In principle the procedure could be applied to any equilibrium if the necessary Raman activity is present in at least one of the species participating, but in practice insufficient precision in relative band intensities can be obtained for acids with equilibrium constants of less than approximately 0.1 mol liter^{-1}. It has been customary for a two-state formalism to be used for acid dissociations, i.e., the equilibrium is represented as

$$HA + H_2O \rightleftharpoons H_3O^+ + A^-$$

inferring that the molecular distribution of species A^- is such that only free ions (see Fig. 1a) and undissociated species (see Fig. 1d) exist in solution. It would seem, however,[140] that the dissociation process is better represented as

$$HA + H_2O \rightleftharpoons H_3O^+ \cdot A^- \rightleftharpoons H_3O^+ + A^-$$

where the first step is a proton transfer step from HA to a water molecule in a suitable orientation and the second step is a disruption of the ionic aggregate. The first step is ionization of the acid and the second step corresponds to dissociation. Investigations of the ν_1 bands arising from the ionic forms of trifluoroacetic and trichloroacetic acids in aqueous solution have been carried out[140] as a function of concentration. These indicate that within experimental error the acetate ions have essentially the same spectral characteristics when present as the ion pair or as the free ionic species. This is important in the present context since it indicates that, at most, only small perturbations in the Raman characteristics of solutes are induced in the formation of solvent-separated ion pairs from free ions.

In this connection mention should also be made of the application of Raman spectroscopy to the kinetics of proton transfer processes.[463,464] Information on such processes can be obtained by a consideration of the variation in linewidth of Raman-active solute species which are able to interact with protons. The basis of the technique is that, if the lifetime of a given vibrational substate is small, then from the uncertainty principle the breadth of the spectral band corresponding to the vibrational level will be correspondingly large. The broadening observed is then given by

$$\beta - \beta_1 = 1/\pi c \tau$$

where β and β_1 are, respectively, the band halfwidths in the presence and absence of exchange, and c and τ denote the velocity of light and the mean lifetime of the species giving rise to the vibrational band. Studies have been made[140,464] of the variation of half-width of the ν_1 band of the anion of trifluoroacetic acid as a function of acid concentration. If it is assumed that the proton transfer may be represented as above and if the species $H_3O^+ \cdot A^-$ and A^- have the same spectroscopic characteristics, then it may be shown[140] that

$$\beta - \beta_1 = (k_1 K_1/\pi c)(1 - \alpha)/\alpha \tag{1}$$

where k_1 is the rate constant for the process

$$H_3O^+ \cdot A^- \rightarrow H_3O^+ + A^-$$

K_1 is the equilibrium constant corresponding to the ionization step,

$$K_1 = [H_3O^+ \cdot A^-]/[HA]$$

and α is given by $([H_3O^+ \cdot A^-] + [A^-])/C$, C being the stoichiometric concentration of acid. Figure 2 shows the results obtained for trifluoroacetic acid solutions and it is apparent that the plot is linear up to $\alpha = 0.6$.

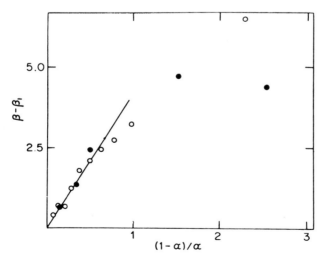

Fig. 2. Test for eqn. (1) for trifluoroacetic acid solutions. (●) Kreevoy and Mead.[464] (○) Covington et al.[140] Reproduced from Ref. 140 with the permission of the copyright holders.

Deviations from linearity at higher concentrations could be accounted for by the neglect of activity corrections in the derivation of the relationship.

A rather more refined approach of the above treatment has been used[99] to interpret line broadening in the SO_4^{2-}–HSO_4^- system. A band observed at 948 cm^{-1} was assumed to arise from SO_4^{2-} which was ion-paired to H_3O^+.[392] The free sulfate ion band occurs at 981 cm^{-1}.

The use of polyatomic, Raman-active ionic species in aqueous solutions as probes into the solute environment has been a particularly active area of research. The most notable and extensive contributions in this field have come from Plane and Irish and their collaborators (see, e.g., Ref. 394 and references quoted therein). Of the polyatomic ions which have been investigated, the nitrate ion is by far the most extensively studied. Rather than discuss in detail all of the work which has been carried out on Raman-active ions in solution, a fairly detailed and comprehensive discussion will be given on the effects which have been observed for various nitrate salts. An indication of the type of conclusions which can be drawn from investigations of this nature will be discussed.

The isolated NO_3^- ion would have D_{3h} symmetry with three allowed Raman bands, the ν_1 symmetric stretch and two (ν_3 and ν_4) degenerate asymmetric stretching bands. The ν_2 band for the ion is Raman forbidden although it is observed in the infrared. It is found experimentally in aqueous systems that rather large perturbations are induced on the ion by solvent and ionic interactions. A thorough and careful investigation[393] has described the variations which are observed in the Raman spectrum of alkali-metal nitrates together with supplementary infrared information. It is found even at low concentrations of solute that the ν_3 band, centered at approximately 1390 cm^{-1}, is split into two components with a peak separation of some 56 cm^{-1}. A similar observation regarding the splitting of this band has been observed for calcium nitrate solutions.[396] Extrapolation of the data on all the salts investigated indicates that this separation would persist in the limit of infinite dilution and hence is a consequence of some nitrate ion–solvent interaction.[393,396] It has been suggested[396] that the band splitting arises from a loss of symmetry of the ion from D_{3h} to C_{2v} through some sort of association between the ion and water molecule(s). This nonequivalence of the oxygen atoms of the nitrate ion could be explained in several ways, although it does appear to be analogous to the loss of equivalence of the oxygen atoms on carboxylate ions in aqueous solution as proposed by Gurney.[302] If this postulated interpretation regarding the loss of symmetry is correct, then a similar band splitting should be observed for the relatively weak ν_4 band, which occurs at 719 cm^{-1}. This band exhibits visually no such band splitting although decomposition of the band into components indicates a broad, weak band centered at about 689 cm^{-1}. It has been suggested[393] that this band arises from a water–nitrate librational mode (see section on water intermolecular bands) but the explanation that it is the second band from the now nondegenerate ν_4 mode appears to be consistent with the suggestion made regarding the hydration of the nitrate ion. Its apparent lack of appearance in D_2O solutions is not consistent with this suggestion, however. It is conceivable that what is being observed here is a breakdown of the selection rules in solutions, as mentioned earlier. Returning now to the discussion on the 1390-cm^{-1} band, it is found that considerable perturbations in the spectral band parameters are observed. The components of the now nondegenerate ν_3 bands show concentration-dependent peak positions and half-widths. Up to concentrations of about 3 M monotonic and approximately linear variations in position and width are observed for solutions in which the counter-ion is Na$^+$, K$^+$, or Cs$^+$. The variations are independent of the cation and would seem to indicate some form of long-range Coulombic effect on the nitrate ion or possibly overlap between the outer fringes of the ionic cospheres. The Li$^+$ ion is exceptional since, although the band positions are concentration-dependent, this dependence is different from that observed for the other ions, indicating a rather more enhanced cation–anion interaction even at comparatively low concentrations. Variations in the spectral parameters for various nitrates of the ν_1 ($\sim$1050 cm^{-1}) symmetric stretching band of NO$_3^-$ have been documented by Vollmar,[804] whose general findings have been confirmed by Irish and Davis.[393] A general increase in half-widths and band positions is observed, the concentration dependence for the lithium salt being different from those of the other alkali metal salts which, within the limits of experimental precision, are independent of the cation at lower concentrations, although above about 1.5 M the widths decrease in the order Li > Na > K $\simeq$ Cs. This is similar to the effects observed for the higher-frequency band in the ν_3 region. It has been suggested[804] that increases in width of the symmetric stretching band are caused by cations breaking down structure in the solution, thus giving a more disordered environment for the nitrate ions. Presumably an effect of this sort is explicable in terms of Fig. 1(b), i.e., overlap of region B in the outer cospheres. A linear correlation between the increase in the bandwidth and the cationic hydrated radius was observed. This correlation with hydrated radius cannot be taken too seriously since, at best, these parameters are reasonable guesses. The band at 719 cm^{-1} was discussed earlier. In more concentrated solutions it is evident from a decomposition of the band into components that a new band appears at 740 cm^{-1}. This band is believed to be due to strong, fairly intimate interactions between the cation and anion. The presence of this band appears to be indicative of contact ion pairing in nitrate solutions.[394]

The occurrence of a new species (e.g., a contact ion pair) in solution confers, perhaps, different symmetry properties and almost certainly different spectroscopic characteristics on the nitrate ion when compared to the infinitely dilute case. This does of course result in rather more complex spectra, since even in the simplest example one would see a superposition of two spectra. It would appear that, although the deconvolution of the concentration dependence of the spectrum of each salt must be considered as a separate problem, certain generalizations may be made. The most striking feature is the splitting of the ν_4 spectral band in liquid water into (at least) two components, observed when cations of high surface charge density are present. This is illustrated in Fig. 3, which shows the Raman spectra of $Cd(NO_3)_2$ and $Zn(NO_3)_2$ solutions at comparable concentrations. The splitting of the band at $\sim$720 cm^{-1} is evident. Decomposition of this band into components indicates the presence of bands at 720

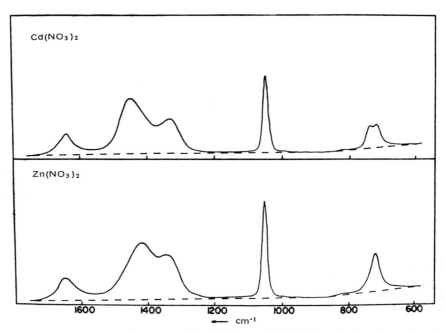

Fig. 3. Raman spectra of concentrated $Zn(NO_3)_2$ and $Cd(NO_3)_2$ solutions. Reproduced from Ref. 394 with the permission of the copyright holders.

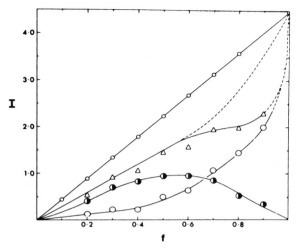

Fig. 4. Integrated intensities against solution composition for solutions in which $[Gd^{3+}] + [NO_3] = 2.75\ M$. $f = [NO_3^-]/[Gd^{3+}] + [NO_3^-]$. ● Intensity of the 718-cm^{-1} NO$_3^-$ line from solutions containing NO$_3^-$. △ At the same concentration as the Gd^{3+}–NO$_3^-$ mixture. Total band intensity between 650 and 800 cm^{-1}. ◐ Intensity of band centered at 750 cm^{-1}. ○ Intensity of band centered at 718 cm^{-1}. Reproduced from Ref. 591 with the permission of the copyright holders.

cm^{-1} and 740 cm^{-1}. The former band is the ν_4 solvated nitrate band which is found in solutions containing the alkali metal cations, whereas the latter is ascribed to contact-ion-paired nitrate ions. Bands in this region (740–760 cm^{-1}) appear to be characteristic of contact cation–anion interactions and have been observed many times[157,159,364,365,367,393,394,396,570,604] and, indeed, the presence of a band in this region is diagnostically useful as an indicator for ionic interactions of this type. Figure 4 shows the variation in relative intensities of the two bands as a function of nitrate concentration for solutions containing gadolinium and nitrate ions.[591] The total concentration of gadolinium and nitrate was fixed at 2.75 M and the ratio of the two ionic compositions was varied by the addition of Gd(ClO$_4$)$_3$ and HClO$_4$. In the diagram $f = [NO_3^-]/([NO_3^-] + [Gd^{3+}])$ and it is apparent that as f increases so does the intensity of the band ascribed to hydrated nitrate, at the expense of the band from Gd$^{3+} \cdot$ NO$_3^-$ contact ion pairs. In the spectral region of about 1200–1600 cm^{-1} quite marked cationic effects are observed. A comparison of the two spectra from Zn and Cd nitrates shown in Fig. 3

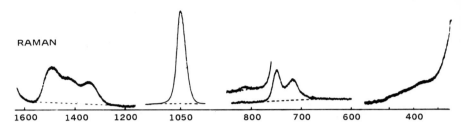

RAMAN

| 1600 | 1400 | 1200 | 1050 | 800 | 700 | 600 | 400 |

Fig. 5. Raman spectrum of 3.06 M Gd(NO$_3$)$_3$ solution. Reproduced from Ref. 591 with the permission of the copyright holders.

exemplifies this point. In the cadmium nitrate solution there is evidently some enhanced splitting and relative intensification of the higher-frequency band compared to the lower-frequency band. Features of this type are also indicative of the presence of some new nitrate species. As an example of the complexity such studies can take, we illustrate in Fig. 5 the Raman spectra obtained in a recent investigation into aqueous Gd(NO$_3$)$_3$ solutions.[591] Considering only the 1200–1600 cm^{-1} region, Fig. 6 shows the computer-analyzed Raman contour. Six bands are necessary to adequately reproduce the spectra obtained in this region. Of these, the bands at 1342 and 1416 cm^{-1} are from the perturbed nitrate band which were discussed earlier and

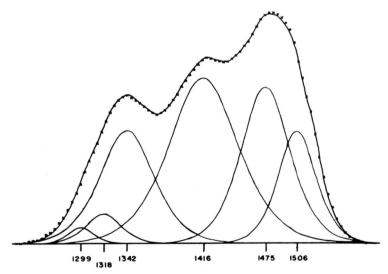

| 1299 | 1342 | 1416 | 1475 | 1506 |
| 1318 | | | | |

Fig. 6. Decomposition of the Raman spectrum of 2.94 M Gd(NO$_3$)$_3$ solution in the region 1200–1600 cm^{-1}. Reproduced from Ref. 591 with the permission of the copyright holders.

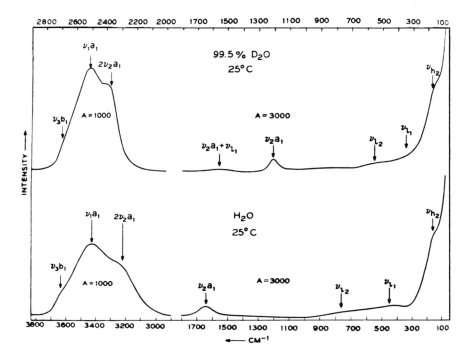

Fig. 7. Raman spectra of liquid H_2O and D_2O at room temperature. Reproduced from Ref. 813 with the permission of the copyright holders.

are due to $H_2O \cdots NO_3^-$ interaction. The bands at 1299 and 1506 cm^{-1} are due to $NO_3^- \cdots Gd^{3+}$ species (i.e., contact ion pairs), and the bands at 1318 and 1475 cm^{-1} are from nitrate-cation triplets ($NO_3^- \cdots Gd^{3+} \cdots NO_3^-$).

3. RAMAN BANDS ARISING FROM THE SOLVENT: LIQUID WATER

The Raman spectra[813] of liquid water (H_2O and D_2O) are shown in Fig. 7.* The most intense bands occur between about 2900 and 3800 cm^{-1} for H_2O and between 2200 and 2700 cm^{-1} for D_2O. These bands arise from O—H and O—D vibrations, respectively. The most striking features of these bands are their breadth and asymmetry. Comparatively sharp sym-

---

* Editor's note: The Raman spectrum of liquid water and its interpretation is discussed in detail in Volume 1, Chapter 5 of this treatise.

metric bands centered at 1645 cm^{-1} and 1210 cm^{-1} are ascribed to H—O—H and D—O—D bending modes. The regions of the spectrum below about 1000 cm^{-1} for liquid H$_2$O and below about 700 cm^{-1} for liquid D$_2$O are believed to arise from intermolecular vibrations of hydrogen-bonded species. The spectrum of liquid D$_2$O shows a very broad, weak band centered at about 1550 cm^{-1}. It has been suggested[190,835] that this band is either composed of overtones of the intermolecular vibrations or some combination of the bending mode with one or more of the intermolecular bands. This band is not visible in the spectrum of liquid H$_2$O, but it is conceivable that if it does exist, then it lies under the envelope of the O—H stretching bands.

In recent work almost all attention has been restricted to investigations on the intramolecular stretching regions of liquid H$_2$O and D$_2$O and their binary mixtures. Some work has been described on the intermolecular region below 1000 cm^{-1}, but very little information is available on the variations induced by temperature and solutes on the bending regions.[92,727,833] The subject has been dealt with in Volume 1, of this treatise (see Chapter 5) and we only summarize here what seem to us the shortcomings of the continuum models (as compared to the two-state mixture model):

1. They assume that the intramolecular stretching region bands arise from a superposition of many narrow Raman bands. Experimental evidence for the presence of such narrow bands in aqueous solution is lacking.

2. The occurrence of isosbestic frequencies would be extremely unlikely from a continuum model.

3. The stimulation of bands by the use of stimulated Raman procedures would likewise not be expected.

Notwithstanding these points, continuum model approaches to the intramolecular region would also have to explain the following conclusions inferred from mixture model approaches:

1. The enthalpy change for hydrogen bond rupture obtained from both the intermolecular and intramolecular regions of the spectrum are the same.

2. There are opposite dependences of component intensities with temperature change.

3. There is agreement between expected and observed H/D frequency ratios for the component band positions.

In view of the doubtfulness of continuum model approaches, throughout the discussions presented here mixture model approaches will be stressed. Further aspects of liquid water in its various forms will be discussed at appropriate points.

3.1. Intermolecular Region Vibrations

The region of the Raman spectrum below 1000 cm^{-1} originates from intermolecular vibrations of hydrogen-bonded units. There is now a considerable body of evidence[821] that the spectral bands which occur in this part of the spectrum correspond to those which would be observed from a five-molecule aggregate of water molecules with four peripheral water molecules disposed tetrahedrally about a central water molecule. The outer water molecules are understood to be hydrogen-bonded to the central molecule and the total structure approximates to C_{2v} symmetry. It is experimentally observed[813,814,818] that increase of temperature or pressure results in a diminution of intensity of the bands in the intermolecular region and this is ascribed to a reduction in the concentration of four hydrogen-bonded units to aggregates containing fewer hydrogen bonds. The application of a thermodynamic method of analysis, assuming that mole fraction statistics can be applied and that all deviations from ideality of the water systems as a whole can be ascribed to hydrogen bonding, yields values of $2.55 \text{ kcal mol}^{-1}$ for the enthalpy corresponding to the rupture of a single hydrogen bonding unit and $8.5 \text{ cal deg}^{-1} \text{ mol}^{-1}$ for the corresponding entropy term. The value derived for the enthalpy term is in agreement with other data.[731,862] If the above method of analysis is accepted, then it might be expected that changes in the intermolecular spectrum of the water by the addition of solutes might similarly be interpreted. In other words, if on addition of a solute, the intensity of the intermolecular region is diminished, then this would correspond to a reduction in the concentration of the four-hydrogen-bonded water species. The actual situation is apparently rather more complex than this. It is found experimentally[812–814] that, depending upon the region of the spectrum considered, different effects are observed. These regions will be considered in turn.

The portion of the spectrum between about 200 and 1000 cm^{-1} corresponds to librational motions in the assumed C_{2v} tetrahedral primary unit. In water three bands centered at about 450, 550, and 720 cm^{-1}[818] contribute to the overall spectral envelope in this region. As was mentioned above, increase in temperature of water results in a decrease in intensity from this region. The addition of salts causes perturbations in this spectral region quite unlike those which would be caused by changes in temperature. As yet no study has been made of the variation in intensity of each component band with salt addition but it has been shown[812,814] that new bands are observed when some salts are added. In solutions containing chloride and bromide ions intensification of the band at 440 cm^{-1}

and the appearance of a new band at about 600 cm^{-1} suggest the formation of a new Raman-active species. This region of the spectrum is a difficult one in which to work (especially with Hg excitation) and some difficulty was encountered in band resolution. The difficulty was circumvented to some extent by consideration of the total intensity in the intermolecular libration region. The total intensity I is given by

$$I = \sum_i J_i c_i$$

where J_i and c_i are the molar intensity and concentration of the ith species giving rise to a Raman line. The summation is made over the entire region considered. It was suggested that since some of the bands have common origins, i.e., they arise from the same species, then the total intensity in this region could be considered to arise from water in combination with salt and uncombined water. The concentration to the intensity from uncombined water would presumably only come from tetrahedrally four-hydrogen-bonded units:

$$I = J_W{}^S [W]_S + J_W{}^F [W]_F$$

where $J_W{}^S$ denotes the sum of the molar intensities of the salt-combined water bands, $J_W{}^F$ the sum of the molar intensities of the "free"-water bands, and $[W]_S$ and $[W]_F$ the molar concentrations of combined and "free" water, respectively. Figure 8 shows the intensity variation of the total intensity with concentration for lithium bromide solutions ranging from 0 to 10.83 M. The observed linearity suggests that a relationship of the form

$$I = K_W{}^S [\text{LiBr}] + J_W{}^F [W]_F{}^\circ$$

holds. Here $[W]_F{}^\circ$ is the concentration of "free" water when no salt is added. If it is assumed in the light of Fig. 1 that two effects occur, namely hydration of the salt and also formation of a disordered region around the solute species, then we can write

$$I = J_W{}^S n_W [\text{LiBr}] + J_W{}^F \{[W]_F{}^\circ - (n_d + n_h)[\text{LiBr}]\}$$

where n_h and n_d are the numbers of moles of hydrating water and "disordered" water per mole of solute. This equation is of the required form. It should be noted that the precise meaning of n_d and n_h are changed if conditions are placed upon the nature of the water giving rise to the "free" water intensity; the form of the equation remains the same, however. It should also be observed that a preferable and more general expression would include cationic and anionic contributions. A molar intensity enhancement (S_{xy} in the original paper[812]) can be defined as $(I - I_W)/[\text{MX}]$,

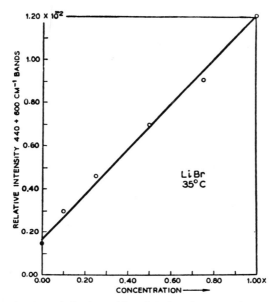

Fig. 8. Relative intensities of the low-frequency inter-molecular bands of aqueous LiBr solutions as a function of concentration $(X = 10.83\ M)$. Reproduced from Ref. 812 with the permission of the copyright holders.

where I and I_W refer to the observed total intensities of solution and water and [MX] denotes the concentration of salt with cation M and anion X. In terms of the above equation we have

$$(I - I_W)/[MX] = [J_W{}^S n_h - (n_d + n_h)J_W{}^F)]$$

If this equation is expressed in terms of ionic contributions, then the molar intensity enhancement should be additive in ionic concentrations. Perusal of the experimental data[812,814] indicates that this is not quite so. Reasons for the departure from additivity could be due to contributions from cosphere overlap or simply from experimental error. As was stated earlier, this region of the spectrum is a difficult one in which to obtain precise data. It is found,[812,814] however, that the molar intensity enhancements increase in the order $NH_4{}^+ < Li^+ < Na^+ \simeq K^+$ for cations and $NO_3{}^- < Cl < Br$ for anions. The magnitude of the anionic effects, especially for the halide ions, are the greater. It is difficult to compare the magnitudes of the effects observed qualitatively since the molar intensities $(J_W{}^S)$ will be dependent on the ion involved in the solute–water interaction. The temperature dependence of the total band intensity of the intermolecular librational region in ionic solutions is small, which indicates that the primary hydration sphere is not very sensitive to thermal changes.

The region of the Raman spectrum below about 300 cm^{-1} corresponds to bending and stretching modes of the hydrogen bond.[818] Two spectral bands are observed, one at 60 cm^{-1} corresponding probably to hydrogen

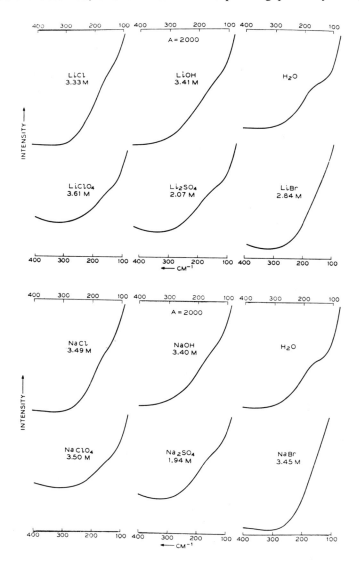

bond bending and the other centered at about 170 cm^{-1} and corresponding to hydrogen bond stretching motions. It is believed that these bands have their origin in the C_{2v} four-hydrogen-bonded unit which was discussed earlier. As yet no quantitative information is available on the effects of added electrolytes on these bands but it has been shown[813] that all of the electrolytes which have been investigated cause a reduction in the intensity in this region of the water spectrum (see Fig. 9). The electrolytes which

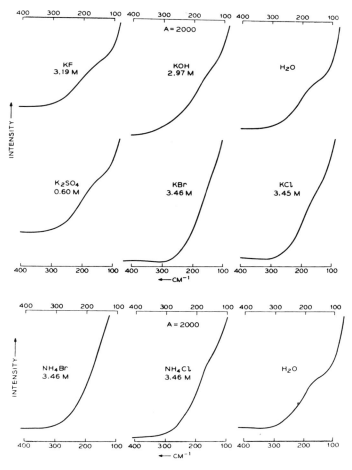

Fig. 9. Effects of electrolytes on the intensity of the 175-cm^{-1} intermolecular band of water. Reproduced from Ref. 813 with the permission of the copyright holders.

have been studied are lithium and sodium chlorides, bromides, hydroxides, perchlorates, and sulfates, potassium fluoride, chloride, bromide, hydroxide, and sulfate, and ammonium chloride and bromide. It is apparent from visual inspection of the spectra that no cationic effects are observed and anionic effects are pronounced and increase in the order $H_2O < SO_4^{2-} < OH^- < ClO_4^- < Cl^- < Br^-$. It would appear that the ion hydrates which contribute to the intensity increase in the librational region of the spectrum do not have Raman bands in this region. Consequently, it can be inferred that the anions which have been investigated have a marked propensity to destroy the four-hydrogen-bonded water units. This removal of hydrogen-bonded species could simply be a perturbation in the distribution of hydrogen-bonded species by the anions, i.e., simple structure breaking as considered for the ClO_4^- intramolecular region, or, in the light of the information from the librational region, it could be that the anion hydrates remove water with a consequent reduction in the concentration of hydrogen-bonded water species. It is possible that for some anions both of these effects occur, but at present no quantative conclusions can be drawn. The apparently small cationic effects are puzzling since one imagines that cations, in some cases at least, have fairly well-defined primary hydration shells and a reduction in intensity should arise simply by removal of water from the bulk environment to the hydration shell, notwithstanding effects induced in the outer cospheres. Negligible cationic effects were also observed in the intramolecular region in solutions containing perchlorate ion (see p. 284). Further experimental work is required on points such as these but it does seem that cations and anions affect liquid water in quite different ways. It would be interesting to investigate systems containing small cations (e.g., Al^{3+}) where it has been shown from PMR measurements[141] that well-defined primary hydration regions exist. To summarize, it can be inferred from the available experimental results that anions interact with the water in a relatively intense way to form partial covalent ion–water bonds in the primary hydration region. This region is apparently fairly extensive or there is an extensive disordered region about a more restricted primary layer of water molecules. The interaction of cations with water molecules in the primary hydration region is less intense and the interaction is probably essentially electrostatic in character. The small or negligible effects of cations on the low-frequency Raman bands of water indicate a fairly restricted primary hydration region and/or only a small disordered region.

The general effects observed for aqueous ionic systems are generally in accordance with observations obtained from neutron inelastic scattering studies.[705]

3.2. Intramolecular Bands of the Solvent

It is now well established[120,731,813–821,862] (also see Volume 1, Chapter 5) that the intramolecular vibration region of the Raman spectrum of liquid water is composed of several overlapping bands. The vibration frequency corresponding to the O—H stretching region of a given oscillator is perturbed by the environment in which it finds itself. Perturbations arise from (a) interactions with the lattice in a noncovalent manner and (b) interaction with the lattice through hydrogen bonds. Further shifts in frequency are induced by intramolecular coupling.[821] The Raman spectrum of the intramolecular regions is consequently rather complex and almost certainly consists of five-component, overlapping bands.[821] Simplification results, however, if isotope substitution is used. For example, if dilute solutions of D_2O in H_2O are investigated in the O—D stretching region, then the intramolecular coupling is diminished since most of the O—D oscillators are present in HOD molecules. Under these conditions the O—D stretching region consists of two overlapping bands, one of which arises from O—D oscillators hydrogen-bonded to water (H_2O) molecules and the other from O—D units which are not hydrogen-bonded although still interacting with the lattice in a nonhydrogen-bonding manner.[818] In most of the investigations which have been carried out[817,821] the binary solvent solutions were sufficiently concentrated in the minor constituent for the coupling vibrational band still to be in evidence, albeit weakly. It is apparent that as the concentration of the minor component diminishes, so, rapidly, do the intramolecular coupling contributions to the overall spectrum.[819]

When considering the addition of ionic solutes to water, then in dilute solutions we would imagine that configurations like those depicted in Fig. 1(a) would prevail. Various subcategories of interaction are possible regarding the primary hydration sphere and the disordered region:

1. The interaction between solute and solvent would be such that a "hydrate" would form. If this hydrate were Raman-active, this would lead to new vibrational bands in the Raman spectrum.

2. A new Raman-active species would not be formed but the solute interacts with the solvent in such a way that the equilibria among the various classes of water molecules in the solvent are perturbed. For example, the solute could induce changes in such a way that from an experimental viewpoint water molecules are being shifted from one category of solvent–solvent interaction to another.

3. Both of the above could occur, i.e., a hydrate is formed and this hydrate then transmits its influence and perturbs the distribution among nonhydrating solvent molecules.

In concentrated solutions the position would be rather more complicated since cosphere overlap and solute–solute interactions generally would need to be considered. In the discussion that follows contributions to the Raman spectrum from interactions of this type will be largely neglected. It is realize that such interactions are of considerable importance and indeed, to a considerable extent, determine the deviations from thermodynamic ideality. However, it would appear that Raman spectroscopy of the solvent motions is a fairly insensitive probe for interactions of this nature. The position is, of course, complicated further when ionic systems are considered because of the fact that one must always have at least two types of primary ionic species present. Another possibility which could arise in ionic solutions is a perturbation of the interaction between solvent molecules outside of the cospheres (i.e., in region C, see Fig. 1a) by long-range Coulombic forces emanating from the ionic species. There is no evidence for such an effect being of importance from the Raman viewpoint.

An extensive study of aqueous binary and ternary solutions of alkali metal perchlorates has been conducted by Walrafen.[821] It has been found that the perchlorate ion affects the solvent intramolecular vibration region in the same way as do temperature changes. Previous infrared investigations[441] indicated a marked splitting of the spectral band in the intramolecular vibration region of the solvent. There was no evidence for the presence of new "hydrate" bands. Figure 10 shows the spectra obtained for the O—D stretching region in dilute solutions of D_2O in H_2O as a function of sodium perchlorate concentration. That no new band appears in the vibrational envelope is apparent both from the deconvolution of the spectra into components and from the observed fact that the same isosbestic points were obtained from the perchlorate solutions as a function of concentration as those obtained from studies on the corresponding aqueous binaries when temperature was used as the perturbing variable.[816,817]* It was also observed[821] that substitution of the counter-ion by Li^+ or K^+ produced no observable change in the stretching region of the solvent spectrum. This latter observation is important and has been observed several times.[92,727,833] It indicates the relative insensitivity of solvent intramolecular vibrations to cations and is no doubt due to the orientation water molecules

---

* Also T. H. Lilley and G. E. Walrafen, unpublished observations.

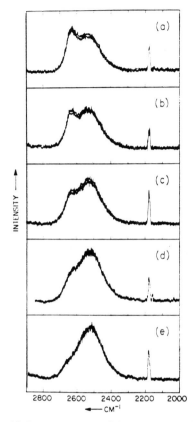

Fig. 10. Raman spectra of the O—D stretching region from solutions containing 5.51 M D_2O in H_2O. Spectrum (e) refers to the solvent and spectra (d)–(a) refer to solutions containing 1–4 M $NaClO_4$. Reproduced from Ref. 821 with the permission of the copyright holders.

tend to adopt when adjacent to cations. Water molecules adjacent to anions would have the H atoms directed toward the ion, whereas water molecules in the primary hydration sphere of cations would have the H atoms directed away from the ion. It would appear that for simple ions at least, the cation–oxygen interaction is predominantly electrostatic in nature and this does not transmit itself to the O—H or O—D bond to any significant extent. Figure 11 shows the component fractional intensities obtained in H_2O/D_2O mixtures dilute in D_2O, as derived from the spectra

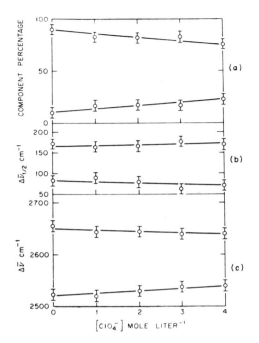

Fig. 11. Spectral parameters obtained from the decomposition of the spectra given in Fig. 9 using two-Gaussian component fits to the spectra. (a)–(c) refer to component percentages, half-widths, and decomposed peak positions. The upper lines refer to the high-frequency component and the lower lines to the low-frequency component. Reproduced from Ref. 821 with the permission of the copyright holders.

shown in Fig. 10. The same procedure was used to obtain these parameters as was used for solutions of D_2O–H_2O binaries. The principal effects are the approximately linear changes with perchlorate concentration in the component fractional band areas from the two spectral bands. In the aqueous binary the band centered at approximately 2645 cm^{-1} contributes some 89% of the total band area, whereas in the 4 M NaClO$_4$ ternary solution the contribution is about 76%. The observed spectroscopic effects are analogous to those which would be expected to be obtained at high temperatures in the aqueous binary solvent. That ClO$_4^-$ ions in some way disrupt the structure of liquid water is indicated by changes in the intermolecular vibration region.[813] As suggested by Walrafen,[821] the spectrum obtained from 4 M NaClO$_4$ solutions at 25°C is equivalent to that which would be obtained from the aqueous binary at about 135°C. Thus it appears that perchlorate solutions behave like high-temperature aqueous systems. In the light of the discussion based on Fig. 1, it appears that from a spectroscopic viewpoint (a) the cation has no observable effect, (b) no hydrate is formed between water molecules and the perchlorate ion, (c) the disordered region consists of water molecules which have the same spectral characteristics as nonhydrogen-bonded water molecules. The experimental results obtained from this study can be placed on a formally more satisfactory basis by assuming that each perchlorate ion has associated with it n disordered water molecules and that the equilibrium constant linking nonhydrogen-bonded and hydrogen-bonded water molecules outside of the disordered region is the same as that in water. Consequently, we may write

$$I_u^S/I_u^W = (K[S]n_{OD}/[HOD]) + 1$$

where I_u^S and I_u^W are the intensities of the nonhydrogen-bonded species (including the disordered species) in the ternary solutions and in the binary D$_2$O–H$_2$O mixture, respectively, K is the equilibrium constant for the process

$$(H_2O)_{non\text{-}H\text{-}bond} \rightleftharpoons (H_2O)_{H\text{-}bond}$$

in water and is assumed the same in the ternary solution, [S] and [HOD] are the molar concentrations of solute (perchlorate ion) and HOD, respectively, and n_{OD} is the number of moles of O—D oscillators in the disordered region per mole of solute. If the HOD species reflect the possible environments of all solvent species, then

$$I_u^S/I_u^W = (K[S]n/[W]) + 1$$

where [W] denotes the concentration of all water species. Similarly for the hydrogen-bonded form of water we have

$$I_b^S/I_b^W = 1 - ([S]n/[W])$$

where I_b^S and I_b^W are the intensities of the hydrogen-bonded species in the ternary solution and the binary solvent. Figure 12 shows the variation of the equilibrium constant K with temperature and was constructed from the data given by Walrafen.[818] Use of the data presented in Figs. 10 and 11 for the aqueous binary and the 4 M solution in conjuction with the appropriate value for K leads to values of n of 2.1 from the intensities of

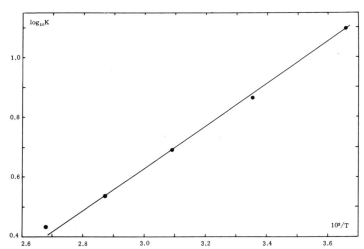

Fig. 12. The temperature variation of the equilibrium constant for the process $(H_2O)_{non\text{-}H\text{-}bond} \rightleftarrows (H_2O)_{H\text{-}bond}$. Redrawn from Ref. 818 with the permission of the copyright holder.

the unbound species and 1.8 from the intensities of the bound species. The agreement is within experimental error. Consequently it appears that each perchlorate ion in solutions at 25°C has associated with it approximately two water molecules the spectroscopic characteristics of which are the same as those for nonhydrogen-bonded water molecules. An alternative procedure is to consider the apparent equilibrium constant K_{app} observed in solution relative to the equilibrium constant in water at the same temperature:

$$\frac{K_{app}}{K} = \frac{1 - ([S]n/[W])}{1 + (K[S]n/[W])}$$

In terms of band intensities this gives

$$\frac{I_b^S/I_u^S}{I_b^W/I_u^W} = \frac{1 - ([S]n/[W])}{1 + (K[S]n/[W])}$$

which may be rearranged to

$$R = 1 - ([S]n/[W])(1 + RK) \qquad (2)$$

where $R = (I_b^S/I_u^S)/(I_b^W/I_u^W)$. Figure 13 is a plot corresponding to eqn. (2) for the experimental data observed and given in Figs. 10 and 11 for sodium perchlorate solutions. It is apparent that the experimental precision

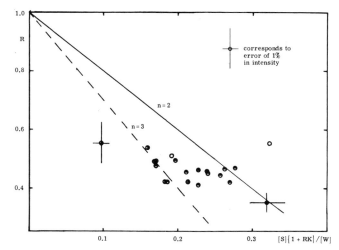

Fig. 13. Plot of eqn. (2) for $NaClO_4$ in D_2O–H_2O at various temperatures: (◖) 2.5 M, (◕) 3 M, (○) at varying $NaClO_4$ concentrations.

is not too high, but nevertheless the earlier conclusion regarding the number of water molecules in the disordered region is confirmed.

This approach can be extended to consider the temperature variation of the O—D spectrum in D_2O–H_2O–$NaClO_4$ mixtures, dilute in D_2O. Combination of eqn. (2) with the equation for the temperature variation of relative band intensities in aqueous binaries[818] gives at a temperature T

$$\log_{10} \frac{I_b^S}{I_u^S} = \frac{546.8_1}{T} - 0.8593_3 + \log_{10} \frac{1 - ([S]n/[W])}{1 + (K[S]n/[W])} \qquad (3)$$

Equation (3) allows the prediction of the temperature variation of the relative intensities of the bands arising from the hydrogen-bonded and nonhydrogen-bonded forms of O—D oscillators in ternary solutions of D_2O in H_2O. Figure 14 shows the results obtained from a study* of 3 M sodium perchlorate–D_2O–H_2O solutions as a function of temperature. Experimental data in these solutions were obtained using argon-ion laser excitation. It was found that a two-Gaussian-component fit adequately represented the spectral data in the O—D stretching region although, as had been found on binary aqueous and ternary perchlorate solutions,[821]

* T. H. Lilley and G. E. Walrafen, unpublished observations.

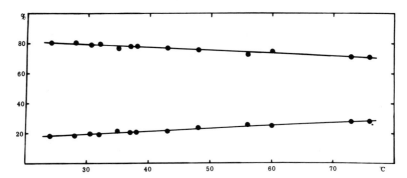

Fig. 14. Gaussian component percentages obtained from 3 M NaClO$_4$–D$_2$O (10%)–H$_2$O solutions in the O—D vibration region as a function of temperature. Data obtained using Ar$^+$ excitation (4880 Å). The upper and lower lines correspond to the component bands centered at $\sim$2525 cm^{-1} and $\sim$2645 cm^{-1}, respectively.

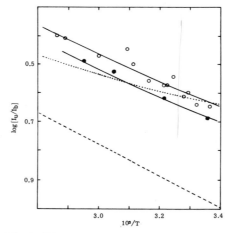

Fig. 15. Plot of eqn. (3) for NaClO$_4$–D$_2$O–H$_2$O mixtures: (○) 3 M, (●) 2.5 M. The solid line passing approximately through the open circles corresponds to the 3 M NaClO$_4$ solution calculated from eqn. (3) assuming 2 mol of disordered water per mole ClO$_4^-$ at 25°C and 3 mol at 75°C. The solid line passing through the filled circles corresponds to the calculated intensity ratio for 2.5 M NaClO$_4$ solution with the same assumptions. The upper broken line is calculated for the 3 M solution assuming 2 mol of disordered water per mole of ClO$_4^-$. The lower dashed line is that observed for the D$_2$O–H$_2$O binary.

a low-intensity high-frequency component was also present. Figure 15 presents the results in the form of eqn. (3). The lower broken line is calculated from eqn. (3) using a value of two for the number of disordered water molecules per perchlorate ion. It is apparent that within experimental error the results are compatible with those obtained earlier[821] where concentration of perchlorate was used as the variable. Deviations between calculated and experimental results appear at higher temperatures and are probably outside of the experimental error. The upper full line was calculated assuming that the number of disordered water molecules associated with each perchlorate ion was a linear function of temperature such that between 25 and 75°C it changed from two to three. Included in Fig. 15 are a few results* (obtained with Hg excitation) from 2.5 M NaClO$_4$ solutions in a D$_2$O–H$_2$O binary solvent. The solid line through the experimental points was calculated in the same way and using the same temperature variation of disordered solvent molecules. It is thus apparent that the variations in intensities of the component bands of O—D oscillators can be reconciled with the observations in HDO by making relatively few assumptions. A consequence of this approach is that if the assumptions are correct regarding the similarity of the spectral characteristics of the nonhydrogen-bonded and disordered water molecules, then it would be expected that at a given temperature the component line parameters (half-width and peak position) would be independent of perchlorate concentration.

It is apparent that within the estimated experimental error the above conditions are satisfied, although in some cases there is a possible tendency for concentration dependence, indicating that the assumptions are not quite correct.* Such small variations would arise if the spectral characteristics of the disordered water species and the nonhydrogen-bonded species differed slightly. Further support for the general approach comes from the observation* that the ternary perchlorate solutions show a temperature isosbestic frequency of 2560 ± 20 cm^{-1} for the O—D region, which is in good agreement with the observed temperature isosbestic points[816,817]† and the concentration isosbestic points reported by Walrafen.[821]

The same sort of approach can be used with sodium meta-periodate (NaIO$_4$) solutions in D$_2$O–H$_2$O mixtures. An investigation* of these solutions was attempted, since it was supposed that this anion would exhibit the same sort of effects on the solvent spectrum as perchlorate but to a

---

* T. H. Lilley and G. E. Walrafen, unpublished observations.
† Also T. H. Lilley and G. E. Walrafen, unpublished observations.

somewhat enhanced degree. This has been verified experimentally. As with the perchlorate investigation, spectra were obtained of the O—D stretching region in ternary salt–D_2O–H_2O mixtures as a function of temperature and salt concentration. The concentration range investigated was somewhat limited because of the lower solubility of this salt in water and generally the investigation was much less exhaustive than the corresponding investigation on perchlorate solutions. Figure 16 is the analogous plot to Fig. 15 and it appears that an effective disordered hydration number of approximately 11 is necessary at the lowest temperature studied (49°C) and this rises to about 15 at 86°C. The data are, however, rather sparse and within the limits of error a single disorder hydration number of approximately 12 will adequately reproduce the experimental results.

It is apparent that solutions containing ClO_4^- and IO_4^- are relatively simple when viewed through the perturbations induced in the intramolecular spectral region (see Addendum, however).

Information on the effects of simple electrolytes on the intramolecular vibration region of the solvent is rather sparse and in some instances of dubious validity. Most of the investigations which have been carried out consider the effects of solutes on the overall vibrational band contours and

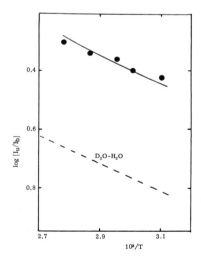

Fig. 16. The dots are for 0.90 M $NaIO_4$ in D_2O–H_2O. The solid line was calculated using eqn. (3) and assuming 11 disordered water molecules per IO_4^- ion at 49°C, rising to 15 at 86°C. The lower line refers to the binary solvent mixture.

discussions have been given on the changes in overall band peak maxima and breadth.[92,500,501,727,810,833] Although such studies are of interest, with the exception of the studies on ClO_4^- and IO_4^-, here no comprehensive investigations have as yet been reported which attempt to explain the observed spectral features in terms of a modified mixture model approach. Preliminary data[821] on ternary solutions of 2 M KCl in 6.2 M D_2O in H_2O and 5 M KBr in 5.5 M D_2O in H_2O have been published. Spectra obtained from the O—D vibrational region indicate quite dissimilar effects to those observed in ClO_4^- and IO_4^- solutions.[821]* No band splitting was observed and visual examination of the spectra indicated that new bands occur which are not present in the binary solvent spectra. It would appear that these bands arise from hydrates of the anions and the comparatively large intensity enhancements indicate that covalent interaction is occurring. The general shapes of the spectra are similar to those observed in earlier studies,[92,500,727,810,833] from which it can be concluded that the iodide ion forms a fairly well-defined hydrate. Comparison of this information with that obtained from the intermolecular region (Section 3.1) indicates that the halide ions form hydrates with adjacent water molecules and at the same time disrupt water structure outside of the primary hydration region.

A recent paper[609] describes the effect of $LiClO_3$ on the vibrational band of liquid water. Measurements were made up to very high concentrations at 25 and 132°C. The spectral band of the solvent decomposed into four components (which disagrees with some other observations) and no new bands occur as solute is added, although changes are observed in the component contributions to the overall intensity. No information is given on component band parameters (other than their relative intensity contributions) but it would appear from the ClO_3^- spectrum that hydration interaction is occurring between ClO_3^- and H_2O and the invariance in component band positions is due to coincidence of hydrate peak frequencies and frequencies arising from water motions.

The effect of salts on the bending region of the water spectrum has not been the subject of many investigations. These effects are comparatively simple. It is found[92,727,833] that the bands are symmetric and remain unchanged in shape and position upon addition of salts. Rather striking intensity changes are observed, however. Figure 17 shows the spectrum of LiI as a function of concentration.[727] Figure 18 compares the relative band intensities for some salt solutions to the intensity of the H_2O bending

---

* Also T. H. Lilley and G. E. Walrafen, unpublished observations.

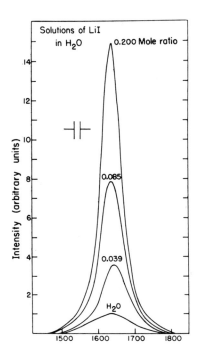

Fig. 17. Raman spectra of the ν_2 bending mode of H_2O in LiI solutions at various concentrations. Reproduced from Ref. 727 with permission of the copyright holders.

band in ternary solutions.[833] The approximately linear effects observed decrease in the order $I^- > Br^- > Cl^- > H_2O > F^-$.[727,833] Figure 19 shows that the same order is observed for the half-widths. It is apparent that little cationic effect is observed, which is similar to the situation in the intramolecular stretching region and the hydrogen bond bending and stretching regions. The observed effects are somewhat puzzling especially with regard to the invariance of band frequency with the addition of solute. Superficially, one would have imagined that the bending mode frequency would be fairly sensitive to the presence of ions, especially anions. The rather large intensity variations for the more polarizable anions indicate considerable covalent character in the water–anion interaction and it is surprising that there is no corresponding perturbation in the band frequency. The experimental results can be rationalized using a similar type of approach as that used earlier (see p. 278) when the effects of salts on the librational region of the water spectrum were discussed. This leads to

$$I/I_W = 1 + [(J_W{}^S - J_W{}^F)/J_W{}^F](n[S]/[W]_F{}^\circ)$$

where I and I_W are the observed band intensities in solution and in water, $J_W{}^F$ and $J_W{}^S$ are the molar intensities of nonhydrating water and hydrating water, $[S]$ and $[W]_F{}^\circ$ are the molar concentrations of solute and pure water, and n denotes the number of water molecules hydrating the solute. It is apparent that the results in Fig. 18 can be represented by an equation of this form. It is not possible to draw many significant conclusions on the data so far available. Further work on the effect of temperature variation on the band spectral parameters and more intense investigations on the effects of solutes are required.

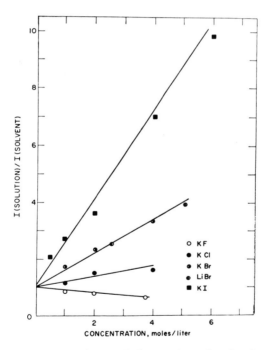

Fig. 18. Intensities (relative to the solvent) of the ν_2 bending mode of H_2O in some electrolyte solutions as a function of concentration. The solvent was 1% D_2O–19% HDO–80% H_2O. Reproduced from Ref. 833 with permission of the copyright holders.

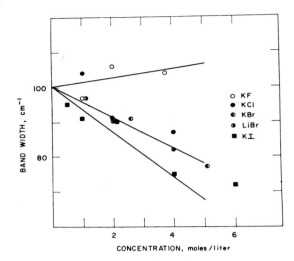

Fig. 19. Half-widths of the ν_2 bending mode of H_2O in some electrolytic solutions as a function of concentration. The solvent was as in Fig. 18. Reproduced from Ref. 833 with permission of the copyright holders.

4. ADDENDUM

A recent paper[822] describes the effects of several ionic solutes on the intramolecular spectrum of ternary mixtures containing H_2O–$D_2O(10\%)$–salt at 25°C. The salts investigated were $NaPF_6$, $NaSbF_6$, $NaClO_4$, $NaBF_4$, $NaReO_4$, $NaIO_4$, $NaNO_3$, $NaClO_3$, $NaSCN$, Na_2SO_4, KI, CsF, and Bu_4NCl. Figure 20 shows the spectra obtained from two $NaPF_6$ solutions at different concentrations. The spectrum obtained from the more concentrated solution shows three features which are new in Raman spectroscopic studies of the solvent species. The O—H stretching region contains two peaks centered at 3590 and 3640 cm^{-1} and in the uncoupled O—D region an intense, narrow peak is evident at 2670 cm^{-1}. Decomposition of the spectra into Gaussian components indicates the presence of six bands under the O—H stretching spectral envelope and three bands in the O—D stretching region. It would appear that what is being observed in any region of the spectrum is a superposition of two spectra, one of which arises from "free" solvent vibrations and the other from vibrations arising from some form of anion–solvent hydrate. It is suggested[822] that the interaction between the anion and solvent peripheral to it is principally electrostatic in nature. The interaction gives rise to relatively narrow O—H vibration bands which occur in the same spectral regions as nonhydrogen-bonded O—H and O—D oscillators. Decomposition of the spectra obtained from solutions containing $NaSbF_6$, $NaClO_4$, and $NaBF_4$ also indicate the presence of six bands in the O—H stretching region. The sharp, partly resolved bands which are apparent in the spectra of $NaPF_6$ solutions are not visually evident. Decomposition of the spectral envelopes in the O—H region indicates that bands with essentially the same characteristics as those obtained from PF_6^- solutions are present. It is suggested that the two relatively sharp bands in the spectral regions 3600–3645 cm^{-1} and 3545–3590 cm^{-1} could arise from weak electrostatic interactions between the anions and O—H vibrators. The presence of a broad band at 3550–3590 cm^{-1},

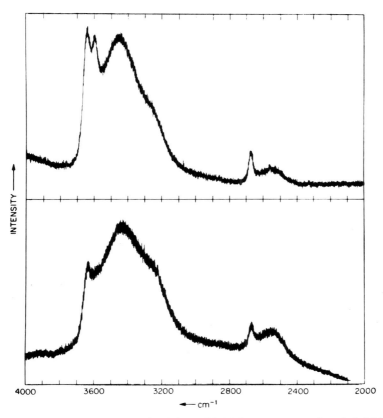

Fig. 20. Raman spectra obtained from 5.3 M (upper spectrum) and 2.6 M (lower spectrum) solutions of $NaPF_6$ in 10 mole % D_2O in H_2O. Reproduced from Ref. 822 with permission of the copyright holders.

common to solutions containing the anions PF_6^-, SbF_6^-, ClO_4^-, and BF_4^-, is assigned to O—H oscillators in the second hydration layers of the anions. In the O—D region only two bands are apparent for the three salts $NaSbF_6$, $NaClO_4$, and $NaBF_4$. The widths of the component high-frequency bands are, however, reduced considerably from the width observed in the pure solvent and it would seem that the interaction between the anion and the O—D species is such that it gives rise to a spectral band which is unresolvable from the band emanating from nonhydrogen-bonded O—D oscillators.

Solutions containing the sodium salts of the anions ReO_4^-, NO_3^-, ClO_3^-, SCN^-, and, to a lesser extent, IO_4^- showed marked spectral differences in the O—H region from those of the solutions containing the anions referred to above. Analyses of the spectral envelopes indicate the absence of the sharp bands previously ascribed to anion–solvent electrostatic interactions. This conclusion is, perhaps, not applicable to the spectrum obtained from the $NaIO_4$ solution. This spectrum was obtained from a solution under pressure at 125°C and although a spectral band in the region 3545–3590 cm^{-1} was absent, a band centered at 3645 cm^{-1} was observed. This latter band was somewhat broader than those obtained for solutions containing anions in the first category, although it is possible that the increase in breadth arises from the increase in temperature. The presence of a broad band in the region 3520–3550 cm^{-1} is believed to arise from O—H oscillators bound in a nonlinear way to the outer atoms of the anion and from O—H$\cdots$O units between the first and second hydration shells. As with the anions SbF_6^-, ClO_4^-, and BF_4^-, the uncoupled O—D region consists of two bands in the ranges 2600–2635 cm^{-1} and 2500–2550 cm^{-1}. The high-frequency bands are somewhat broader than those observed for anions in the first category but again are unresolved from the nonhydrogen-bonded O—D oscillators outside of the primary solvation region.

It was suggested that for the polyatomic anions investigated there may be a correlation between the strength of the acid corresponding to the anion and changes in the Raman spectrum of the solvent. This may well be so, although, as mentioned by Walrafen, the validity of acidic equilibrium constants for acids of moderate strength ($K_A > \sim 10^{-1}$ mol liter^{-1}) are usually very questionable (see also Ref. 140). A subdivision was made between those anions which gave relatively sharp bands between 3600–3645 cm^{-1} and 3545–3590 cm^{-1} in the O—H stretching region and contain splitting in the O—D region and those anions which showed no such spectral features. Anions in the first category were considered to be conjugate to strong or very strong acids and those in the second category to be conjugate to moderately strong acids. It was suggested that within a given class the ratio of the intensity of the band in the nonhydrogen-bonding region to that of the band in the hydrogen-bonding region per mole of added anion could be used to give an order of conjugate acid strength. The correlation is, however, only qualitative and in one example at least breaks down completely. Recent Raman and PMR measurements on perrhenic acid[139] indicate that this acid is as strong as perchloric acid[6] in aqueous solution and consequently it would be expected that the anions ClO_4^- and ReO_4^- would fall in the same category, namely very strong acids. This is not found to be so and it would appear that if a correlation between acid strength and effects on the intramolecular vibration spectrum of the solvent does exist, then it certainly is not a simple one.

The O—D stretching bands in ternary solutions indicate the presence of bands at 2580, 2545, and 2525 cm^{-1} for the anions I$^-$, Br$^-$, and Cl$^-$, respectively. The intensities of these bands increase in the order Cl$^-$ to I$^-$. It is suggested that these bands arise from hydrate species of the anions and the ordering of effects is similar to that observed from studies on the ν_2 bending band. In addition to these bands, components which are ascribed to nonhydrogen-bonded and hydrogen-bonded solvent are observed near 2640 cm^{-1} and 2510 cm^{-1}. The CsF solution shows in addition to these latter two bands, a broad, low-frequency component centered at 2325 cm^{-1} which, it is suggested, arises from a special interaction between the F$^-$ anion and water molecules where the F$^-$ substitutes for H_2O on the water lattice. The O—D region of the Bu_4NCl solutions is quite dissimilar from that obtained from any of the other solutions investigated. Two bands are observed of approximately equal intensity, one centered at 2553 and the other at 2483 cm^{-1}. This latter band is close to the frequency observed in HDO–H_2O ice and it would appear that it arises from a pronounced hydrogen-bonded structure induced by the large cation. (This structure may well be of the clathrate type.)

The application of approaches similar to that given earlier for ClO_4^- and IO_4^- solutions would be of considerable interest. It is apparent, however, that to invoke any process involving the existence of possible solvent subspecies would require a great deal of very careful experimental work for establishing its credibility.

CHAPTER 7

Nuclear Magnetic Relaxation Spectroscopy

H. G. Hertz

Institut für Physikalische Chemie und Elektrochemie
der Universität Karlsruhe
Karlsruhe, Germany

1. INTRODUCTION

As in all liquids, so, too, in water the molecules undergo thermal motion. The self-diffusion coefficient D describes the nature of the erratic translational motion:

$$\langle r^2 \rangle = 6Dt \tag{1}$$

where $\langle r^2 \rangle$ is the mean square displacement of the water molecule after a time t. For water at 25°C, $D = 2.31 \times 10^{-5}$ cm² sec⁻¹,[580] so if we consider a mean square displacement of $\sim a^2$, where a is the diameter of the water molecule, the time necessary for such a displacement is 6×10^{-12} sec. Equation (1) only holds if the time is long enough so that we actually have random motion, that is, the instantaneous velocity of the water molecule must have suffered a large number of changes in magnitude and direction. For times $t \gtrsim 6 \times 10^{-12}$ sec this condition is fulfilled.

But the water molecule does not only perform translational diffusive motion. If we observe the vector connecting the two intermolecular protons, we will of course find that its orientation relative to a laboratory coordinate system changes in time, due to thermal motion. There is no reason why the orientation of this vector should be preserved for a time longer than the time just mentioned, i.e., a time after which the water molecule has suffered an appreciable displacement. Indeed, from dielectric relaxation time measurements we know that the reorientation time of a water molecule is $\sim 10^{-11}$ sec. Again, if the rotational motion is diffusive, it consists of many small rotational steps which are necessary until the molecule, that is, the proton–proton vector or any other intramolecular vector, has made an overall reorientation of 180°, say. The rotational diffusion coefficient D_r of pure water at 25°C is about 7×10^{10} sec^{-1}. Thus, in analogy to eqn. (1) we could write for the mean square angular displacement

$$\langle \Delta \Omega^2 \rangle = 4Dt$$

where the time t has to be properly chosen. But the definition of D_r in water is not so clear as in the translational case, where observation times are much longer (see below). Therefore it is more convenient to describe the rotational motion directly by a time which at the present stage we can only define very loosely: Assume the instantaneous orientation of the vector to be θ; then the mean time after which $\cos \theta$ has changed sign is denoted as the reorientation time τ_r. Alternatively, the mean time after which $\cos \theta \sin \theta$ has changed sign may be denoted as the rotational correlation time τ_c. We have $\tau_c \leq \tau_r$, and for water at 25 °C, $\tau_r \approx 7 \times 10^{-12}$ sec and $\tau_c = 2.5 \times 10^{-12}$ sec. Since the numerical value of τ_c is better known at present, we shall use this as the representative time to describe reorientational motion.

It is well known that at room temperature the coordination number of water is ~ 4.4,[190] this is apparently preserved in liquid water from the open four-coordinate structure of ice. So, if the water molecule suffers a displacement or if it reorients, then for a certain fraction of the time it will be connected with or bound to one or more water neighbors, or in other words, there will exist a certain amount of correlated motion of water molecules. The mean lifetime τ_b of such a bound state in pure water where neighboring molecules perform trajectories which, apart from vibrations, are those of different points of a rigid body is not known precisely; $\tau_b \approx 10^{-12}$ sec seems to be a reasonable estimate (see below).

Now suppose we dissolve ions in water. Then all three time constants just mentioned will change. More than this: The liquid becomes more complex. There appear to be a number of different classes of water molecules: those that have only other water molecules as nearest neighbors, those that have a cation as one of their nearest neighbors, those that have an anion as one of their nearest neighbors, finally, those that have two ions as nearest neighbors, etc. If we now observe the motion of the water molecule, taking into account that they belong to one of the different classes during the entire relevant observation time, then we would find numerical values other than those of pure water for the self-diffusion coefficient D, the rotational correlation time τ_c, and the time of attachment to a neighbor τ_b; the last fact in certain cases implies that the degree of correlated motion of different and equivalent particles is modified in a drastic way. We denote the water molecules of the first class as free water, those of the second class as belonging to the hydration sphere of the cation, and those of the third class as belonging to the hydration sphere of the anions, in this way defining the term hydration sphere of an ion; the fourth and higher classes are less convenient to handle and we shall omit explicit discussion of these configurations.

Starting from the three quantities characteristic of pure water at 25°C,

$$D = 2.31 \times 10^{-5} \text{ cm}^2 \text{ sec}^{-1}, \qquad \tau_c = 2.5 \times 10^{-12} \text{ sec}, \qquad \tau_b \approx 10^{-12} \text{ sec}$$

we shall consider the changes in these quantities when ions are dissolved in water. It may be anticipated that the first two quantities vary by not more than one order of magnitude, but τ_b in bound configurations may change by many orders of magnitude.

In our treatment, as the tool with which the desired information is obtained, we shall exclusively consider the study of nuclear magnetic relaxation times. More precisely, the self-diffusion coefficients are obtained from a modification of the nuclear magnetic relaxation process caused by diffusion of the molecules carrying the nuclear magnets.

Residence times in bound states close to the ion can also be considered in terms of conventional kinetics of fast reactions; additional information concerning rotational correlation times in electrolyte solutions are to be expected in the near future from neutron scattering and light scattering experiments. Tracer diffusion measurements and similar methods are well known as sources for self-diffusion coefficients.

Each of the three dynamical parameters of water, D, τ_c, and τ_b, in a simple electrolyte solution splits into three local parameters characteristic of the respective region: free water and plus and minus hydration spheres. The result thus obtained may be denoted as the microdynamic structure of the electrolyte solution. As interesting as these results may be, however, they give rise to the question: What is the relation between the microdynamic structure of the electrolyte solution and the real structure of the solution? The real structure of a liquid is to be expressed in terms of the depth of effective intermolecular potentials, e.g., of pairs of particles in the liquid, in the presence of all other particles. We shall try to provide a rough translation of our results obtained for the microdynamic structure into the language of the real structure of electrolyte solutions. There have been a number of review articles about NMR studies on electrolyte solutions.[176, 344,352,376]

2. NUCLEAR MAGNETIC RELAXATION

2.1. Principles

2.1.1. Definition of Relaxation Times and Some Experimental Aspects

At equilibrium the nuclear magnetization forms the vector $\mathscr{M}^{\circ}$ = $\{\mathscr{M}_x^{\circ}, \mathscr{M}_y^{\circ}, \mathscr{M}_z^{\circ}\}$,[194,330,640,737] with the components

$$\mathscr{M}_x^{\circ} = \mathscr{M}_y^{\circ} = 0, \qquad \mathscr{M}_z^{\circ} = [\mathscr{N}\gamma_I^2\hbar^2 I(I+1)/3kT]H_0 \qquad (2)$$

where $\mathscr{N}$ is the number of nuclei per cm² with spin I and gyromagnetic ratio γ_I, H_0 is the z component of the static magnetic field, $\mathbf{H} = \{0, 0, H_0\}$, k is the Boltzmann constant, and T is the absolute temperature. Suppose, however, that at a given time $t = 0$ the nuclear magnetization $\mathscr{M} \neq \mathscr{M}^{\circ}$, $\mathscr{M} = \{\mathscr{M}_x, \mathscr{M}_y, \mathscr{M}_z\}$ and that for $t > 0$, H_0 is the only magnetic field present at the sample. Then the vector $\mathscr{M}$ will vary in time and will "move" toward the equilibrium value $\mathscr{M}^{\circ}$. The change in time of the three components is different. Let us first consider the motion of $\mathscr{M}_z$. Very often one observes

$$d\mathscr{M}_z/dt = (1/T_1)(\mathscr{M}_z^{\circ} - \mathscr{M}_z) \qquad (3)$$

or in the integrated form,

$$\mathscr{M}_z = \mathscr{M}_z^{\circ} - [\mathscr{M}_z^{\circ} - \mathscr{M}_z(0)]e^{-t/T_1} \qquad (4)$$

where T_1 is the longitudinal (spin–lattice) relaxation time. $\mathscr{M}_z(0) = \mathscr{M}_z$ at $t = 0$ Equation (3) is a generally valid equation, but there are cases where it is only approximately valid. This is also the case for a number of examples to be described in this chapter. However, for the purpose of our treatment it is permissible to neglect small deviations and to consider eqn. (3) to be a general definition of T_1. We generally call $1/T_1$ the longitudinal relaxation rate. The motion of the components $\mathscr{M}_x$ and $\mathscr{M}_y$ is more complicated at first sight because the transverse component $\mathscr{M}_\perp$ performs a precessional motion about the z axis with (angular) frequency

$$2\pi v = \omega_I = \gamma_I H_0 \qquad (5)$$

ω_I is the Larmor frequency or the nuclear magnetic resonance frequency of the nucleus considered. We give one typical example: For protons $\gamma_I = 2.675 \times 10^4$ rad sec^{-1} G^{-1}; thus the resonance frequency is $v = 42.57$ MHz when the magnetic field is 10 kG.

However, the motion of the transverse component becomes simpler if we relate it to a coordinate system which itself rotates with the resonance frequency, the z axis of the rotating system being fixed and coincident with our former z axis. In this rotating coordinate system one observes

$$d\mathcal{M}_x'/dt = -(1/T_2)\mathcal{M}_x', \qquad d\mathcal{M}_y'/dt = -(1/T_2)\mathcal{M}_y' \qquad (6)$$

T_2 is the transverse (spin–spin) relaxation time and $1/T_2$ is the transverse relaxation rate. The integrated form of eqn. (6) is

$$\mathcal{M}_{x,y}' = \mathcal{M}_{x,y}'(0)e^{-t/T_2} \qquad (7)$$

We now deal briefly with the experimental determination of T_1 and T_2. If the transverse magnetization precesses about the z axis, it induces an ac voltage in a coil which contains the sample and the axis of which is perpendicular to the z axis; of course, the frequency of the ac voltage is equal to ω_I and the amplitude is proportional to $\mathcal{M}_x'$. Assume that at $t = 0$, $\mathcal{M}_z(0) = 0$ and $\mathcal{M}_x = \mathcal{M}_z^\circ$. This situation can be produced by a so-called 90° rf pulse which again has the frequency ω_I and which is of appropriate length so as to rotate $\mathcal{M}_z^\circ$ by 90°[305] into the x axis, say. Then the maximum transverse ac signal induced in the coil is $\sim\mathcal{M}_z^\circ$. From eqn. (4) we see that with $\mathcal{M}_z(0) = 0$

$$\mathcal{M}_z = \mathcal{M}_z^\circ(1 - e^{-t_1/T_1})$$

at a time $t_1 > 0$. At t_1 this gradually redeveloping magnetization is again rotated by 90° through the application of a proper rf pulse; the maximum value of the transverse component is again $\sim\mathcal{M}_z(t)$. Repeating this procedure and observing the gradual regrowth of $\mathcal{M}_z$ as a function of time allows determination of T_1.

The time T_1 in our systems ranges from about 10^{-3} to about 10^2 sec. It is therefore an important property of our nuclear magnetic relaxation times that they are essentially macroscopic times, in contrast to the dielectric relaxation times, which directly represent microscopic times of the order of 10^{-11} sec, as described in the introduction.

In principle T_2 may be measured by direct application of eqn. (7), but field inhomogeneity, diffusion of molecules in the liquid, and electrical damping preclude a simple observation of the transverse magnetization as a function of time, as required for the determination of T_2. More complicated methods utilizing the so-called spin echo and the application of series of rf pulses have been developed to measure T_2; details cannot be given here.[94,565]

We add some remarks on the use of NMR methods to measure the self-diffusion coefficient: Assume that the magnetic field H_0 is not strictly homogeneous but varies over the sample. The molecules in the liquid perform translational diffusion and thus the nucleus "sees" varying magnetic field strengths as time proceeds. This is equivalent to a slight change of the Larmor frequency of the individual diffusing spins, which in turn destroys phase coherence of the precessional motion. The result is an additional decay of the transverse magnetization; it may be shown[782] that the variation of the transverse magnetization with time is now given by

$$\mathscr{M}_x' = \mathscr{M}_x'(0)[\exp(-t/T_2)] \exp(-1/12\gamma_I^2 G^2 D t^3)$$

where G is the field gradient in the z direction. Observation of the spin echo eliminates another (reversible) influence of the field inhomogeneity and allows measurement of D.

There exists another method to measure T_2 which has been applied for all the ^{17}O studies which we shall describe. If eqns. (4) and (6) are combined with the equations governing the motion of the nuclear magnetization under the influence of an rf magnetic field, one arrives at the so-called Bloch equations.[194,330,640,737] The solution of the Bloch equations under steady-state conditions yields a connection between the width of the resonance line and the transverse relaxation rate:

$$1/T_2 = \tfrac{1}{2} \Delta\omega_{1/2} \tag{8}$$

or

$$1/T_2 = \tfrac{1}{2}\gamma \Delta H_{1/2} \tag{9}$$

where $\Delta\omega_{1/2}$ is the linewidth measured as (angular) frequency. Linewidth is understood here as the half-width. Equation (9) relates to the case when the linewidth is measured in gauss. If the linewidth $\Delta\omega$ or ΔH is defined by the separation of points of maximal slope on the absorption curve, the corresponding equations read

$$1/T_2 = \tfrac{1}{2}\sqrt{3}\,\Delta\omega \tag{8a}$$

$$1/T_2 = \tfrac{1}{2}\sqrt{3}\,\gamma\,\Delta H \tag{9a}$$

Linewidth measurements can always be applied when $1/T_2$ is much greater than the linewidth caused by field inhomogeneity. The most reliable technique for linewidth measurements is the side band technique.[328]

2.1.2. Time Correlation Functions; Some Introductory Remarks

In what way are the macroscopic parameters $1/T_1$ and $1/T_2$ connected with the erratic thermal motion in the liquid? An important link in this relationship is the time correlation function of a certain quantity. We describe the time correlation functions as follows.

Assume one molecule selected at random in the liquid is observed. Let the molecule be at $r_0 = r(0)$ relative to an arbitrary origin at time $t = 0$, r_0 being given in polar coordinates: $r_0 = r(0) = \{\theta(0), \phi(0), r(0)\}$. From these three quantities we calculate the function

$$y(0) = Y_2^1(\Omega_0)/r(0)^3 = [1/2r(0)^3](15/2\pi)^{1/2} \sin\theta(0) \cos\theta(0)\, e^{i\phi(0)} \quad (10)$$

where $Y_2^1(\Omega_0) = Y_2^1(\theta(0), \phi(0))$ is one of the normalized spherical harmonics of second order $Y_2^m(\Omega_0)$ $(m = \pm 2, \pm 1, 0)$; we have put $m = +1$. Our choice of the special form of $y(0)$ as given in eqn. (10) has no physical reason so far; we could have chosen any other function. We are going to construct the time correlation function of the function y and it will be seen that our special choice as an example is useful.

As time goes on, the selected molecule undergoes translational diffusion. At time $t_1 > 0$ the same particle is at $r = r(t_1) = \{\theta(t_1), \phi(t_1), r(t_1)\}$. We again form the function, the only difference being that we now take the conjugate complex of Y_2^1, i.e., Y_2^{-1}:

$$y(t_1) = Y_2^{-1}(\Omega)/r(t_1)^3 = [1/2r(t_1)^3](15/2\pi)^{1/2} \sin\theta(t_1) \cos\theta(t_1)\, e^{-i\phi(t_1)} \quad (11)$$

We now obtain the product of the two numbers $y(0)$ and $y(t_1)$:

$$y(0)y(t_1) = Y_2^1(\Omega_0) Y_2^{-1}(\Omega)/r(0)^3 r(t_1)^3 \quad (12)$$

Consider that as the observation leading to this result was being made other observers were following the motion of other molecules. Assume the special result in eqn. (12) was obtained by the first observer,

$$(y(0)y(t_1))_1$$

There are N other observers, N being a large number. All these observers perform the "same" measurement. The "same" means that the time separation between the two observations of r_0 and r is t_1 in all cases. Their results are

$$(y(0)y(t_1))_k, \qquad k = 2, \ldots, N$$

These N observers may follow the motion of different molecules in 1 cm³ of liquid, say, or each observer may select one molecule from one component of an ensemble of N systems, each system being 1 cm³ of liquid. Of course, the N results $(y(0)y(t_1))_k, k = 1, \ldots, N$, differ from one another. However, we may form the mean value of the products obtained by all observers:

$$(1/N) \sum_k (y(0)y(t_1))_k = \overline{y(0)y(t_1)}$$

$$= \overline{Y_2^1(\Omega_0) Y_2^{-1}(\Omega)/r(0)^3 r(t_1)^3} \qquad (13)$$

Pairs of measurements are then repeated for other time separations t and the mean value of the corresponding product $y(0)y(t)$ for these time separations is evaluated. The result is the time correlation function of the function $y \equiv Y_2^1(\Omega)/r^3$:

$$(1/N) \sum_k (y(0)y(t))_k = \overline{y(0)y(t)} \equiv g(t) \qquad (14)$$

We note that the same time correlation function could have been obtained by k repeated measurements of pairs $y(0)y(t)$ along the path of *one* single particle. For sufficiently long total observation time the ensemble average (14) and the time average become equal.

We now rearrange the members in the sum on the left-hand side of eqn. (14). For this purpose we group the r's into classes $r^{(i)}$ defined by finite volume elements Δv_i; thus Δv_i contains all $r^{(i)}$ and the sum of all Δv_i covers the entire volume of the system. Investigating the sum $(1/N) \sum_k (y(0)y(t))_k$, we find that it contains n_i of the $y(0)$ formed from r_0 which fall in the class i, and we find n_{ij} products which connect a $y(0)$ formed from r_0 of class i with a $y(t)$ formed from r of class j. As a consequence we may rewrite eqn. (14) as

$$g(t) = \overline{y(0)y(t)} = (1/N) \sum_i n_i y_i \sum_j (n_{ij}/n_i)y_j \qquad (15)$$

where $y_i \equiv y(r^{(i)})$. At $t = 0$ we must have $n_{ij}/n_i = 1$ for $i = j$, and $n_{ij}/n_i = 0$ for $i \neq j$ since products connect quantities which evolve dynamically from one another. Thus $g(0)$ must be a positive quantity, since our choice of $t = 0$ is arbitrary:

$$g(0) = \overline{|y|^2} \qquad (16a)$$

At sufficiently large t all n_{ij} become equal because the particle may be anywhere in the system with equal probability irrespective of our choice of $\mathbf{r}(0)$, but the mean value of $y(\mathbf{r})$ over the entire system vanishes:

$$g(t) = 0 \quad \text{as} \quad t \to \text{very large} \tag{16b}$$

So our time correlation function decays to zero as the time goes on, i.e., correlations of positions of a molecule at times $t = 0$ and $t > 0$ are gradually lost.

The reasoning leading up to eqns. (16a) and (16b) implied that n_{ij}/n_i is the only quantity in eqn. (15) which is time-dependent. Obviously this ratio expresses the probability of finding a molecule in class j at time t if we know that it belonged to class i at time 0. If this probability changes with time, it can only be the microscopic dynamic processes which cause this change. The dynamic process, here translational diffusion, brings the particle from one class (position) to another class (position). The faster the dynamic process, i.e., the faster the transport from one class to another, the faster is the change of the probability n_{ij}/n_i and the faster the time correlation function decays.

We denote n_{ij}/n_i as the propagator and we see that it is exclusively the propagator which represents the dynamical nature of the systems under study. The time correlation function of $y(\mathbf{r})$ reflects the behavior of the propagator in a particular way; the time correlation function of another arbitrary function $x(\mathbf{r})$ for the same liquid would reflect the same behavior of the propagator in a different way.

In eqn. (15) n_i/N is the probability of finding a vector of class i as the position of the molecule at time $t = 0$. Converting the sum in eqn. (15) to an integral, we have

$$g(t) = \overline{y(0)y(t)} = \int\int y(\mathbf{r}_0)p(\mathbf{r}_0)P(\mathbf{r}_0, \mathbf{r}, t)y(\mathbf{r}) \, d\mathbf{r}_0 \, d\mathbf{r} \tag{17}$$

$dp = p(\mathbf{r}_0) \, d\mathbf{r}_0$ is the probability of finding the molecule in $d\mathbf{r}_0$ at $\mathbf{r}_0$, $P(\mathbf{r}_0, \mathbf{r}, t)$ is again the propagator, and

$$dP = P(\mathbf{r}_0, \mathbf{r}, t) \, d\mathbf{r}$$

is the probability of finding the molecule at time t in $d\mathbf{r}$ at $\mathbf{r}$ if it was at $\mathbf{r}_0$ at time $t = 0$.

So far we have established the connection between the fundamental quantity characterizing the microdynamical behavior in the liquid, namely the propagator $P(\mathbf{r}_0, \mathbf{r}, t)$, and the time correlation function of a certain function, namely $y(\mathbf{r})$. The next step is the conversion of the time dependence given by $g(t)$ to a frequency dependence. This is done by the Fourier transform

$$J(\omega) = \int_{-\infty}^{+\infty} g(t)e^{-i\omega t}\,dt \tag{18}$$

It may be shown that $g(t)$ is an even function as concerns $t < 0$; thus there are no difficulties in evaluating the integral (18). We do not want to discuss any details; we give only two limiting cases: at very low frequencies, where the periodic function in the integrand is practically constant for times during which $g(t)$ decays to zero, $J(0)$ is the time integral of $g(t)$:

$$J(0) = 2 \int_{0}^{\infty} g(t)\,dt \tag{19}$$

At very high frequencies $g(t)$ is practically constant during the time of many oscillations; thus

$$J(\omega) = 0 \quad \text{as} \quad \omega^{-1} \ll \text{decay time of correlation function}$$

Thus $J(\omega)$ is a bell-shaped function centered at $\omega = 0$, the wings extending the further out to higher frequencies the faster the decay of the correlation function; the area under $J(\omega)$ is a constant quantity,

$$g(0) = \overline{|y|^2} = (1/2\pi) \int_{-\infty}^{+\infty} J(\omega)\,d\omega \tag{20}$$

$J(\omega)$ is sometimes denoted the spectral density.

2.1.3. *Relaxation Rates, Two Fundamental Formulas*

The change with time of the components of the nuclear magnetization toward equilibrium as given by eqns. (3) and (6) is caused by the interaction of the relaxing nuclei with the nuclei and electrons in their surroundings. Three kinds of interactions are important for the present discussion: (1) the magnetic dipole–dipole interaction between the nucleus studied and other nuclei or unpaired electrons; (2) the interaction of the electric quadrupole moment of the nucleus studied with the electric field gradient at the nucleus, the field gradient being produced by all kinds of surrounding charges—by electrons in particular; and (3) the scalar interaction or contact interaction between the nucleus studied and an unpaired electron. This interaction comes about when the unpaired electron has a finite probability of being at the nucleus.

The strength of interaction in all cases depends on the positions of the other nuclei or other electrons relative to the nucleus under study (the reference nucleus). Since all these other particles undergo rapid thermal motion in liquids, these interaction energies fluctuate rapidly. The fluctuations cause the magnetic relaxation of the nuclei.

It may be shown[1,709] that the relaxation rates are generally given by linear combination of spectral densities $J(\omega_k)$ at a number of distinct frequencies ω_k:

$$1/T_1 = K[l_{11}J(\omega_1) + l_{12}J(\omega_2) + l_{14}J(\omega_4)] \tag{21}$$

$$1/T_2 = K[l_{20}J(0) + l_{21}J(\omega_1) + l_{22}J(\omega_2) + l_{23}J(\omega_3) + l_{24}J(\omega_4)] \tag{22}$$

The constants K depend on the type of interaction. Let ω_S be the resonance frequency of the other (not relaxing) spin (spin S) with which spin I interacts; then the frequencies occurring as arguments of the spectral densities in eqns. (21) and (22) are the following combinations of the resonance frequencies ω_I and ω_S:

$$\omega_1 = \omega_I - \omega_S, \qquad \omega_2 = \omega_I, \qquad \omega_3 = \omega_S, \qquad \omega_4 = \omega_I + \omega_S$$

The spectral densities $J(\omega_k)$ in eqns. (21) and (22) are the Fourier transforms

$$J(\omega_k) = \int_{-\infty}^{+\infty} g(t)e^{-i\omega_k t}\, dt \tag{23}$$

of the time correlation function

$$g(t) = \overline{f(0)f(t)} \tag{24}$$

where $f(t)$, apart from a constant factor, is the interaction energy mentioned above which fluctuates with time as a consequence of thermal motion in the liquid.

In the case of magnetic dipole–dipole interaction it may be shown that $f(t)$ is proportional to $y(t)$ as given in eqn. (10). For convenience we put

$$f(t) = y(t)$$

This was the reason for choosing $y(t)$ as an example for the construction of a time correlation function of a given function. Only one slight modification is necessary: formerly the vectors $\mathbf{r}_0$ and $\mathbf{r}$ were defined relative to an origin fixed in the laboratory system. We must now understand these vectors as giving the position of a selected particle (spin) relative to the reference spin whose relaxation we are calculating. The z axis is the direction of the static magnetic field H_0. The dipole–dipole interaction contains as well terms $\sim Y_2{}^m$ with $m = \pm 2, 0$; however, for isotropic systems the correlation function does not depend on the m's. For magnetic dipole–dipole interaction between unlike spins I and S the theory yields for the constants in eqn. (21)

$$K = (8\pi/15)\gamma_I{}^2\gamma_S{}^2\hbar^2 S(S + 1) \tag{21a}$$

and

$$l_{11} = \tfrac{1}{2},\ l_{12} = \tfrac{3}{2},\ l_{14} = 3,\ l_{20} = 1,\ l_{21} = \tfrac{1}{4},\ l_{22} = \tfrac{3}{4},\ l_{23} = \tfrac{3}{2},\ l_{24} = \tfrac{3}{2} \tag{21b}$$

For like spins we put $\omega_I = \omega_S$ and the theory gives[1]

$$K = (4\pi/5)\gamma^4\hbar^2 I(I + 1) \tag{21c}$$

$$l_{11} = 0,\ l_{12} = 1,\ l_{14} = 4,\ l_{20} = \tfrac{3}{2},\ l_{21} = 0,\ l_{22} = \tfrac{5}{2},\ l_{23} = 0,\ l_{24} = 1 \tag{21d}$$

Details concerning the scalar and quadrupole interaction will be given below; here we only give the respective constants for reasons of completeness. For scalar interaction with coupling constant A [see eqn. (56)]

$$K = S(S + 1)/3 \tag{21e}$$

$$l_{11} = 1,\quad l_{12} = 0,\quad l_{14} = 0,\quad l_{20} = \tfrac{1}{2},\quad l_{21} = \tfrac{1}{2},\ l_{22} = l_{23} = l_{24} = 0 \tag{21f}$$

Likewise, the nature of the quadrupolar relaxation will be described below. We quote here the constants to be used in eqns. (21) and (22), $\omega_I = \omega_S$. If the spin of the relaxing nucleus $I = 1$, then*

$$K = (3\pi/20)(eQV'_{zz}/\hbar)^2(1 + \tfrac{1}{3}\tilde{\eta}^2) \tag{21g}$$

$$l_{11} = 0,\ l_{12} = 1,\ l_{14} = 4,\ l_{20} = \tfrac{3}{2},\ l_{21} = 0,\ l_{22} = \tfrac{5}{2},\ l_{23} = 0,\ l_{24} = 1;$$
$$V'_{zz} = \partial^2 V/\partial z'^2;\quad V'_{yy} = \partial^2 V/\partial y'^2;\quad V'_{xx} = \partial^2 V/\partial x'^2 \tag{21h}$$
$$\tilde{\eta} = (V'_{xx} - V'_{yy})/V'_{zz}$$

Here V'_{xx}, V'_{yy}, and V'_{zz} are the components of the electrical field gradient tensor in a coordinate system fixed in the molecule. Therefore eqns. (21g) and (21h) describe the intramolecular relaxation rate caused by quadrupole interaction. Q is the electric quadrupole moment of the nucleus. For $I > 1$, T_1 and T_2 are not defined in general,[1] but only for the situation of extreme narrowing; then

$$\frac{1}{T_1} = \frac{1}{T_2} = \frac{3\pi}{20}\cdot\frac{2I + 3}{I^2(2I - 1)}\left(\frac{eQV'_{zz}}{\hbar}\right)^2\left(1 + \frac{\tilde{\eta}^2}{3}\right)J(0) \tag{21i}$$

* Usually in the literature the factor $K/2\pi$ is quoted.

2.2. Proton Relaxation Times and Correlation Times of Water in Paramagnetic Electrolyte Solutions

2.2.1. *Magnetic Dipole–Dipole Interaction*; *Theoretical Background*

The proton relaxation time in water is of the order of seconds. However, in the presence of paramagnetic salts the relaxation time becomes very much shorter. For example at 20 MHz one finds $T_1 \approx 10$ msec for a 10^{-2} M solution of Mn^{2+} at room temperature. Other ions like Ni^{2+} and Co^{2+} are less effective by one order of magnitude. In general the interpretation of the proton relaxation rate in the presence of paramagnetic ions is a difficult matter and not very much work has so far been done to elucidate the structure and molecular motion of the hydration sphere concerned. A simple treatment is possible only in the case where the interaction of the electron spin with the static magnetic field is very much stronger than any other of its interactions with the molecules, the electron spins, and the nuclei of its surroundings. This situation applies particularly well to the ion Mn^{2+}, which is in an S state, if the static magnetic field is not too low.[633,710] The ions Fe^{3+}, V^{2+}, Cr^{3+}, Cu^{2+}, and Gd^{3+} also represent favorable cases. The former condition is not at all fulfilled for ions like Ni^{2+}, Co^{2+}, and Fe^{2+}.[522,584] Thus the treatment to be given is not applicable to solutions of the latter ions.

We select one water proton and one paramagnetic ion at random from the solution. The spins of the proton and of the electron we denote as I and S, respectively. Then according to our presupposition the interaction of this spin pair with the static magnetic field is given by

$$\mathscr{H}_0 = \omega_I I_z + \omega_S S_z \tag{25}$$

where I_z and S_z are the z components of the angular momentum vector (in units of $\hbar$) of the proton and the ion, respectively. The z direction is the direction of the static magnetic field

$$\mathbf{H} = \{0, 0, H_0\}$$

ω_I is the nuclear magnetic resonance frequency:

$$\omega_I = \gamma_I H_0$$

ω_S is the electron paramagnetic resonance frequency:

$$\omega_S = \gamma_S H_0$$

as introduced in the preceding section. The magnetic dipole–dipole interaction between the spins I and S causes the relaxation of the nuclear magnetization. On the other hand, the magnetic dipole–dipole interaction between I and S is so weak as compared with other interactions of S with its surroundings that the relaxation of S is not due to the interactions with the protons. Note that although these interactions are effective in inducing relaxation of S, they are much less than $\omega_S S_z$.

In order to connect the experimental nuclear magnetic relaxation time with microdynamic data, we need the time correlation function of the functions

$$Y_2{}^m(\theta, \phi)/r^3, \qquad m = 0, \pm 1, \pm 2$$

as has been explained in Section 2.1.3. Here θ and ϕ are the polar and azimuthal angles of the vector connecting the proton with the center of the paramagnetic ion, both selected at random; r is the length of this vector. As a consequence of the thermal molecular motion θ, ϕ, and r vary with time, i.e., $\theta = \theta(t)$, $\phi = \phi(t)$, $r = r(t)$; of course $\mathbf{r}(t) = \{\theta(t), \phi(t), r(t)\}$ follows a complicated, erratic path. The time correlation function which we need is, according to eqn. (17),

$$g_m(t) = \overline{[Y_2{}^m(0)/r(0)^3]Y_2^{m*}(t)/r(t)^3}$$

$$= \iint [Y_2^{m*}(\theta, \phi)Y_2{}^m(\theta_0, \phi_0)/r^3 r_0{}^3]p(\mathbf{r}_0)P(\mathbf{r}_0, \mathbf{r}, t)\, d\mathbf{r}_0\, d\mathbf{r} \qquad (26)$$

At $t = 0$ the vector connecting our two spins (the proton spin and the electron spin of the ion) is $\mathbf{r}_0 = \{\theta_0, \phi_0, r_0\}$; at time t this vector has developed to $\mathbf{r} = \{\theta, \phi, r\}$; $p(\mathbf{r}_0)\, d\mathbf{r}_0$ is the probability of finding the center of the ion in $d\mathbf{r}_0$ at $\mathbf{r}_0$ as seen from the proton spin; $P(\mathbf{r}_0, \mathbf{r}, t)\, d\mathbf{r}$ is the probability of finding the ion in $d\mathbf{r}$ at $\mathbf{r}$ relative to spin I at time t if we know that at time $t = 0$ it was at $\mathbf{r}_0$ (relative to spin I).

As we have remarked in Section 2.1.2, the propagator $P(\mathbf{r}_0, \mathbf{r}, t)$ contains in a concentrated, and of course truncated, form all dynamical peculiarities of the system. Thus we have to construct the correct propagator from a proper model of the solutions. Since the interconnection between the propagator and the frequency dependence of $1/T_{1,2}$ occurs via the integral (26), there is no strictly unique relationship between these two quantities. To construct the propagator $P(\mathbf{r}_0, \mathbf{r}, t)$ in question for our paramagnetic electrolyte solution, we proceed as follows: Our first statement is, $P(\mathbf{r}_0, \mathbf{r}, t)$ depends strongly on $\mathbf{r}_0$. We consider a vector $\mathbf{r}_0$ for which $|\mathbf{r}_0| = r_0$ is very small, $r_0 = r_{ion} + r_{H_2O}$, say. In this event the water molecule on which our reference proton sits is a member of the first hydration sphere of the ion. If we ask for the probability $P(\mathbf{r}_0, \mathbf{r}, t)\, d\mathbf{r}$ with $|\mathbf{r}_0| = r_{ion} + r_{H_2O} \equiv R_0$ and an $\mathbf{r}$ for which $|\mathbf{r}| > R_0$, then we shall find $P(R_0, \mathbf{r}, t) = 0$ for times of the order 10^{-11} sec. This is simply a consequence of the fact that the water is bound to the ion in the first hydration layer and that it cannot leave it. In other words we have

$$\int_{|\mathbf{r}| > R_0} P(R_0, \mathbf{r}, t)\, d\mathbf{r} = 1 - e^{-t/\tau_b} \tag{27}$$

where τ_b is the mean life time of a water molecule in the first hydration sphere, e^{-t/τ_b} being the probability that a given water molecule resides in the hydration layer at time t if it was there at time zero. The fact that eqn. (27) vanishes for times $t \approx 10^{-11}$ sec is not a general property of ionic hydration, but experimentally we know that for the ions Mn^{2+}, Fe^{3+}, Cu^{2+}, Cd^{3+}, and Cr^{3+} this condition is fulfilled. It implies that $\tau_b \gg 10^{-11}$ sec. Later we shall discuss cases where $\tau_b \lesssim 10^{-11}$ sec.

We thus see that for a proton which is always in the first hydration sphere R_0 is a constant quantity and for this situation we may write the propagator as

$$P(R_0, \theta_0, \phi_0, R_0, \theta, \phi, t)$$

which now describes the isotropic rotational diffusion of a vector of length R_0. We may also write

$$P(R_0, \theta_0, \phi_0, \theta, \phi, t) = P_{R_0}(\Omega_0, \Omega, t)$$

where Ω_0 and Ω describe the directions of the vector connecting the two spins. We stated that $P_{R_0}(\Omega_0, \Omega, t)$ describes the rotational diffusion process of the vector I–S. Thus one would expect that $P_{R_0}(\Omega_0, \Omega, t)$ is a solution of the rotational diffusion equation

$$\partial P(\Omega, t)/\partial t = D_r^* \, \Delta_r P(\Omega, t) \tag{28}$$

Here D_r^* is the rotational diffusion coefficient and Δ_r is the Laplacian operator on the surface of a sphere. A solution of eqn. (28) with the desired properties is

$$P_{R_0}(\Omega_0, \Omega, t) = \sum_{lk} Y_l^{k*}(\Omega_0) Y_l^k(\Omega) \exp[-tl(l+1)D_r^*] \tag{29}$$

whence at $t = 0$

$$P_{R_0}(\Omega_0, \Omega, t) = \delta(\Omega - \Omega_0) \tag{30}$$

i.e., at any time P is the distribution function of the orientation which is a δ-function at Ω_0 for $t = 0$. On the right-hand side of eqn. (29) the subscript R_0 has been dropped. This has the following meaning: $D_r{}^*$ depends in some manner on the distance R_0. The larger R_0 is, the smaller $D_r{}^*$ will be. However, the precise connection between R_0 and $D_r{}^*$ is unknown; in any case it depends on the surroundings of the hydration complex. Were we allowed to use the ideal hydrodynamic limit, then the surroundings would be fully characterized by the viscosity η and

$$D_r{}^* = kT/8\pi R^3 \eta \tag{31}$$

where R is the radius of the hydration sphere, $R \approx R_0$. But of course the situation is more complex in real systems.

Some remarks are in place to justify the use of the diffusion equation (28). We observe the hydration complex at $t = 0$. Apart from the fact that the vector we selected has an orientation Ω_0, the entire complex possesses an instantaneous angular velocity $\boldsymbol{\omega}(0) = \{\omega_x(0), \omega_y(0), \omega_z(0)\}$. We neglect vibrations of the complex. Later at time t we again observe the complex and find $\boldsymbol{\omega}(t) = \{\omega_x(t), \omega_y(t), \omega_z(t)\}$. If t is sufficiently long, then, e.g., for the x component of $\boldsymbol{\omega}$,

$$\overline{\omega_x(0)\omega_x(t)} = 0$$

but of course

$$\overline{\omega_x(0)^2} = (kT/I)^{1/2} \neq 0$$

where I is the moment of inertia of the hydration complex including the ion. Thus we see that the angular velocity correlation function $\overline{\omega_x(0)\omega_x(t)}$ must decay to zero as the time proceeds:

$$\overline{\omega_x(0)\omega_x(t)} = (kT/I)^{1/2} f(t); \qquad f(t) \to 0 \qquad \text{as } t \text{ becomes long} \tag{32}$$

Now, clearly, as ω_x has any nonzero value, the complex rotates; thus Ω also changes with time. The crucial point is, how large is the change of Ω during a time after which $f(t) \to 0$? For an almost freely rotating system $f(t)$ will be almost constant for a long time before it finally decays toward zero. During this time the system of particles has performed many rotations. But if the damping forces become stronger and stronger, then eqn. (32) will be increasingly well approximated by the relation

$$\overline{\omega_x(0)\omega_x(t)} = (kT/I)^{1/2} e^{-t/\tau_\omega} \tag{33}$$

where τ_ω is the angular velocity correlation time of the hydration complex (a quantity usually not accessible experimentally). Roughly speaking, τ_ω is the time during which the complex follows a dynamically coherent rotational path. For times much greater than τ_ω the path is of random nature. Now, if the time after which Ω has changed appreciably, say by 2π, is very much longer than τ_ω—or better, very much longer than the time after which correlations in angular velocity have died out—then eqn. (28) may be applied. In the situation of a strongly hydrated ion with high electric field strength we may safely assume that the damping of the angular velocity correlation function caused by the intermolecular forces is sufficiently strong so that eqn. (28) may be applied. The experimental results to be presented later give full *a posteriori* justification of this assumption. Of course at very short times ($\sim 10^{-13}$ sec) eqns. (28) and (29) are not valid.

Now let us consider the situation where $|\mathbf{r}_0| = 2r_{H_2O} + R_0$, where r_{H_2O} is the radius of the water molecule. Clearly, the reference water now sits in the second hydration sphere of the ion and the chance that the water moves out of this sphere is greater than for the first hydration sphere. In principle one could write the same propagator for the second hydration sphere as was given by eqn. (29); one would have to take account of the fact that residence times in the second sphere are shorter.[346] However, since computational results which would allow a test of the model by measurement of the frequency dependence of the relaxation rate are not yet available, we shall neglect the particular influences which come from the second coordination sphere of the ion. Rather we consider the other limiting situation: The reference water is far away from the ion; it is a member of the undisturbed water structure of pure water; we assume that the ionic concentration is small enough so that such regions exist. Now, since $\mathbf{r}_0$ is assumed to be large, the translational diffusion equation may be used to supply us with an expression for the propagator $P(\mathbf{r}_0, \mathbf{r}, t)$ analogously to eqn. (28):

$$\partial P(\mathbf{r}, t)/\partial t = D\, \varDelta P(\mathbf{r}, t) \tag{34}$$

We have supposed that at $t = 0$ the spin S is at $\mathbf{r}_0$; thus we take the solution of eqn. (34) which is a δ-function $\delta(\mathbf{r} - \mathbf{r}_0)$ at $t = 0$:

$$P_{tr}(\mathbf{r}_0, \mathbf{r}, t) = (4\pi D_{ion}t)^{-3/2} \exp[-(\mathbf{r} - \mathbf{r}_0)^2/4D_{ion}t] \tag{35}$$

where D_{ion} is the self-diffusion coefficient of the ion. The subscript tr means "translational motion." But so far eqn. (35) is valid relative to an origin defining $\mathbf{r}_0$ which is fixed in space. If $\mathbf{r}_0$ refers to an origin fixed in the water molecule, the probability of finding the ion at $\mathbf{r}$ in $d\mathbf{r}$, at a given time, changes faster when the water diffuses as well. For large distances $\mathbf{r}_0$ and $\mathbf{r}$ we may modify eqn. (35) to the form

$$P_{tr}(\mathbf{r}_0, \mathbf{r}, t) = [4\pi(D_{ion} + D)t]^{-3/2} \exp[-(\mathbf{r} - \mathbf{r}_0)^2/4(D + D_{ion})t] \quad (36)$$

where now D denotes the self-diffusion coefficient of the water molecule. Again at small distances $\mathbf{r}_0$ and $\mathbf{r}$ there may appear complication as the motions of the two particles are not strictly uncorrelated and the use of the simple sum of the two self-diffusion coefficients may be questionable. Equation (29) has been written as a sum of exponentials; it is convenient for further application to write eqn. (36) in the analogous form

$$P_{tr}(\mathbf{r}_0, \mathbf{r}, t) = (2\pi)^{-3} \int \{\exp[-(D + D_{ion})t\varrho^2]\} \exp[i\boldsymbol{\varrho} \cdot (\mathbf{r} - \mathbf{r}_0)] \, d\boldsymbol{\varrho} \quad (36a)$$

The integration has to be extended over the whole vector space $\boldsymbol{\rho}$. We see that in eqn. (36a) the propagator is formulated as a Fourier transform of an exponential

$$\exp[-(D + D_{ion})t\varrho^2]$$

We summarize our final result for the propagator of the motion of the vector $I\!-\!S$ in the electrolyte solution as

$$
\begin{array}{ll}
P_{R_0}(\Omega_0, \Omega, t) & \text{for} \quad |\mathbf{r}_0| = R_0 \\
P_{tr}(\mathbf{r}_0, \mathbf{r}, t) & \text{for} \quad |\mathbf{r}_0| > R_0
\end{array}
\quad (37)
$$

This propagator contains only two simple types of dynamics: (a) (isotropic) rotational motion in the hydration complex and (b) relative translational motion of water and paramagnetic ion.

The propagator does *not* contain the following dynamical details: motion in the second coordination sphere, capture of the water molecule by the first or second coordination sphere, release of the water molecule from one of these spheres, anisotropic rotational motion of the vector $I\!-\!S$ in the hydration sphere, which also leaves $|\mathbf{r}_0| = $ const (see below).

Next, with the propagator thus developed we form the correlation function eqn. (26). First we seek the quantity $p(\mathbf{r}_0)$ and according to the "two-region" form of the propagator (37) we ask, what is the probability of finding an interparticle vector $\mathbf{r}_0$ with $|\mathbf{r}_0| = R_0$? The answer is: x_h^+, the mole fraction of water molecules that are members of the first hydration sphere of an ion, divided by N_{ion},

$$x_h^+ = n_h^+ c^*/55.5 \quad (38)$$

where n_h^+ is the hydration number and c^* is the ion concentration in mol $(kg\ H_2O)^{-1}$. N_{ion} is the number of (paramagnetic) ions present in the solution. The factor N_{ion}^{-1} appears because the vector r_0 connects the proton with a specified ion. Thus we obtain for the probability $p(r_0)$, at r_0 with $|\ r_0\ | = R_0$, as occurs in eqn. (26),

$$\{p(r_0)\}_{|r|=R_0} = (1/4\pi N_{ion})n_h^+c^*/55.5 \tag{39}$$

This equation implies that the concentration c^* is sufficiently low so that a given water molecule never belongs to two hydration spheres (of paramagnetic cations) at the same time.

We have to take account of the requirement that the water molecule is a member of the first hydration sphere not for an infinitely long time, but only for the mean time τ_b. To do this, we multiply the rotational propagator eqn. (29) by a factor

$$P_{R_0,\tau_b}(\Omega_0, \Omega, t) = e^{-t/\tau_b}P_{R_0}(\Omega_0, \Omega, t)$$

The probability of finding a specified cation in dr_0 at r_0 with $|\ r_0\ | > R_0$ is

$$dp = dr_0/V$$

that is, for $|\ r_0\ | > R_0$ we have simply

$$\{p(r_0)\}_{|r_0|>R_0} = 1/V$$

where V is the volume of the liquid. With these results the time correlation function eqn. (26) is

$$g_m(t) = \frac{n_h^+c^*e^{-t/\tau_b}}{4\pi 55,5 R_0^6 N_{ion}} \int Y_2^{m*}(\Omega_0)Y_2^{m}(\Omega)P_{R_0}(\Omega_0, \Omega, t)\ d\Omega_0\ d\Omega$$

$$+ \frac{1}{V}\int \frac{Y_2^{m*}(\Omega_0)Y_2^{m}(\Omega)}{r_0^3 r^3}\ P_{tr}(r_0, r, t)\ dr_0\ dr \tag{40}$$

We note that this is the time correlation function which corresponds to one spin pair $I–S$. Of course, the correlation function appearing in eqns. (21) and (22) regards the total interaction of the spin I with all other spins S. So the right-hand side of eqn. (40) has to be multiplied by a factor N_{ion}, the number of ions present in the solution. All the ions are supposed to move independently, which gives the justification for this multiplication by N_{ion}.

One further comment is necessary. The correlation function as written in eqn. (40) by virtue of the two propagators $P_{R_0}(\Omega_0, \Omega, t)$ and $P_{tr}(\mathbf{r}_0, \mathbf{r}, t)$ contains information about two microdynamic quantities: D_r^* and $D+D_{ion}$, both corresponding to the respective region of the solution, namely hydration water and free water. The experimental quantity, however, according to eqn. (21) is a single quantity, the longitudinal relaxation time. (The transverse relaxation time does not contain any additional information in this situation.) So one sees that a condition for T_1 to give information regarding both the regions of the solution is that τ_b, the mean lifetime of the proton in the hydration sphere, is very much shorter than T_1 itself. Only in this way can T_1 tell us something of the average environment the protons sees during the time the magnetization approaches the equilibrium value. If $\tau_b > T_1$, then we would just have two nuclear magnetic relaxation times, one for the hydration sphere and one for the free state and the change of the magnetization in time has to be described by a superposition of two exponentials. We shall reconsider this point below.

The rotational part of the correlation function eqn. (40) can be easily integrated. One has to apply the orthogonality relations of the spherical harmonics and finds after insertion of eqn. (29) in eqn. (40) and multiplication by N_{ion} that

$$g_m^{(r)}(t) = (n_h^+ c^*/4\pi 55.5 R_0^6) \exp\{-t[(1/\tau_c^*) + (1/\tau_b)]\} \qquad (41)$$

where we have written $g_m(t)$ as

$$g_m(t) = g_m^{(r)}(t) + g_m^{(tr)}(t) \qquad (42)$$

The rotational correlation time τ_c^* is given by the relation

$$1/\tau_c^* = 6D_r^* \qquad (43)$$

and is the rotational correlation time of the vector connecting a proton in the first hydration sphere with the center of the ion which forms the hydration sphere. More generally, one may write

$$1/\tau_c^* = l(l + 1)D_r^*; \qquad l = 2 \qquad (44a)$$

which indicates that we are concerned with the correlation time of the second-order spherical harmonics determined by the orientation of the vector $I \to S$. For the reorientation time τ_r^* which describes the decay of correlation of the vector components themselves (i.e., of the first-order spherical harmonics) one has to put $l = 1$; thus

$$\tau_r^* = 3\tau_c^* \qquad (44b)$$

The introduction of one rotational diffusion coefficient D_r* of the hydration complex implies that the latter is a rigid body performing isotropic rotational diffusion. The word rigid has no meaning in itself. It is obvious from this treatment that rigid must mean that all point distances within the complex are constant for a time much longer than τ_c*. Vibrational motion need not be considered. Elsewhere this type of rigidity of an intermolecular aggregate has been denoted by θ-rigid.[349] Of course, it can occur that the hydration complex is not θ-rigid. The most probable type of motion is a rotation of the water molecules about the O—M axis, where O is the oxygen of the water and M the metal ion. In this event our vector I–S rotates on a cone about the O–M axis. Considering one water–ion pair, reorientation times water and M the metal ion. In this event our vector I–S° rotates on a cone about the O–M axis. Considering one water–ion pair, reorientation times of different vectors belonging to this pair are different, so that the rotational motion becomes anisotropic. Equation (41) would only be valid for the O–M axis itself, but if the spin–spin vector does not coincide with the O–M direction, the correlation function no longer decays exponentially. Rather it is a sum of exponentials[797,811,843,844,846]*

$$g_m{}^r(t) = \{\tfrac{1}{4}(3\cos^2\theta - 1)^2 \exp(-t/\tau_c*)$$
$$+\tfrac{3}{4}(\sin^2 2\theta)\exp\{-t[(1/\tau_c*) + (1/\tau_i)]\}$$
$$+\tfrac{3}{4}(\sin^4\theta)\exp\{-t[(1/\tau_c*) + (4/\tau_i)]\}\}$$
$$\times n_h c*[\exp(-t/\tau_b)]/4\pi 55.5 R_0{}^6 \qquad (45)$$

if the rotational diffusion of the axis O–M and that about this axis are assumed to be independent. Here τ_c* is the orientational correlation time of the axis O–M and τ_i is the time constant for the stochastic rotation process about the axis O—M. As $\tau_i \to \infty$ eqn. (45) reduces to eqn. (41). In the following discussions for paramagnetic ions only the case $\tau_i \to \infty$ shall be considered; the extension to finite τ_i will be obvious.

The integration of the translational part of eqn. (40) is more complicated; using eqn. (36a) and multiplying by N_{ion} yields

$$g_m{}^{(tr)}(t) = \frac{N_{ion}}{Va^3} \int_0^\infty [J_{3/2}(\varrho a)]^2 \exp(-2\bar{D}\varrho^2 t)\frac{d\varrho}{\varrho}, \qquad \bar{D} = \tfrac{1}{2}(D+D_{ion}) \qquad (46)$$

$J_{3/2}$ is the Bessel function of order 3/2 and a is the closest distance of ap-

---

* We omit here the presentation of the propagator, being aware that we are thereby conflicting with a logical manner of treatment.

proach between the proton and the ion if we only count those configurations where the proton is not in the first hydration sphere. $J_{3/2}(x)$ is a damped oscillatory function approaching zero as $x \to \infty$, $J_{3/2}(x) = 0$ for $x = 0$.

The important result from eqns. (41) and (46) is the fact that $g_m(t)$ is a sum of exponentials. There is one single term, as given by eqn. (41) [or a finite number if eqn. (45) is taken] and an infinite sum over a continuous distribution of exponentials $\exp(-2\bar{D}\varrho^2 t)$ represented by eqn. (46). The time constants for these exponentials range from zero towards infinity. Large time constants $(2\bar{D}\varrho^2)^{-1}$ correspond to pair partners being very far apart, so that the relative motion is extremely slow, and vice versa. Since the rotational part of the correlation function is proportional to $1/R_0^6$ and the translational part is proportional to $1/a^3$, the former term dominates in the case of small ions because a is about equal to twice the radius of the water molecule. Thus for strongly hydrated paramagnetic ions the rotational part is more important and the translational part amounts to $\sim 10\%$ of the rotational part for $N_{\text{ion}}/V = 10^{-3}$ mol liter^{-1}.

The Fourier transformation of an exponential is a Lorentzian function, so that we find for the spectral density of the rotational part (see Fig. 1)

$$J_{\text{rot}}(\omega) = (n_h^+ c^*/4\pi 55.5 R_0^6) 2\tau_c^{**}/[1 + \omega^2(\tau_c^{**})^2] \qquad (47)$$

where we have set

$$1/\tau_c^{**} = (1/\tau_c^*) + (1/\tau_b)$$

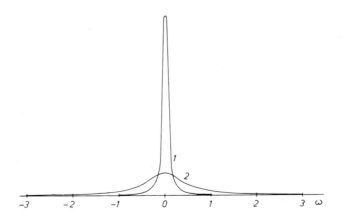

Fig. 1. Spectral densities for two different correlation times τ_c. τ_c of curve 2 is 1/8 of the correlation time of curve 1. Frequency is given in arbitrary units.

and for the translational part

$$J_{\text{tr}}(\omega) = \frac{\tilde{c}}{\bar{D}a^3} \int_0^\infty [J_{3/2}(\varrho a)]^2 \frac{1}{\varrho^3} \frac{1}{1 + (\omega^2/4\bar{D}^2\varrho^4)} \, d\varrho \tag{48}$$

$$= \frac{\tilde{c}}{\bar{D}a^3} \int_0^\infty [J_{3/2}(\varrho a)]^2 \frac{1}{\varrho^3} \frac{d\varrho}{1 + (\omega^2\tau_t^2/a^4\varrho^4)}$$

$$= \frac{\tilde{c}}{\bar{D}a} \int_0^\infty [J_{3/2}(\varrho a)]^2 \frac{u \, du}{u^4 + \omega^2\tau_t^2} \tag{49}$$

where $u = \varrho a$ and the abbreviation $\tau_t = a^2/2\bar{D}$ have been introduced in the second line ($\tilde{c} = N_{\text{ion}}/V$). We note that sometimes τ_t (or $\tau_t/3$) is denoted as the translational correlation time; this may be useful but it should be kept in mind that actually we have a distribution of correlation times $(2\bar{D}\varrho^2)^{-1}$, the distribution being given by $[J_{3/2}(a\varrho)]^2/\varrho$, and τ_t is the correlation time at $\varrho = 1/a$.

We do not give the details of the integration over ϱ in eqn. (49); these may be found in Refs. 383, 632, and 781; the result is

$$J_{\text{tr}}(\omega) = (4\tilde{c}/15a^3)\tau_t \bar{f}(\omega\tau_t) \tag{50}$$

The function $\bar{f}$ has the properties $\bar{f}(\omega\tau_t) \to 1$ for $\omega \to 0$ and $\bar{f}(\omega\tau_t) \to 0$ as $\omega \to \infty$, which is the same limiting behavior as that of the Lorentzian curve

$$1/(1 + \omega^2\tau_c^2)$$

appearing in eqn. (47). But apart from this coincidence at $\omega \to 0$, $\omega = \pm\infty$, $\bar{f}(\omega\tau_t)$ is "sharper" around $\omega = 0$, i.e., there is a stronger decrease of $\bar{f}(\omega\tau_t)$ for small ω, so that we have $\bar{f}(\omega\tau_t) < (1 + \omega^2\tau_t^2)^{-1}$ at $\omega\tau_t \ll 1$; but on the other hand, the wings of $\bar{f}(\omega\tau_t)$ for $\omega\tau_t \gg 1$ are extended further out than is found for the Lorentzian curve.

Figure 1 shows two examples of spectral densities representing Lorentzian behavior.

The total spectral density is [see eqn. (42)]

$$J(\omega) = J_{\text{rot}}(\omega) + J_{\text{tr}}(\omega) \tag{50a}$$

According to eqns. (21) and (22), $1/T_1$ and $1/T_2$ are proportional to linear combinations of spectral densities at various frequencies. We see from eqn. (21) that for the computation of $1/T_1$ we need the spectral densities at the resonance frequency of the proton, ω_I, and at the two frequencies $\omega_I - \omega_S$ and $\omega_I + \omega_S$. For $1/T_2$ we need the Fourier transforms of the correlation function at frequencies 0, ω_I, ω_S, $\omega_I - \omega_S$, and $\omega_I + \omega_S$. But the situation simplifies because $\omega_S \gg \omega_I$, e.g., for Mn^{2+}, $\omega_S \approx 700\omega_I$. Furthermore, we anticipate that $\tau_c^{**} \approx 10^{-11}$ sec, whereas $\omega_I \approx 10^9$ sec^{-1}, i.e., $\omega_I\tau_c^{**} \ll 1$, which gives a further simplification. Taking account of these facts, one finds after introduction of eqns. (47) and (50) into eqn. (50a) and using the results of eqns. (21), (22), (21a), and (21b)*

$$\left(\frac{1}{T_1}\right)_p = \left(\frac{1}{T_1}\right)_{tr} + \frac{2}{5}\frac{\gamma_I^2\gamma_S^2\hbar^2 S(S+1)c^*n_h^+}{55.5R_0^6}\left[\tau_c^{**} + \frac{(7/3)\tau_c^{**}}{1+\omega_S^2(\tau_c^{**})^2}\right] \quad (51)$$

$$\left(\frac{1}{T_2}\right)_p = \left(\frac{1}{T_2}\right)_{tr} + \frac{1}{15}\frac{\gamma_I^2\gamma_S^2\hbar^2 S(S+1)c^*n_h^+}{55.5R_0^6}\left[7\tau_c^{**} + \frac{13\tau_c^{**}}{1+\omega_S^2(\tau_c^{**})^2}\right] \quad (52)$$

with

$$(1/T_1)_{tr} = (16/75)\pi\gamma_I^2\gamma_S^2\hbar^2 S(S+1)(\tilde{c}/a^3)[\tau_t + (7/3)\tau_t\tilde{f}(\omega_S\tau_t)] \quad (53)$$

$$(1/T_2)_{tr} = (8/225)\pi\gamma_I^2\gamma_S^2\hbar^2 S(S+1)(\tilde{c}/a^3)[7\tau_t + 13\tau_t\tilde{f}(\omega_S\tau_t)] \quad (54)$$

The important result which is contained in eqns. (51)–(54) is the following: $(1/T_1)_p$ and $(1/T_2)_p$ both depend on the magnetic field strength. The field dependence is via the electron spin resonance frequency or Larmor frequency ω_S which in turn related to the static magnetic field by

$$\omega_S = \gamma_S H_0$$

The Larmor frequency of the electron spin is of the order of the "rate" of the molecular motions which occur in the liquid and thus indirectly causes field and frequency dependences of the proton relaxation rate. The Larmor frequency of the proton spin is always much lower than the molecular reorientation rates and the rates of other molecular motions which contribute to the relaxation. Thus, relative to the Larmor frequency of the nuclei we have always the situation of "extreme narrowing," $\tau_c\omega_I \ll 1$. According to eqns. (51)–(54), the relaxation rates $(1/T_1)_p$ and $(1/T_2)_p$ decrease with increasing frequency. This is the behavior given by the bell-shaped curves shown in Fig. 1. According to eqns. (51)–(54), via $(1/T_1)_p$ and $(1/T_2)_p$ we "see" two contributions to the bell-shaped function $J(\omega)$: One is always at the extreme center, $\omega = 0$, and the other varies along the "wings" according to the frequency ω_S which is determined by the magnetic field H_0.

* The index p stands for paramagnetic dipole–dipole interaction.

The ratio of $(1/T_1)_p$ at very low and very high frequencies is 10/3 in the limiting case. The corresponding ratio for $(1/T_2)_p$ is 20/7. At low frequencies we have $(T_1)_p = (T_2)_p$ from eqns. (51)–(54); the limiting value of $(T_1/T_2)_p$ at high frequencies is 7/6.

For the rotational contribution to $(1/T_1)_p$ and $(1/T_2)_p$ which is written explicitly in eqns. (51) and (52) the drop of the relaxation rate with increasing frequency is more localized around the "critical" frequency ω_c, with $\omega_c \tau_c^{**} = 1$. For the translational contribution the decrease in $(1/T_1)_p$ and $(1/T_2)_p$ is more spread out over a wider frequency range. Still, at very low and very high frequencies both relaxation rates are practically independent of the frequency. The frequency range where $(1/T_1)_p$ and $(1/T_2)_p$ vary markedly with frequency is called the "dispersion range." As already mentioned, the rotational contribution is the dominant part of the relaxation rate for strongly hydrated ions. The rigorous separation of the translational contribution from the total relaxation rate is a difficult matter. This is understandable, because the propagator implicit in this contribution does not take account of partly correlated and rotational motions in the second hydration sphere of the paramagnetic ion and other dynamical details. In spite of this, eqns. (51)–(52) and the experimental results concerning these equations (to be described below) are of great importance. They are the only cases where we have direct information about τ_c^* ($\tau_p \gg \tau_c^*$, thus $\tau_c^{**} = \tau_c^*$), the rotational correlation time of relatively small molecular aggregates, i.e., hydration complexes in electrolyte solution. Here, by studying the dispersion behavior of the relaxation rate, the correlation time can be determined in the most direct way, within the limit of the accuracy given by the concept of separate rotational and translational propagators. With the knowledge of τ_c^* the interaction factor in front of the brackets in eqns. (51)–(54) can also be found. If it were not for the term containing $\omega_S \tau_c^{**}$ in eqns. (51)–(54), *independent* measurement of both these quantities would be impossible. In this latter situation, which happens for all highly fluid diamagnetic liquids, we would only measure the product of τ_c^{**} and a quantity which characterizes the strength of nuclear interaction, here $1/R_0^6$, and the separation of these factors may cause difficulties or at least may not be sufficiently certain in some cases. Still, much of the present chapter is devoted to diamagnetic liquids, and as we shall see the results for paramagnetic electrolyte solutions give favorable support to the data obtained for these liquids.

It should be mentioned here that eqns. (51)–(54) are not yet the complete expressions to be applied in many practical situations. Two additions are necessary:

1. According to our formulas, $(1/T_1)_p \to 0$ if c^* and $\tilde{c} \to 0$. Of course for very dilute solutions the relaxation times do not become infinitely long. In fact, more precisely, the equations have to be rewritten in the form

$$1/T_1 = (1/T_1)_{\mathrm{H_2O}} + (1/T_1)_p, \qquad 1/T_2 = (1/T_2)_{\mathrm{H_2O}} + (1/T_2)_p \qquad (55)$$

where $(1/T_1)_{\mathrm{H_2O}}$ and $(1/T_2)_{\mathrm{H_2O}}$ are the relaxation rates caused by the magnetic dipole–dipole interaction between the water protons. Now

$$1/T_1 = (1/T_1)^{\circ}_{\mathrm{H_2O}} \qquad \text{and} \qquad 1/T_2 = (1/T_2)^{\circ}_{\mathrm{H_2O}} \qquad \text{as} \quad c^* \to 0$$

and $(1/T_{1,2})^{\circ}_{\mathrm{H_2O}}$ is the relaxation rate of pure water. At 25°C, $(1/T_1)^{\circ}_{\mathrm{H_2O}} \approx (1/T_2)^{\circ}_{\mathrm{H_2O}} = 0.28 \ \mathrm{sec}^{-1}$; as the ionic concentration increases the increase of $(1/T_{1,2})_{\mathrm{H_2O}}$ is usually less than 100% per mole per liter. On the other hand, $(1/T_1)_p c^{*-1}$ is of the order $10^4 \ \mathrm{sec}^{-1} \ \mathrm{mol}^{-1} \ (\mathrm{kg} \ \mathrm{H_2O})$. Thus for concentrations $c^* \gtrsim 10^{-3} \ m$ the diamagnetic contribution to the relaxation rate may be neglected.

2. So far we considered two interactions of the water protons: the magnetic dipole–dipole interaction with the unpaired electrons on the ion and the magnetic dipole–dipole interaction with other water protons. Very often, however, there is a third important interaction which also causes relaxation. This is the so-called scalar interaction which we introduced briefly in Section 2.1.4. The name derives from the form of the interaction energy, which is proportional to the scalar product of the two angular momenta:

$$\hbar A \mathbf{I} \cdot \mathbf{S}$$

where A is the coupling constant characterizing the strength of interaction. We shall discuss details of this relaxation mechanism below. It is useful first to describe two systems where relaxation is entirely due to magnetic dipole–dipole interaction.

2.2.2. Special Cases: Cu^{2+} and Gd^{3+}

There are only two ions in aqueous solution, Cu^{2+} and Gd^{3+}, for which eqns. (51)–(54) represent the correct relaxation behavior, i.e., dipole–dipole interaction is the only source of relaxation. Figure 2 gives the proton relaxation rates in a $6.65 \times 10^{-2} \ M$ solution of $CuCl_2$ at three different temperatures.[326] The arrows indicate the center of the dispersion range; here $\omega_S \tau_c^{**} = 1$. Similar results have also been obtained by Morgan and

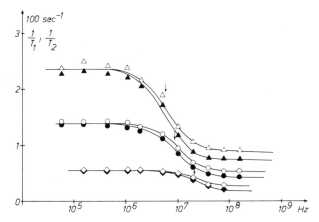

Fig. 2. $1/T_2$ and $1/T_1$ (filled symbols) of water protons in 6.65 $\times 10^{-2}\ M$ aqueous $CuCl_2$ solution as a function of the proton resonance frequency. Triangles, 5°C; circles, 25°C; diamonds, 65°C. The arrows indicate centers of the dispersion range.[326]

Nolle[584] and Bernheim et al.[49] (see also Ref. 60). The experimental data have been evaluated by using only the rotational contribution to the relaxation rate; the results are listed in Table I. Recently, Held and Noack[334] have attempted to fit the experimental data to the complete system of eqns. (51)–(54) including the translational contribution. The result is $\tau_c^{**} = 2.95 \times 10^{-11}$ sec at 25°C, but the translational correlation time $\tau_t' = a^2/6D = 2.6 \times 10^{-10}$ sec, which the authors find is much too long, because the effective self-diffusion coefficient would be much too small. We see, therefore, that the detailed understanding of the form of the dispersion curve is still an open question. Of course, the accuracy of the measurements presents another problem.

For Gd^{3+} the experimental data cover a somewhat less extended frequency range and the data are comparatively old, but the authors[584] find the ratio of the low- to high-frequency relaxation rates is somewhat less than 10/3, the value expected according to eqn. (51). A correlation time $\tau_c^{**} = 2 \times 10^{-11}$ sec is derived from the experimental data.[60]

2.2.3. Relaxation by Scalar Interaction

As has already been mentioned, the scalar interaction between two magnetic dipoles $\gamma_N \hbar \mathbf{I}$ and $\gamma_S \hbar \mathbf{S}$ may be written as

$$\mathscr{H}_{sc} = \hbar A \mathbf{I} \cdot \mathbf{S} \tag{56}$$

TABLE I. Orientational Correlation Times τ_c^* (at 25°C), Lifetimes τ_b, and Scalar Coupling Constants A for the Hydration Spheres of a Number of Transition Metal Ions[a]

Ion	Nucleus	$\tau_c^* \times 10^{11}$, sec	$E_A^{\ddagger}$, kcal mol^{-1}	τ_b,[b] sec	E_b, kcal mol^{-1}	E_b^*, kcal mol^{-1}	A, MHz	Ref.
Ti^{3+}	$H_2^{17}O$	—	—	1.0×10^{-5}	—	6.2	4.4	101
V^{2+}	H_2O	2.1	4.3	—	—	—	2.0	60
V^{2+}	$H_2^{17}O$	—	—	1.1×10^{-2}	—	16.4	—	610
VO^{2+}	$H_2^{17}O$	—	—	2×10^{-3}	—	13.7	3.8	680, 868
Cr^{3+}	H_2O	4.5	3.0(?)	1.8×10^{-6}	3.5(?)	—	1.3	326
Cr^{3+}	H_2O	8.3(?)	—	4.5×10^{-6}	10	—	2.0	60
Cr^{3+}	$H_2^{17}O$	—	—	Very long	—	—	—	122, 610
Mn^{2+}	H_2O	3.1	4.5	2.7×10^{-8}	8.4	—	1.0	49
Mn^{2+}	H_2O	3.1	4.3	2.5×10^{-8}	8.1	—	1.0	60
Mn^{2+}	H_2O	3.2	4.3	—	—	—	—	632
Mn^{2+}	$H_2^{17}O$	—	—	2.0×10^{-8}	—	—	0.82	326
Fe^{3+}	$H_2^{17}O$	—	—	3.2×10^{-8}	—	8.1	9.2	767
Fe^{3+}	$H_2^{17}O$	—	—	3.3×10^{-4}[c]	9.5	8.9	>0.047	263
Fe^{2+}	$H_2^{17}O$	—	—	3.1×10^{-7}	—	7.7	11.9	767
Co^{2+}	$H_2^{17}O$	—	—	7.4×10^{-7}	—	8.0	14.8	767
Co^{2+}	$H_2^{17}O$	—	—	4.2×10^{-7}[c]	—	10.4	17.0	100
Ni^{2+}	$H_2^{17}O$	—	—	3.7×10^{-5}	—	11.6	43.0	767
Cu^{2+}	H_2O	2.6 (2.95?)	4.4	—	—	—	—	326
Cu^{2+}	$H_2^{17}O$	—	—	5×10^{-9}	~4	5	4	767
Cu^{2+}	$H_2^{17}O$	—	—	1.4×10^{-10}[d]	—	—	50	512

[a] Definition of activation energy: 1) $\tau_c^* = \text{const. exp. } E_A^+/RT$, 2) $\tau_b = \text{const. exp } E_b/RT$, 3) $\tau_b = \text{const. } T^{-1} \exp E_b^*/RT$.
[b] At 298°K if not indicated otherwise.
[c] At 300°K.
[d] At 297°K.

The coupling constant A is given by the expression

$$A(\mathbf{r}) = (8\pi/3)\hbar \, | \, \gamma_N \, | \, | \, \gamma_S \, | \, | \, \psi(\mathbf{r}) \, |^2 \qquad (57)$$

$| \, \psi(\mathbf{r}) \, |^2$ is the probability that one of the unpaired electrons of the ion is at the nucleus to be considered when the ion center has the position $\mathbf{r}$ relative to the nucleus. We write γ_N as the gyromagnetic ratio of the nucleus here because the above formula is not confined to the water proton; we shall also apply it to the nucleus ^{17}O. Since this interaction is due to the direct contact between the nucleus and the electron, expressed by the factor $| \, \psi(\mathbf{r}) \, |^2$, it is also sometimes denoted as contact interaction.

In principle $A(\mathbf{r})$ represents the delocalization of the unpaired electrons; however, the functional dependence of $A(\mathbf{r})$ on $\mathbf{r}$ is not known. But it is natural that the dominant effect will be observed when the water molecule is in the first hydration sphere of the ion. Further away from the ion $A(\mathbf{r})$ will probably not be exactly zero, but it will decay to zero rapidly as the water–ion distance increases. We shall assume here that

$$
\begin{aligned}
A(\mathbf{r}) &= A &\quad \text{for} \quad | \, \mathbf{r} \, | \leq R_0 \\
A(\mathbf{r}) &= 0 &\quad \text{for} \quad | \, \mathbf{r} \, | > R_0
\end{aligned}
\qquad (58)
$$

where, as before, R_0 is the equilibrium distance between the protons and the ion in the first hydration sphere. A more general treatment where the delocalization of the electron spin is spread further out from the ion (or radical) has been published by Hubbard.[383] The important difference between the scalar interaction and the magnetic dipole–dipole interaction is given by the fact that $A = A(\mathbf{r})$ does not vary when the hydration complex performs rotational thermal motion relative to the laboratory system which determines the direction of the magnetic field, whereas the latter interaction fluctuates as a result of this motion. Thus rotational diffusion or similar rotational motion cannot cause a relaxation mechanism via scalar interaction. For the scalar interaction to cause relaxation, the water molecule must be removed from the hydration sphere. This process is usually called the exchange of water in the hydration sphere. This is the only possibility for the scalar interaction to fluctuate, the fluctuation occurring by the change of A with time; essentially we have a fluctuation between the value $A = A(R_0)$ and $A = 0$. The relaxation rate due to the scalar interaction is given by eqns. (21), (21e), (21f), and (22) as

$$(1/T_1)_{\text{sc}} = \tfrac{2}{3}S(S+1)J_{\text{sc}}(\omega_I - \omega_S) \qquad (59)$$

$$(1/T_2)_{\text{sc}} = \tfrac{1}{6}S(S+1)[J_{\text{sc}}(0) + J_{\text{sc}}(\omega_I - \omega_S)] \qquad (60)$$

with

$$J_{sc}(\omega) = \int_{-\infty}^{+\infty} e^{-i\omega t} \overline{A(0)A(t)} \, dt \tag{61}$$

For the correlation function $\overline{A(0)A(t)}$ we write as before (allowing for the interaction with any of the N_{ion} ions):

$$\overline{A(0)A(t)} = N_{ion} \int\int p(\mathbf{r_0})A(\mathbf{r_0})P(\mathbf{r_0}, \mathbf{r}, t)A(\mathbf{r}) \, d\mathbf{r_0} \, d\mathbf{r} \tag{62}$$

The propagator $P(\mathbf{r_0}, \mathbf{r}, t)$ is exactly the same as that introduced in eqn. (26), but we shall see that now the only dynamic process of importance is the passage of the H_2O molecule outward through the boundary of the hydration sphere.

Due to the assumed behavior of $A(\mathbf{r})$, according to eqn. (58), integration over $\mathbf{r_0}$ outside $|\mathbf{r_0}| = R_0$ gives no contribution to eqn. (62) and we can write

$$\overline{A(0)A(t)} = N_{ion} \int \tilde{p}(R_0)A(R_0)P(R_0, \mathbf{r}, t)A(\mathbf{r}) \, d\mathbf{r} \tag{63}$$

and after integration over $\mathbf{r}$ with eqn. (58) we get

$$\overline{A(0)A(t)} = N_{ion}\tilde{p}(R_0)A(R_0)^2 \int_{|\mathbf{r}|\leq R_0} P(R_0, \mathbf{r}, t) \, d\mathbf{r} \tag{64}$$

Considering eqn. (27), we have

$$\overline{A(0)A(t)} = N_{ion}\tilde{p}(R_0)A(R_0)^2 e^{-t/\tau_b} \tag{65}$$

$\tilde{p}(R_0)$ is the integral of eqn. (39) over all orientations of $\mathbf{r_0}$:

$$\tilde{p}(R_0) = n_h^+ c^*/N_{ion}55.5 = x_h^+/N_{ion} \tag{66}$$

Thus the Fourier transform of eqn. (65) is $[A \equiv A(R_0)]$

$$J_{sc}(\omega) = [x_h^+ 2\tau_b/(1 + \omega^2\tau_b^2)]A^2 \tag{67}$$

and the relaxation rates are, according to eqns. (60) and (61),

$$\left(\frac{1}{T_1}\right)_{sc} = \frac{2S(S+1)}{3} A^2 \frac{\tau_b}{1 + (\omega_I - \omega_S)^2\tau_b} x_h^+ \tag{68}$$

$$\left(\frac{1}{T_2}\right)_{sc} = \frac{S(S+1)}{3} A^2\left[\tau_b + \frac{\tau_b}{1 + (\omega_I - \omega_S)^2\tau_b^2}\right]x_h^+ \tag{69}$$

We see that again $T_1 = T_2$ for this relaxation mechanism, when $(\omega_I - \omega_S)\tau_b \ll 1$; however, for high frequencies T_1/T_2 can have any value, provided that $\tau_b \ll T_2$, because otherwise the theory is no longer valid.

There is, however, another effect which limits the ratio T_1/T_2 at high frequencies. This is due to the fact that the lifetime of a spin state of the spin S is not infinitely long. Rather, the spin S itself fluctuates among its various possible quantum states, which also causes fluctuation of the phase of the precessing spins S. For instance, for the z component of $\mathbf{S}$ we have the correlation function

$$\overline{S_z(0)S_z(t)} = \tfrac{1}{3}S(S + 1)e^{-t/\tau_{s1}} \tag{70}$$

where τ_{s1} is the longitudinal electron spin relaxation time. Omitting any further detail of the derivation, we only give the result[1]:

$$\left(\frac{1}{T_1}\right)_{\text{sc}} = \frac{2S(S + 1)}{3} A^2 \frac{\tau_{e2}}{1 + (\omega_I - \omega_S)^2\tau_{e2}^2} x_h^+ \tag{71}$$

$$\left(\frac{1}{T_2}\right)_{\text{sc}} = \frac{A^2 S(S + 1)}{3} \left[\tau_{e1} + \frac{\tau_{e2}}{1 + (\omega_I - \omega_S)^2\tau_{e2}^2}\right] x_h^+ \tag{72}$$

with

$$1/\tau_{ei} = (1/\tau_b) + (1/\tau_{si}), \qquad i = 1, 2 \tag{73}$$

τ_{s2} is the transverse electron spin relaxation time. Thus we see from eqns. (71)–(73) that the scalar relaxation measures the shorter of the two time constants τ_b and $\tau_{s2} \approx \tau_{s1} = \tau_s$. Usually $\tau_s = 10^{-10}$–10^{-8} sec, so the residence time τ_b may only be "seen" by the scalar relaxation if $\tau_b \ll \tau_s$, which quite often is not the case. Examples will be given shortly. It should be mentioned that the correlation time for the dipole–dipole interaction τ_c^{**} has also to be replaced by

$$1/\tau_e^{**} = (1/\tau_c^*) + (1/\tau_b) + (1/\tau_s)$$

if the electron spin relaxation time $\tau_s = \tau_{s1} \approx \tau_{s2}$ is not very much longer than the reorientational correlation time.[178,709] The same holds true for the translational correlation time.[632]

Summarizing, we obtain for the entire relaxation rate of a proton in solution

$$1/T_1 = (1/T_1)_{\text{H}_2\text{O}} + (1/T_1)_p + (1/T_1)_{\text{sc}} \tag{74}$$

$$1/T_1 = (1/T_2)_{\text{H}_2\text{O}} + (1/T_2)_p + (1/T_2)_{\text{sc}} \tag{75}$$

So far we have only considered the proton to be the relaxing nucleus, but eqns. (74) and (75) are more general and can also be used for other nuclei, including those of the solvent water as well as those of other diamagnetic solutes present in the paramagnetic electrolyte solution.

2.2.4. Solutions of V^{2+}, Mn^{2+}, and Cr^{3+}

V^{2+}: Here $1/T_1$ is due to dipole–dipole interaction and behaves essentially as predicted by eqn. (51).[60] The ratio of $1/T_1$ for low and high frequencies has been found to be somewhat less than $10/3$. This may be explained by a partial contribution of the translational term in eqn. (51); the frequency range studied has not been sufficiently wide. However, it was not within the scope of the relatively early paper cited to get detailed information concerning the form of $1/T_1$ in the dispersion range. The reorientational correlation time for the hydration complex was found to be 2×10^{-11} sec, as is listed in Table I. For $1/T_2$ the scalar part as given by eqns. (75), (69), and (72) is much more important than the dipole–dipole contribution $(1/T_2)_p$, at least above room temperature. However, the gross frequency dependence is not given by eqns. (69) and (72). This means that we must have $\omega_S \tau_{s2} \gg 1$ ($\omega_S \gg \omega_I$). The second term in the parentheses of eqn. (72) may be neglected. A lifetime, however defined, should certainly decrease as the temperature increases. However, $(1/T_2)_{sc}$ increases as the temperature increases. In many cases the electron spin relaxation time increases (as does the nuclear magnetic relaxation time) with increasing temperature.[60] Consequently in eqn. (73) $\tau_b \gg \tau_{s1}$ and thus $\tau_{e1} = \tau_{s1}$, i.e. the electron spin relaxation time is the correlation time for the transverse proton relaxation. It may also be shown that the electron spin relaxation time increases as the frequency increases.[60] This dependence of the correlation time on the frequency is reflected in the frequency dependence of $1/T_2$: $(1/T_2)_{sc}$ increases with increasing frequency. We see that the dynamic quantity which is of interest here, namely the residence time of the proton in the hydration sphere, cannot be obtained from proton relaxation time studies because the electron spin relaxation time is too short; it acts as the correlation time of the proton relaxation.

Mn^{2+}: Figure 3 shows the proton relaxation rates in aqueous solutions of Mn^{2+} as a function of the proton resonance frequency at two temperatures. Let us first consider the behavior of $1/T_1$. We see two dispersion steps. At low frequencies we have $\omega_S \tau_{e2} = 1$, according to eqn. (71), i.e., the decrease of T_1^{-1} is due to the decrease and finally the disappearance of the scalar contribution to the longitudinal relaxation rate.[49,59,115] At low

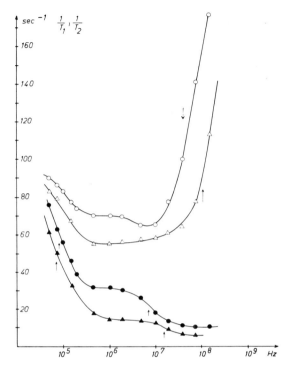

Fig. 3. $1/T_2$ and $1/T_1$ (filled symbols) of water protons in 1.53×10^{-3} M aqueous $MnCl_2$ solution as a function of the proton resonance frequency. Circles, 25°C; triangles, 65°C. The arrows indicate centers of the dispersion ranges.[326]

temperatures τ_{e2} is determined by the electron spin relaxation time [see eqn. (73)]. With increasing temperature the electron spin relaxation time increases[49,60,197,258,374,545,779] (regarding the concentration dependence of τ_s, see also Refs. 329 and 590). Finally, at sufficiently high temperatures the residence time becomes shorter than the electron spin relaxation time. However, in this event eqns. (71) and (72) are no longer valid at low frequencies which is a consequence of the hyperfine interaction of the unpaired electrons with the nuclear spin of ^{55}Mn.[633]

At higher frequencies the second dispersion step occurs, which is due to the magnetic dipole–dipole interaction, eqn. (51). The correlation time τ_c^* for the reorientation of the hydration complex of Mn^{2+} is 3.2×10^{-11} sec at 25°C $(\tau_b \gg \tau_c^*)$. One sees clearly that τ_c^* decreases with increasing temperature, i.e., the dispersion step $(\omega_S \tau_c^* = 1)$ is shifted to higher frequencies.

Now let us consider the transverse relaxation rate. Again at low frequency $1/T_2$ shows a dispersion step which is caused by the frequency dependence of the scalar contribution as given by eqns. (69) and (72). However, the scalar contribution does not vanish, it only decreases to one-half of the original value. The dipolar dispersion step at higher frequencies is hardly detectable from the transverse relaxation rate (Fig. 3); the latter is almost entirely caused by scalar interaction. At an NMR frequency concerned here the electron spin relaxation time is $\tau_{s1} \approx 3 \times 10^{-9}$ sec.[60,779] We see from Fig. 3 that with increasing frequency the transverse relaxation rate increases. On the other hand, according to eqn. (72), $T_2^{-1} \sim \tau_{e1}$. Thus the frequency dependence can only derive from τ_{e1}. Clearly τ_b is independent of ω_S; however, the electron spin relaxation time depends on the frequency, again in a form corresponding to a Lorentzian behavior,

$$\frac{1}{\tau_{s1}} \sim \left(\frac{\tau_v}{1 + \omega_S^2 \tau_v^2} + \frac{4\tau_v}{1 + 4\omega_S^2 \tau_v^2} \right) \tag{76}$$

where τ_v is an appropriate correlation time, given by the respective theory.[60,599,780] Thus the increase of the nuclear magnetic relaxation rate with increasing frequency is caused by an increase of the electron spin relaxation time. At higher temperature the electron spin relaxation time becomes longer since τ_v, the correlation time which determines τ_{s1} [eqn. (76)], decreases according to an ordinary activation law[60]

$$\tau_v = \tau_v^\circ e^{E_v/RT}$$

but, of course, the residence time τ_b becomes shorter as the temperature increases, again according to

$$\tau_b = \tau_b^\circ e^{E_b/RT}$$

where E_b is the activation energy of the water exchange in the first hydration sphere of Mn^{2+}. As a consequence, the increase of the relaxation rate $1/T_2$ with increasing frequency is less at higher temperature, since here the contribution of τ_b to the correlation time τ_{e1} according to eqn. (73) has become greater. The fitting of the necessary parameters for the water exchange and electron spin relaxation yields the dynamic data presented in Table I.

It should be mentioned that the data reported by Pfeifer[632] are obtained from a full analysis of the longitudinal relaxation rate, including the translational contribution which accounts for about 10–20% of the total relaxation rate. Again the diffusion coefficient of the water relative to the Mn^{2+} ion comes out somewhat too small.

Cr^{3+}: It is well known that water exchange in the hydration sphere of Cr^{3+} is very slow.[122,389] However, the protons of H_2O exchange faster than the oxygen does (as will be seen below), but still the condition which underlies eqns. (74) and (75), namely $\tau_b \ll T_{1,2}^+$, where $T_{1,2}^+$ is the relaxation time of the proton if it were permanently sitting in the hydration sphere, is no longer valid at low temperatures ($\sim 0°C$).

Anticipating the results derived below in eqns. (85) and (98), we put

$$1/T_1 = (1/T_1)_{H_2O} + (1/T_1)_{tr} + [x_h^+/(T_1^+ + \tau_b)] \tag{77}$$

$$1/T_2 = (1/T_2)_{H_2O} + (1/T_2)_{tr} + [x_h^+/(T_2^+ + \tau_b)] \tag{78}$$

where

$$\frac{1}{T_1^+} = \frac{2}{5} \frac{\gamma_I^2 \gamma_S^2 \hbar^2 S(S+1)}{R_0^6} \left[\tau_c^* + \frac{(7/3)\tau_c^*}{1 + \omega_S^2(\tau_c^*)^2} \right]$$
$$+ \frac{2S(S+1)A^2}{3} \frac{\tau_{s2}}{1 + \omega_S^2 \tau_{S2}^2} \tag{79}$$

$$\frac{1}{T_2^+} = \frac{1}{15} \frac{\gamma_I^2 \gamma_S^2 \hbar^2 S(S+1)}{R_0^6} \left[7\tau_c^* + \frac{13\tau_c^*}{1 + \omega_S^2(\tau_c^*)^2} \right]$$
$$+ \frac{S(S+1)A^2}{3} \left(\tau_{s1} + \frac{\tau_{s2}}{1 + \omega_s^2 \tau_{s2}^2} \right) \tag{80}$$

As already mentioned, $T_{1,2}^+$ is the relaxation time of a proton in the hydration sphere, where this complex is considered as a "stable" molecular system; any lifetimes of constituents do not enter. Evidently, $1/T_1^+$ and $1/T_2^+$ are the sums of eqns. (71) and (72) and of the rotational part of (51) and (52) for $x_h^+ = c^* n_h^+/55.5 = 1$ and $\tau_b \to \infty$. We see from eqns. (77) and (78) that the condition $\tau_b \gtrsim T_{1,2}^+$ has two typical consequences: (1) The total variation of $1/T_{1,2}$ over the entire frequency range is reduced; in the limit $\tau_b \gg T_{1,2}^+$ the frequency dependence of $1/T_{1,2}$ is only determined by that of $(1/T_{1,2})_{tr}$, which may be a weak contribution to the total relaxation rate. (2) For sufficiently long τ_b as compared with $T_{1,2}^+$ both relaxation rates $1/T_1$ and $1/T_2$ increase with increasing temperature because τ_b occurs in the dominant contribution, as given by

$$\tau_b = \tau_b^° e^{E_b/RT}$$

In such a situation we say that the relaxation rate is governed by exchange. For higher temperature τ_b gets shorter until $\tau_b \approx T_{1,2}^+$; then the temperature dependence of $1/T_1$ becomes normal, i.e., $1/T_1$ decreases with increasing temperature, and the behavior of $1/T_1$ is governed by relaxation in the strict sense.

It is also possible for $1/T_2$ to decrease with increasing temperature; whether it does so or not depends on the strength of scalar interaction relative to dipolar interaction and on the temperature dependence of τ_{s1}. Very often τ_{s1} increases with increasing temperature, but there are cases where τ_{s1} decreases with increasing temperature (see below).

It has indeed been observed that in Cr^{3+} solution near $0°C$ both $1/T_1$ and $1/T_2$ increase as the temperature increases.[79,325] This allows an estimate to be made of τ_b (see Table I). At higher temperatures, $\tau_b \ll T_1^+$ and the temperature dependence of $1/T_1$ is reversed (see Fig. 4). Two dispersion steps may be clearly recognized, of which one is due to the electron spin relaxation time ($\tau_{s1} \approx 5 \times 10^{-10}$ sec at $300°K$) as the correlation time of the scalar term in $1/T_1$, and the other one to the rotational motion of the hydration complex, i.e., it gives τ_c^*. With increasing temperature the two steps move in opposite directions, as shown in Fig. 4.[326] At higher temperatures $1/T_2$ is determined by the scalar contribution which is proportional

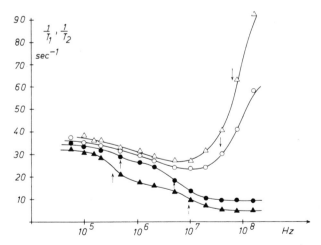

Fig. 4. $1/T_2$ and $1/T_1$ (filled symbols) of water protons in 1.41×10^{-3} M aqueous $Cr(NO_3)_3$ solution as a function of the proton resonance frequency. Circles, $25°C$; triangles, $65°C$. The arrows indicate centers of dispersion ranges (according to the authors).[326]

to τ_{s1}; the last quantity increases with increasing frequency and temperature, and thus $1/T_2$ continues to increase with increasing temperature and frequency, contrary to the behavior of $1/T_1$. At about $80°C$ there is also a reversal of the $1/T_2$ temperature dependence;[79] probably τ_{s1} is shortened due to exchange processes.

2.2.5. *Summary*

We see from Table I that the rotational correlation time $\tau_c{}^*$ in the hydration sphere of a doubly charged cation is $\sim 3 \times 10^{-11}$ sec; the ionic radius varies between 0.72 and 0.88 Å. For V^{2+} the correlation time is somewhat shorter than 3×10^{-11} sec; however, more refined measurements for this ion, which is chemically difficult to handle, may bring better agreement. For higher charge of the cation one expects a longer correlation time; this is confirmed for Cr^{3+} ($r_{ion} = 0.61$ Å). However, for Gd^{3+} a relatively short correlation time has been reported, which should not be justifiable in terms of a larger ionic radius (0.94 Å). As already mentioned, a re-examination of the experimental results is desirable. To what extent the smaller correlation times may be ascribed to a rotation of the water about the dipole axis has not yet been analyzed; this possibility may exist for Gd^{3+}

We emphasize once again that the importance of all these measurements lies in the fact that they give a direct and independent determination of the correlation time and, as already mentioned, this is not possible for highly fluid diamagnetic liquids containing small dissolved particles.

The further advantage of the use of paramagnetic ions may be seen from eqns. (51) and (52). If we have determined $\tau_c{}^* = \tau_c{}^{**}$ by an analysis of the dispersion behavior, then the terms in brackets in these formulas may be calculated and, neglecting $(1/T_1)_{tr}$, $R_0{}^6$ can be found if all other quantities in front of the brackets are known. The only quantity for which experimental data are not available in all cases is the hydration number $n_h{}^+$, but there should be hardly any doubt that for the cations considered here $n_h{}^+ = 6$ is the correct assumption. The agreement of the R_0 thus calculated with the value expected from crystal radii and water geometry is satisfactory. At least, all these results confirm the (almost self-evident) expectation that the protons are directed away from the cation, according to the stable orientation of an electric dipole in the ionic field. Moreover, the fact that we have independent information about $\tau_c{}^*$ and the radius of the complex serves as a check on the results which are obtained for diamagnetic, highly fluid liquids, in particular for water. The radius of the hydration complex composed of the ion and the first shell is about 3 Å. The radius of the water molecule is 1.4 Å; thus if we use the "hydrodynamic" Debye formula

$$\tau_c = 4\pi r^3 \eta / 3kT \tag{81}$$

we estimate for the correlation time τ_c of a water molecule in pure water

$$\tau_c \approx (1.4/3)^3 \times 3 \times 10^{-11} \text{ sec}$$
$$= 3 \times 10^{-12} \text{ sec}$$

This is indeed close to the figure one finds for pure water, as shown below.

We consider it to be outside the scope of this chapter to treat the motion and relaxation properties of complexes where one or more of the hydration waters are replaced by other ligands. Although some work has been done on complexes, thorough and systematic data are sparse. The same is true for the relaxation studies of nonaqueous solutions of paramagnetic ions.

2.3. Oxygen-17 Relaxation in Aqueous Solutions of Paramagnetic Ions

2.3.1. *Additonal Theory for the Situation of Fast Relaxation and Slow Exchange*

Besides $H_2^{16}O$, there are two other water species the nuclear magnetic relaxation behavior of which may be studied: $D_2^{16}O$ and $H_2^{17}O$. Both nuclei, $D \equiv {}^2H$, and ${}^{17}O$, have electric quadrupole moments; thus relaxation due to quadrupole interaction is the dominating process in pure water. Returning to eqns. (74) and (75), $(1/T_{1,2})_{H_2O}$ is no longer caused by magnetic dipole–dipole interaction between the water protons but is determined by quadrupolar relaxation and $(1/T_1)_{H_2O} \cong (1/T_2)_{H_2O}$ is by one and three orders of magnitude greater in $\underline{D}_2^{16}O$ and $H_2^{17}\underline{O}$, respectively, than in $\underline{H}_2^{16}O$ (the nucleus of interest is underlined).

The ratio of gyromagnetic ratio for the deuteron $\gamma_{I'}$ and proton γ_I, is $\gamma_{I'}/\gamma_I = 0.154$, so the term $(1/T_1)_p$ becomes very small [see eqns. (51) and (52) with γ_I replaced by $\gamma_{I'}$]; the same is true for the scalar part of relaxation [see eqns. (57), (68), and (69)]. We thus understand that it is hardly worthwhile to perform deuteron relaxation studies in paramagnetic solutions; the precision of deuteron relaxation measurements is usually worse because the signal is weaker. The situation is different if we consider $H_2^{17}\underline{O}$. Again the magnetic dipole–dipole contribution is very small due to the small gyromagnetic ratio of the ^{17}O nucleus. But the coupling constant A which determines the scalar interaction is larger here, because according to eqn. (57), $\psi(\mathbf{r})^2$, the probability that one of the unpaired electrons is at the nucleus, is larger for the water oxygen than for the water hydrogen. So the last term on the right-hand side of eqns. (74) and (75) acquires great importance in solutions of paramagnetic ions in water when the ^{17}O magnetic resonance is studied.

An important new point now enters: Equations (74) and (75) were derived under the condition that the lifetime in both environments, in the bulk or free water and in the hydration sphere of the ion, are very much shorter than the relaxation times in these two environments. Due to the strong relaxation of ^{17}O in water of the bulk phase (three orders of magnitude stronger relaxation than that of the water protons) and the short relaxation time in the hydration complex, this condition is not fulfilled in many cases. In this event, i.e., $\tau_b > T_{1,2}$, we actually have two relaxation times, one in the bulk water and another in the hydration sphere. Let us denote the relaxation rates in the bulk or free water as $1/T_{1a}$ and $1/T_{2a}$. The relaxation rates in the hydration complex we denote as $1/T_{1b}$ and $1/T_{2b}$. We note that $1/T_{1b}$ and $1/T_{2b}$ are given by the relations

$$1/T_{ib} = (1/T_i)_Q + (1/T_i)_p, \qquad i = 1, 2 \tag{82}$$

where $(1/T_{1,2})_Q$ is the quadrupolar contribution and $(1/T_{1,2})_p$ is the dipolar contribution to the relaxation rate in the hydration complex. The $(1/T_{1,2})_Q$ is determined by the correlation time in the hydration complex and by the quadrupole coupling constant in the hydration complex; the latter may in principle be much different from that in pure water. It seems, though, that the effect of variation of the coupling constant is not large for water. However, if we consider the relaxation rate of an anion, then $(1/T_{1,2})_Q$ may be very large: Formally, the equation to be discussed may be also applied to the relaxation of anions; then $1/T_{ib}$ would be the relaxation rate of the "ligand" bound to the metal ion.

The water molecule now fluctuates between two environments characterized by these two relaxation rates. The lifetime in the hydration sphere is, as before, τ_b, and the one in the free water is $\tau_a = \tau_b(1 - x_h^+)/x_h^+$, $x_h^+ = n_h^+ c^*/55.5$. The total magnetization $\mathscr{M}$ is the sum of $\mathscr{M}_a = (1-x_h^+)\mathscr{M}$ and $\mathscr{M}_b = x_h^+\mathscr{M}$. The time dependence of the z component (component in the H_0 direction) of the total magnetization may be calculated from a set of Bloch equations taking proper account of the exchange of water between the two environments.[543] Of course the limiting case $\tau_b \gg T_{1a}$,

T_{1b} gives the approach of the z component from zero to the equilibrium value $\mathscr{M}_z{}^\circ$:

$$\mathscr{M}_z(t) = \mathscr{M}_z{}^\circ[x_h{}^+(1 - e^{-t/T_{1b}}) + (1 - x_h{}^+)(1 - e^{-t/T_{1a}})] \qquad (83)$$

i.e., the motion of $\mathscr{M}_z$ is no longer characterized by one relaxation time. However, very often the situation $x_h{}^+ \ll 1$, $T_{1a} \gg T_{1b}$ occurs; then we once again have *one* relaxation time:

$$\mathscr{M}_z(t) = \mathscr{M}_z{}^\circ(1 - e^{-t/T_1}) \qquad (84)$$

and it may be shown from the formulas of Zimmerman and Brittin[879] that the following relation holds:

$$1/T_1 = (1/T_{1a}) + [x_h{}^+/(T_{1b} + \tau_b)] \qquad (85)$$

an equation which we already applied [eqn. (77)].

Now, if we consider the behavior of the transverse magnetization, we find that the situation is more complicated. This is so because not only do the water molecules exchange between two environments with different T_2, but the Larmor frequency is also different in the two environments. Moreover, in the hydration state the ^{17}O nucleus may find a number of different Larmor frequencies, the actual instantaneous Larmor frequency being dependent on the orientation of the unpaired electrons at the transition metal ion. Eisenstadt and Friedman[191] have given the corresponding calculation of T_2 for a system where the unpaired electron has spin 1/2. Thus the ^{17}O nucleus, in the course of time, passes through three different environments: (I) State a is the free state, the water not being a member of the transition metal ion; (II) in state b_+ the water molecule is in the hydration sphere of an ion with electron-spin z component of $+1/2$; (III) in state b_- the water is in the hydration sphere of an ion with electron-spin z component of $-1/2$. We shall reproduce the formulas of Eisenstadt and Friedman here with the only modification that we distinguish the two b states in a more general way as having a positive or a negative z component of the electron spin, but not necessarily $\pm 1/2$. Then in the free state the Larmor frequency of the nucleus is ω_I; in the b_+ state it is

$$\omega_I + \delta^+ = \omega_I + A[S(S + 1)]^{1/2} \cos\theta \qquad \text{with} \quad \cos\theta > 0 \qquad (86)$$

and in the b_- state it is

$$\omega_I + \delta^- = \omega_I + A[S(S + 1)]^{1/2} \cos\theta \qquad \text{with} \quad \cos\theta < 0 \qquad (87)$$

$\cos\theta$ is the angle between the electron spin and the magnetic field. The energies $\mathscr{H}^{\pm}$ of the electron spin in the two states b_+ and b_- are not equal, and are given by

$$\mathscr{H}^+ = -[S(S+1)]^{1/2}(\cos\theta)\gamma_S\hbar H_0 \qquad \text{with } \cos\theta > 0 \qquad (88)$$

$$\mathscr{H}^- = -[S(S+1)]^{1/2}(\cos\theta)\gamma_S\hbar H_0 \qquad \text{with } \cos\theta < 0 \qquad (89)$$

which is important, since due to the Boltzmann factor, both states are not of exactly equal population and this causes the occurrence of a net chemical shift of the nucleus studied (here ^{17}O). Eisenstadt and Friedman applied the McConnel–Bloch equation formalism to an exchange between the three environments a, b_+, and b_-; the time constant for the transition of the water molecule from b_+ to b_- and vice versa is the electron spin relaxation time, $\tau_{s1} = \tau_s$. One finds that the decay of the transverse magnetization is an exponential characterized by the transverse relaxation time if x_h^+, the mole fraction of hydration sphere water, is sufficiently small, $x_h^+ \ll 1$; then the result is

$$\frac{1}{T_2} = \frac{1}{T_{2a}}$$

$$+ x_h^+ \left\{ \left[\frac{1}{T_{2b}} \left(\frac{1}{T_{2b}} + \frac{1}{\tau_b} + \frac{1}{\tau_s} \right) + \overline{(\delta^{\pm})^2} \right] \right.$$

$$\times \left[\left(\frac{1}{T_{2b}} + \frac{1}{\tau_b} \right) \left(\frac{1}{T_{2b}} + \frac{1}{\tau_b} + \frac{1}{\tau_s} \right) + \overline{(\delta^{\pm})^2} \right] + \overline{(\delta^{\pm}\mathscr{H}^{\pm})^2} \frac{1}{\tau_s^2} \right\}$$

$$\times \tau_b \left\{ \left[\left(\frac{1}{T_{2b}} + \frac{1}{\tau_b} \right) \left(\frac{1}{T_{2b}} + \frac{1}{\tau_b} + \frac{1}{\tau_s} \right) + \overline{(\delta^{\pm})^2} \right]^2 \right.$$

$$\left. + \overline{(\delta^{\pm}\mathscr{H}^{\pm})^2} \frac{1}{\tau_s^2} \right\}^{-1} \qquad (90)$$

where $\overline{(\delta^{\pm})^2}$ is the mean value of the square of $\delta^{\pm}$ according to eqns. (86) and (87):

$$\overline{(\delta^{\pm})^2} = A^2 S(S+1)/3 \qquad (91)$$

and $\overline{\delta^{\pm}\mathscr{H}^{\pm}}$ is the thermal average of $\delta^{\pm}\mathscr{H}^{\pm}$ according to eqns. (86)–(89):

$$\overline{\delta^{\pm}\mathscr{H}^{\pm}} = [AS(S+1)/3kT]\omega_s\hbar$$

$$= [AS(S+1)/3kT]\hbar\gamma_s\omega_I/\gamma_I, \qquad \omega_S = \gamma_S H_0 \qquad (92)$$

Equation (90) is of quite general validity but of such complex appearence that it is very useful to demonstrate the various limiting cases for which it leads to simpler formulas. Let us first consider the case that $\tau_{s1} = \tau_s$, the electron spin relaxation time is so long and the scalar coupling constant so small that the thermal average term $\overline{(\delta^\pm \mathscr{H}^\pm)^2 \tau_s^{-2}}$ in eqn. (90) may be neglected. However, we shall assume that $\tau_s, \tau_b \ll T_{2b}$. In this event eqn. (90) together with eqn. (91) yields

$$\frac{1}{T_2} = \frac{1}{T_{2a}} + \frac{x_h^+}{T_{2b}} + \frac{x_h^+ A^2 S(S+1)}{3} \frac{\tau_b \tau_s}{\tau_b + \tau_s} \tag{93}$$

We see that the third term on the right-hand side of this formula is identical with eqn. (72) when $(\omega_S - \omega_I)\tau_e \gg 1$, i.e., the usual scalar contribution to the relaxation rate, but we note that it will not be possible to derive the frequency dependence of eqn. (72) from eqn. (90). Thus eqn. (90) is no longer valid when the residence time becomes so short that it approaches the reciprocal electron spin resonance frequency, which is $\sim 10^{-11}$ sec. If $\tau_s \ll \tau_b$, then the influence of τ_b disappears in eqn. (93) and we may now write

$$1/T_2 = (1/T_{2a}) + (x_h^+/T_2^+) \tag{94}$$

where

$$1/T_2^+ = (1/T_{2b}) + \tfrac{1}{3} A^2 S(S+1)\tau_s \tag{95}$$

is the relaxation rate of the nucleus when it is a member of the paramagnetic hydration complex for a very long time $\tau_b \gg \tau_s$.

Now, if τ_s becomes shorter and shorter, the term $\overline{(\delta^\pm \mathscr{H}^\pm)^2 \tau_s^{-2}}$, which we may call the "first-order" term because it takes account of the first-order perturbation of the Zeeman energy of the nucleus due to the preferential electron spin orientation in the magnetic field, increases in importance, particularly if the coupling constant is large. Now we may derive from eqn. (90) that

$$\frac{1}{T_2} = \frac{1}{T_{2a}} + \frac{x_h^+}{\tau_b} \frac{(1/T_2^+)[(1/T_2^+) + (1/\tau_b)] + \Delta\omega_b^2}{[(1/T_2^+) + (1/\tau_b)]^2 + \Delta\omega_b^2} \tag{96}$$

for $\tau_s \ll \tau_b$. This equation was originally derived by Swift and Connick.[767] In (96)

$$\Delta\omega_b = \overline{\delta^\pm \mathscr{H}^\pm} = [AS(S+1)\hbar/3kT]\gamma_s \omega_s/\gamma_I \tag{97}$$

is the chemical shift of the nucleus relative to the free state in the hydration sphere due to the scalar interaction with the unpaired electrons. However, it should be noted that eqn. (96) is more general; $\Delta\omega_b$ may be the chemical shift of the nucleus in the hydration sphere caused by any other effect, e.g., the electrostatic field of the ion. Since the derivation of eqn. (96) by Swift and Connick was based on a two-site process, $\Delta\omega_b$ necessarily had this more general meaning in their treatment. We see, when $\tau_b \ll T_2^+$ and $\Delta\omega_b$ is small, that we obtain eqn. (94) from eqn. (96), but for eqn. (96) to hold it is by no means necessary that $1/T_2^+$ is caused by scalar interaction; $1/T_2^+$ in eqn. (96) may be caused by any mechanism, and it has been derived by Swift and Connick for quite arbitrary $1/T_2^+$, but $1/T_2^+$ must describe a relaxation rate of the nucleus in the bound state. Equation (96) just describes the nature of T_2 when the lifetime τ_b can assume any value as compared with the relaxation time in the complex. When the chemical shift $\Delta\omega_b$ is sufficiently small we obtain from eqn. (96)

$$1/T_2 = (1/T_{2a}) + [x_h^+/(T_2^+ + \tau_b)] \qquad (98)$$

which demonstrates that T_1^{-1} and T_2^{-1} behave in the same way [see eqn. (85)] when we are allowed to neglect the chemical shift or first-order effect. The only difference between eqns. (85) and (98) is the occurrence of T_{1b} in the former and of T_2^+ in the latter case. According to eqs. (95), $1/T_2^+$ is the transverse relaxation rate in the hydration water state, including the effect of scalar interaction—apart from quadrupole and dipole contributions. Of course, an equation of the type (85) is more general, and if $\tau_{s2} \ll \tau_b$, we may also include

$$(1/T_1)_{sc} = \tfrac{1}{3}[2A^2 S(S+1)]\tau_{s2}/[1 + (\omega_I - \omega_S)^2\tau_{s2}^2]$$

in $1/T_1^+$:

$$1/T_1^+ = (1/T_{1b}) + (1/T_1)_{sc}$$

Thus

$$1/T_1 = (1/T_{1a}) + [x_h/(T_1^+ + \tau_b)]$$

but unless the scalar coupling constant is very large, the scalar contribution to $1/T_1^+$ is negligible ($\omega_S \tau_{s2} \gg 1$), as may be seen from a comparison of eqns. (71) and (72).

Equation (98) is valid for the situation where the increase of $1/T_2$ is caused by the strong relaxation in the hydration sphere, the chemical shift effect being unimportant. Now, the reverse situation may happen, namely that the entire additional relaxation effect is due to a large chemical shift in the hydration sphere. The first limiting case which can occur here is

$\Delta\omega_b \gg 1/T_2^+, \ 1/\tau_b$; then from eqn. (96)

$$1/T_2 = (1/T_{2a}) + (x_h^+/\tau_b) = (1/T_{2a}) + (1/\tau_a) \tag{99}$$

which is the same result as eqn. (98) in the limit $\tau_b \gg T_2^+$.

It may also happen that $\Delta\omega_b \gg 1/T_2^+$ but that the second condition $\Delta\omega_b \gg 1/\tau_b$ is no longer fulfilled; on the contrary, we have $\Delta\omega_b \ll 1/\tau_b$. We now derive from eqn. (96)

$$1/T_2 = (1/T_{2a}) + (x_h^+/T_2^+) + x_h^+(\Delta\omega_b)^2\tau_b \tag{100}$$

Finally, it may be useful to add two formulas which give the chemical shift observed in the solution in terms of the shift $\Delta\omega_b$ for a nucleus existing for an infinitely long period in the hydration sphere. In the general case (note, however, that all our formulas apply to the situation $x_h^+ \ll 1$), Swift and Connick derived the formula[767]

$$\Delta\omega = x_h^+ \frac{\Delta\omega_b}{\tau_b^2\{[(1/T_2^+) + (1/\tau_b)]^2 + \Delta\omega_b^2\}} \tag{101}$$

and for the fast exchange limit $1/\tau_b \gg \Delta\omega_b, \ \Delta\omega_b \gg 1/T_2^+$,

$$\Delta\omega = x_h^+ \Delta\omega_b \tag{102}$$

We have reported all these formulas in order to demonstrate the different ways in which the residence time τ_b occurs in them. We therefore expect considerable amounts of information regarding τ_b to be obtained from ^{17}O transverse relaxation time measurements. As already indicated, these T_2 measurements are linewidth measurements. A number of applications to transition metal ions of the iron group are reviewed in the next section.

2.3.2. Residence Time of $H_2^{17}O_h$ in the Hydration Sphere of First Group Transition Metal Ions*

Ti^{3+}: The electron spin relaxation time in this ion is rather short, $\tau_s \lesssim 3.2 \times 10^{-11}$ sec at room temperature.[101] This causes the relaxation rate of ^{17}O in the hydration sphere of Ti^{3+} to be small. From the chemical shift of $H_2^{17}O$ in the presence of Ti^{3+} Chmelnick and Fiat[101] determined the coupling constant A [eqns. (97) and (102)] as tabulated in Table I.

* For results concerning rare earth ions see Ref. 681.

n_h^+ is assumed to be six. The transverse relaxation rate of $H_2^{17}O$ in Ti^{3+} solutions is given by eqns. (99) and (100) for lower and higher temperatures, respectively. We have $1/T_2^+ < \Delta\omega_b$, τ_b; with the knowledge of A, τ_b has been obtained from the high-temperature portion of $1/T_2(T)$; the result is listed in Table I.

V^{2+}: Above room temperature $1/T_2$ of $H_2^{17}O$ in the presence of V^{2+} increases with increasing temperature.[610] Thus the transverse relaxation is governed by exchange and formula (98) or (99) is applied:

$$1/T_2 = (1/T_{2a}) + (x_h/\tau_b) \qquad (103)$$

Whether this relaxation behavior is due to a relaxation effect—small T_2^+—or to a $\Delta\omega_b$ effect—large $\Delta\omega_b$—is not discussed. Probably we have a T_2^+ effect, since the electron spin relaxation time is longer than for Ti^{3+}. The kinetic parameters obtained are given in Table I. At lower temperatures $\tau_b^{-1} \ll T_{2a}^{-1}$, and $d(1/T_2)/dT < 0$. The $1/T_{2a}$ contains a dipolar contribution according to eqn. (54), a second-sphere scalar contribution according to eqn. (69), which, however, may also be more of a translational character,[383] and possibly a contribution due to the first-order effect, eqn. (100); Olson *et al.*[610] were unable to draw definite conclusions from their experimental results; the temperature dependence of $1/T_2$ below room temperature was found to be unexpectedly large.

VO^{2+}:[680,868,869] Above room temperature and below $\sim120°C$ $H_2^{17}O$ again relaxes according to eqn. (103). Above $120°C$ τ_b becomes comparable to T_2^+ and eqn. (98) will be obeyed. The correlation time for T_2^+ is the electron spin relaxation time, the latter being rather long and practically temperature-independent. The τ_b which determines the transverse relaxation in this range is ascribed to the four equatorial water molecules bound to the vanadyl ion; at and below room temperature the influence of this exchange becomes negligibly small, and the remaining relaxation is ascribed to the axial water, which exchanges very much faster. Dipole–dipole and quadrupole mechanisms are discussed.

Cr^{3+}: τ_b for the oxygen of water cannot be determined from ^{17}O relaxation because the exchange is much too slow.[122,123a,389,610] Relaxation of $H_2^{17}O$ occurs by translational dipole–dipole, scalar, and quadrupole effects, but a separation of the contributions has not yet been achieved.

Mn^{2+}: At and below room temperature the order of the three time constants which are relevant is $\tau_b > T_2^+ > \tau_{s1}$, but all three times are of a

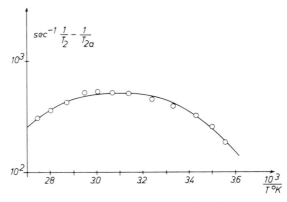

Fig. 5. $(1/T_2) - (1/T_{2a})$ of ^{17}O in $1.82 \times 10^{-4} M$ aqueous $MnSO_4$ solution as a function of the temperature. $\nu = 5.43$ MHz.[767]

similar magnitude. The time τ_{s1} is sufficiently short, thus defining T_2^+ [see eqn. (95)]. With increasing temperature τ_{s1} increases slightly;[60] thus T_2^+ (neglecting the influence of $1/T_{2b}$) and τ_b decrease. Thus, at low temperatures eqn. (103) is valid and τ_b may be determined. At temperatures above 100°C the order of the time constants is as follows: $T_2^+ > \tau_{s1} > \tau_b$, and with the aid of eqn. (90) we arrive at eqn. (93), with τ_b becoming smaller as compared to τ_{s1} as the temperature increases. Thus at high temperatures one expects for plots of $\log[(1/T_2) - (1/T_{2a})]$ versus $1/T$ a limiting slope equal in magnitude but opposite in sign to that observed for low temperatures. This behavior has been found experimentally (and is depicted in Fig. 5). For the entire temperature range Swift and Connick write

$$\frac{1}{T_2} - \frac{1}{T_{2a}} = \frac{x_h^+}{\tau_b + C^{-1}[(1/\tau_{s1}) + (1/\tau_b)]} \tag{104}$$

where

$$1/T_2^+ = C\tau_{e1}; \qquad 1/\tau_{e1} = (1/\tau_{s1}) + (1/\tau_b)$$

It should be noted, however, that the more correct and general description is given by eqn. (90), and that eqn. (104) is only correct provided that

$$\tau_{s1} \ll \tau_b \qquad \text{when} \qquad \tau_b \approx C^{-1}\tau_{e1}^{-1}$$

i.e., in eqn. (104) τ_b must not occur twice.

Fe^{3+}: Equation (103) is valid over the entire temperature range which has been studied experimentally, 25° to 80°C,[263] τ_b can therefore be determined. The validity of eqn. (103) is probably related to strong relaxation in the hydration complex.

Fe^{2+}: Here (see Fig. 6) the electron spin relaxation time is very short, which considerably reduces the relaxation rate of ^{17}O in the hydration complex. Thus we are left with a $\Delta\omega_b$ effect, producing relaxation of $H_2^{17}O$. With increasing temperature $1/T_2 - 1/T_{2a}$ decreases.[767] This excludes the slow exchange situation [eqn. (103)] and one therefore has to use eqn. (100). We can obtain $\Delta\omega_b$ from chemical shift measurements, according to eqn. (102). Results are given in Table I.

Co^{2+}: This ion again has a very short electron spin relaxation time and T_2^+ effects are not to be expected (see Fig. 6); indeed one finds relaxation caused by differences of the Larmor frequency in the hydration and free states; this has been checked by a study of the frequency dependence of $1/T_2$.[767] Below 0°C one observes the "$\Delta\omega_b$-governed" form of eqn. (103) and at temperatures $\gtrsim 25$°C eqn. (100) becomes valid.[100,767] Chemical

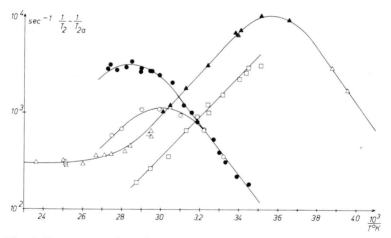

Fig. 6. Temperature dependences of $(1/T_2) - (1/T_{2a})$ of ^{17}O in aqueous Fe^{2+}, Co^{2+}, and Ni^{2+} solutions. ($\bullet$) 0.100 M $Ni(NO_3)_2$ at 5.43 MHz.[767] ($\bigcirc$) 0.100 M $Ni(NO_3)_2$ at 2.00 MHz.[767] ($\blacktriangle$) 0.205 $\bar{m}$ $Co(ClO_4)_2$ ($+0.34\,\bar{m}$ $HClO_4$) at 8.13 MHz.[100] ($\triangle$) 1.64 $\bar{m}$ $Co(ClO_4)_2$ ($+0.37\,\bar{m}$ $HClO_4$) at 8.13 MHz.[100] The data for the latter solution are converted to 0.205 $\bar{m}$ ($\bar{m}$ = moles solute per 55.5 mol water). ($\square$) $[(1/T_2) - (1/T_{2a})]10$ for $1.44 \times 10^{-2}\,M$ $Fe(NH_4)_2(SO_4)_2$ solution at 5.43 MHz.[767]

shift measurements yield $\Delta\omega_b$ which, in turn, allows determination of τ_b. The time τ_b decreases steadily with increasing temperature, so finally at about 140°C one observes only the relaxation rate[100]

$$1/T_2 = (1/T_{2a}) + [x_h^+/(T_2^+)]$$

Here we have put $1/T_2^+$ in parenthesis because the relaxation rate of ^{17}O in the hydration sphere of Co^{2+} cannot be expressed by any of our previous formulas because of the invalidity of the theory so far presented in the case of strong interactions (other than the Zeeman interaction with the static field). Results are given in Table I.

Ni^{2+}: In Ni^{2+} solutions the situation is very similar to that in Fe^{2+} and Co^{2+} solutions (see Fig. 6). Again τ_{s1} is very short. Thus the chemical shift $\Delta\omega_b$ causes the $1/T_2$ effect and one again observes a transition from the slow exchange behavior to the fast exchange behavior (eqn. (100)); this transition, however, is at rather high temperature ($\sim$80°C, depending on the resonance frequency) due to the slower exchange of water from the hydration sphere of the nickel ion. The frequency dependence of $1/T_2$ is as expected; at higher temperatures the chemical shift becomes detectable [see eqns. (101) and (102)] and allows the determination of the scalar coupling constant. Results are given in Table I.

Cu^{2+}: This is another example of an ion with a long electron spin relaxation time: $\tau_{s1} = 9.3 \times 10^{-10}$ sec at 24°C.[511,512] As regards the $H_2^{17}O$ relaxation, $(1/T_2) - (1/T_{2a})$ decreases strongly as the temperature is increased[512,767] and even at the lowest temperatures ($\sim$0°C) no indication of exchange-governed relaxation has been found $[d(1/T_2)/dT < 0]$ in the most recent investigation of Lewis et al.[512] Thus, our eqn. (93) or eqn. (94) has to be applied. With Cu^{2+} the electron spin relaxation time τ_{s1} also decreases as the temperature increases[247,452,512] and since the activation energies are rather alike, one would suspect that τ_{s1} is the correlation time for the relaxation process. However, the measurement of the chemical shift gives the coupling constant A and indicates that τ_{e1} must be smaller than τ_{s1} by a factor of 6–8 and thus $1/T_2$ is determined by the lifetime τ_b, i.e., by the exchange process of the water in the first hydration sphere of Cu^{2+}. The Jahn–Teller theorem predicts a tetragonal distortion in the hexa-aquo copper(II) ion. Hence, it was proposed by Lewis et al.[512] that scalar coupling exists only for the equatorial positions (see also Ref. 598) of the distorted octahedron, while at the axial positions there is no scalar coupling. Thus by an inversion process of the hydration complex the scalar interaction is interrupted and the correlation time for $1/T_2{}^+$ might be given by the frequency of inversion of the complex. However, it is not the interruption of the scalar interaction that is important for relaxation, but the loss of correlation in the interaction between a given nucleus and the unpaired electron, and this loss of correlation can only occur if the nuclear spin actually loses its interaction partner irreversibly as has been shown on p. 330 (τ_{s1} being sufficiently long).

Inspection of the data collected in Table I shows that in some cases there is good agreement between results found by different authors; in other cases, however, the data show obvious deviations from expected numerical values, in some cases a question mark is added to these numbers, and in some cases values are omitted. It should be kept in mind that some of the papers quoted aimed at a principal understanding of new experimental facts and not at a presentation of precise kinetic numerical results. We see from Table I that residence times τ_b are equal for the water protons and the water oxygen for Mn^{2+}, but for Cr^{3+} the protons exchange faster than the oxygen; the same is true for $Fe^{3+(537)}$ and for Al^{3+} (see below). For a discussion of the τ_b results presented in Table I in connection with the binding mechanism to the ion, the reader is referred to the literature cited, in particular to Ref. 767.

2.4. Proton Relaxation Times and Correlation Times of Water in Diamagnetic Electrolyte Solutions

2.4.1. *Pure Water; The Propagator*

In the preceding section we have described a number of formulas which permit direct determination of rotational correlation times and residence times of water molecules in ionic hydration complexes from nuclear magnetic relaxation times. In all these cases the fundamental source of information was the interaction of the nucleus residing in the water molecule with the magnetic moment of the unpaired electron on the cation of interest. We shall now turn to the question, What information can be obtained from nuclear magnetic relaxation time studies if unpaired electrons at the ions are absent, i.e., if we wish to study diamagnetic electrolyte solutions? Indeed, the amount of information available is more limited, but nevertheless some nuclear magnetic relaxation work has also been done with diamagnetic electrolyte solutions and this will be described in the following sections.

We shall begin our discussion with the proton relaxation rate of the solvent water in diamagnetic electrolyte solutions. We recall that the large relaxation rate of the proton in paramagnetic electrolyte solutions was caused by the magnetic dipole–dipole interaction of the water proton with the point dipoles represented by the unpaired electrons in the corresponding ions. If these electron paramagnetic point dipoles are absent, then magnetic dipole–dipole interaction occurs with those point dipoles with which we are necessarily left after the former have been removed, namely the nuclear magnetic moments of the protons in the surrounding water. Apart from the fact that we now have interaction between like spins, whereas formerly we had interaction between unlike spins, and that now the interaction partner has a magnetic moment which is roughly three orders of magnitude smaller than that of the electron, the theoretical situation is not at all changed and the starting point is again eqn. (26). We note that magnetic dipole–dipole interaction with nuclear magnetic moments of diamagnetic ions need not be considered; in most cases the magnetic moments of these nuclei are small compared with that of the proton. Furthermore, the concentration of ions is usually not high enough to render the ionic effect to be significant.

Assume the aqueous system we are studying contains $\mathscr{N}$ water molecules. Then the reference proton selected at random interacts with $2(\mathscr{N} - 1) + 1$ other water protons. One of these protons sits on the same water molecule as the reference proton; the $2(\mathscr{N} - 1)$ other protons belong to other water molecules. Thus for one proton neighbor the distance r_0 occurring in the time correlation function, eqn. (26), is constant and thus the propagator $P(\mathbf{r}_0, \mathbf{r}, t)$ is of rotational type; for the remaining $2(\mathscr{N} - 1)$ protons $\mathbf{r}$ varies, due to translational (and rotational) diffusion. The former type of correlation function leads to the intramolecular relaxation rate, the latter type leads to the intermolecular relaxation rate.

The rotational propagator is given by eqns. (29) and (30) when we assume that the rotational diffusion equation is the correct equation to describe the reorientational motion. Now the quantity D_r^* which occurs in the expression for $P(\Omega_0, \Omega, t)$ [eqn. (29)] is denoted as D_r; it is the rotational diffusion coefficient of the water molecule in the pure liquid. Likewise, according to eqn. (43) a reorientational or rotational correlation time of the water molecule is defined by

$$1/\tau_c = 6D_r \qquad \text{(rotational diffusion)}.$$

Whereas for the large hydration complex the strict diffusion model is certainly appropriate, for the small, single water molecule this picture may in principle be questionable. As was pointed out on p. 317, the validity of the diffusion model implies that the entire turning over of the molecule (or molecular system, e.g., a complex) is composed of a very large number of small rotational steps. The other extreme is the free rotation of a molecule in the gas, where we know that the turning over of the molecule is accomplished by a coherent dynamic process, a process which cannot be described by the diffusion equation. Thus we do not know where in the range between these two limits water has to be placed. Clearly, the strong hydrogen bonding of water at room temperature excludes almost-free rotation of the water molecule. On the contrary, the existence of hydrogen bonding means that the water molecule is a member of an aggregate formed from two or more water molecules at least for a certain fraction of the time. This makes the situation resemble that of a water molecule bound in the hydration complex, but, as we saw, the associated binding energies and thus the binding times are usually far beyond the values to be expected for the H-bond coupling in pure water. Without going further into the details at the moment, we see that, whereas the long lifetimes in the hydration complexes gave us a safe basis for applying the diffusion equation, the shorter lifetimes in H-bonded water just remove this basis. In this connection, one model has been discussed very often which leads to a type of motion which is not correctly described by the diffusion equation. This is the jump model: A given water molecule is locked in a quasicrystalline environment for a certain time τ and performs torsional oscillations, then after this period it changes its orientation during a very short time, i.e., it undergoes a rotational jump, then again it rests for a certain time in the bound state "expecting" the next rotational jump, and so on. It is obvious that this model is taken from the solid state, and its extension to a highly fluid liquid like water is very questionable. Certainly, in the bound state the water molecule, apart from oscillations, is not at rest, in the strict sense of the word, relative to the laboratory system and certainly the jump does not occur in an infinitely short time. Where, then, has the distinction to be made between jump and nonjump?

We see now that in order properly to describe the rotational thermal motion of the water molecule, we have to construct an improved rotational or intramolecular propagator. The propagator thus obtained will later also turn out to be appropriate for diamagnetic electrolyte solutions. We construct the intramolecular propagator as follows.

First we must define "bound" and "unbound" molecules in pure water. For the unbound molecule, during a mean time interval τ_a, all nearest neighbors possess a high total energy $E > \Delta E$ relative to a coordinate system in the molecule to be considered. We choose $\Delta E = -kT$ to be the energy defining our classification. We denote a water molecule as bound if at least one of its nearest neighbors has a total relative energy E for which $E < \Delta E = -kT$. Total energy means the sum of kinetic and potential energies. We could have taken any other energy ΔE for the classification of bound and unbound molecules. The reason for our choice of $\Delta E = -kT$ will be seen below. The probability of finding the molecule in an unbound (or "free") state is $p_0(\Delta E) = p_0(kT)$; kT in parentheses indicates that bound and unbound are defined by the energy $-kT$ (relative to an appropriate zero[349–351]). Then the lifetime in the bound state, $\tau_{b'}$, is given by

$$\tau_{b'} = \tau_a(1 - p_0)/p_0$$

We note here that a clear distinction of the concepts used is important. In Sections 2.2 and 2.3 we distinguished between water molecules being members of the hydration sphere (of a paramagnetic cation) and those being in the bulk phase. But we never stated that the water was bound to the ion. Thus our definition was given purely in terms of geometry: $|\mathbf{r}_0| > R_0$ for the bulk water, $|\mathbf{r}_0| = R_0$ for hydration water. Later we learned that the water molecule may indeed be considered as bound to the cation in the energetic sense. This is because the propagator was found to be of rotational character for $|\mathbf{r}_0| = R_0$, hence

$$\tau_b \gg \tau_c{}^* = \tau_0 e^{E/RT}$$

which gives us $E \gtrsim 4.5$ kcal with $\tau_c{}^* \approx 2 \times 10^{-11}$ sec and $\tau_0 = 10^{-14}$ sec; and the water is certainly bound *to the ion* when we again define binding in terms of an energy $< -kT$. Likewise, since the coupling constant $A = A(R_0) \neq 0$ for $|\mathbf{r}_0| = R_0$ and $A = 0$ for $|\mathbf{r}_0| > R_0$ we found appreciable binding energies between H_2O *and the cation* holding the water in the position $|\mathbf{r}_0| = R_0$. But we bear in mind that the distinction between hydration water and bulk water was primarily a geometric one and we shall discuss cases below where grouping of molecules in hydration spheres has not an *a priori* implication of binding in the energetic sense. To indicate that the lifetime $\tau_{b'}$ in a bound state has primarily an energetic meaning we add the prime to τ_b.

In the unbound state intermolecular forces are still acting and the reorientational motion may in fair approximation be considered to be diffusive. Let $D_r{}^\circ$ and $D_r^{b'}$ be the rotational diffusion coefficient in the unbound and bound states, respectively. In a way, one may describe this situation as a "jump" mechanism if $D_r{}^\circ \gg D_r^{b'}$, since the water molecule fluctuates between two states with fast and slow reorientational motion (the diffusion coefficient being different, e.g., by a factor ten).

It may be shown from a number of independent starting points that the rotational propagator for this situation has the form[42,345,879]*

$$P(\Omega_0, \Omega, t) = \sum_{l,k} Y_l^{k*}(\Omega_0) Y_l^k(\Omega) P_l(t) \tag{105}$$

where

$$P_l(t) = p_a{}'[\exp(-t/\tau_{al}')] + p_{b'}' \exp(-t/\tau_{b'l}') \tag{106}$$

with

$$\frac{1}{\tau_{al}'} = \frac{1}{2}\left\{\frac{1}{\tau_{al}} + \frac{1}{\tau_{b'l}} + \frac{1}{\tau_a} + \frac{1}{\tau_{b'}}\right.$$
$$\left. -\left[\left(\frac{1}{\tau_{b'l}} - \frac{1}{\tau_{al}} + \frac{1}{\tau_{b'}} - \frac{1}{\tau_a}\right)^2 + \frac{4}{\tau_a\tau_{b'}}\right]^{1/2}\right\} \tag{107}$$

$$\frac{1}{\tau_{b'l}'} = \frac{1}{2}\left\{\frac{1}{\tau_{al}} + \frac{1}{\tau_{b'l}} + \frac{1}{\tau_a} + \frac{1}{\tau_{b'}}\right.$$
$$\left. +\left[\left(\frac{1}{\tau_{b'l}} - \frac{1}{\tau_{al}} + \frac{1}{\tau_{b'}} - \frac{1}{\tau_a}\right)^2 + \frac{4}{\tau_a\tau_{b'}}\right]^{1/2}\right\} \tag{108}$$

$$p_{b'}' = \frac{(p_0/\tau_{al}) + [(1-p_0)/\tau_{b'l}] - (1/\tau_{al}')}{(1/\tau_{b'l}') - (1/\tau_{al}')} \tag{109}$$

$$p_a' = 1 - p_{b'}' \tag{110}$$

and

$$1/\tau_{al} = l(l+1)D_r{}^\circ \tag{111}$$

$$1/\tau_{b'l} = l(l+1)D_r^{b'} \tag{112}$$

We quote the two limiting values of $P_l(t)$: For $\tau_a, \tau_{b'} \ll \tau_{al}, \tau_{b'l}$ we obtain

$$P_l(t) = \exp\{-t[(1/\tau_{al})p_0 + (1-p_0)(1/\tau_{b'l})]\}$$
$$= \exp\{-tl(l+1)[p_0 D_r{}^\circ + (1-p_0)D_r^{b'}]\} \tag{113}$$

On the other hand for $\tau_a, \tau_{b'} \gg \tau_{al}, \tau_{b'l}$ we obtain

$$P_l(t) = p_0[\exp(-t/\tau_{al})] + (1-p_0)\exp(-t/\tau_{b'l}) \tag{114}$$

We see that whereas eqn. (113) yields an exponential decay of $P_l(t)$, eqn. (114) gives the superposition of two exponentials. Assuming that $\tau_a, \tau_{b'} \ll \tau_{al}, \tau_{b'l}$, then as a consequence of eqns. (111) and (112) there will be

* See also Refs. 15 and 16.

always τ_{al}, $\tau_{b'l}$ for which

$$\tau_{al} \approx \tau_a; \qquad \tau_{b'l} \approx \tau_{b'}$$

This means that for sufficiently long times the diffusion equation is approximately obeyed with $P_l(t)$ given by the mean diffusion coefficient, eqn. (113); but for short times the propagator is no longer a solution of the diffusion equation; this occurs when the observation time is of the same order as the exchange times. Now the dynamic details become important. As already mentioned, dynamic coherence effects also contribute to make the diffusion equation a poor description for short times. We have here presented the formulas for a two-state model. More general models with more states may be similarly treated[879] but the resulting formulas for any intermediate case become very complicated.

We recall that if in our two-state treatment the rotational diffusion coefficients of the bound and unbound water are very different, then we may consider this situation as "jump" model. However, frequently the meaning of the jump model is given by the fact that the unbound molecule undergoes dynamically coherent motion for which no diffusion coefficient can be defined. Whatever the justification of such a model may be, we wish to add some comments concerning this model.

A treatment whose main merit is its mathematical elegance is that of Ivanov[398] (see also Ref. 384), who derived the propagator which results from a rotational random-walk process. Given a certain probability distribution for the change of orientation caused by one infinitely short jump, the probability is calculated that after a time t the change of orientation has a certain value due to the sequence of many jumps. Between the jumps orientation is preserved. We do not reproduce Ivanov's result here because the formulas are complicated. Rather we shall present a much simpler expression which also contains all the important features characteristic for a jump model.[345] Suppose that the molecule resides in a bound state for a certain time $0 < t' < t_1$. Within this time interval $P(\Omega_0, \Omega, t)$ is determined by the solution of the diffusion equation (28). At $t = t_1$ the molecule performs a finite rotational jump, and immediately after this jump $P(\Omega_0, \Omega, t)$ is supposed to be given by

$$P(\Omega_0, \Omega, t_1) = \sum_{l,m} Y_l^{m*}(\Omega_0) Y_l^m(\Omega) e^{-D_r l(l+1)(t_1+\xi)} \qquad (115)$$

$$= \sum_{l,m} Y_l^{m*}(\Omega_0) Y_l^m(\Omega) C_l e^{-D_r l(l+1)t_1} \qquad (116)$$

i.e.,

$$C_l = e^{-D_r l(l+1)\xi}, \qquad \xi \geq 0 \qquad (117)$$

Thus the rotational jump at time t_1 is assumed to cause a probability distribution that would have been developed at the time $t_1 + \xi$ if only true infinitesimal step diffusion had been effective all the time. In the special case $C_l = 0$ for $l \geq 1$ we have at $t > t_1$

$$P(\Omega_0, \Omega, t) = 1/4\pi$$

For times $t > t_1$ the water molecule is assumed to continue a diffusive motion in the bound state according to eqn. (28), until after a certain time it performs another finite rotational jump, etc. It may be shown that the whole effect of these finite rotational jumps is a propagator[345]

$$P(\Omega_0, \Omega, t) = \sum_{l,m} Y_l^{m*}(\Omega_0) Y_l^m(\Omega) \exp\{-t[l(l+1)D_r + (1-C_l)/\tau_{b'}]\}$$

$$= \sum_{l,m} Y_l^{m*}(\Omega_0) Y_l^m(\Omega) \exp\{-t[(1/\tau_l) + (1-C_l)/\tau_{b'}]\} \qquad (118)$$

with $1/\tau_l = l(l+1)D_r$. We see again that $P(\Omega_0, \Omega, t)$ as given by eqn. (118) is no longer a solution of the diffusion equation, but according to eqn. (117), the propagator approaches to the infinitesimal step propagator as $C_l \to 1$.

In the deeper physical significance the two propagators of eqns. (105) and (118) are equivalent. The only difference is that in the former case the diffusionlike behavior is realized for sufficiently long observation times if the lifetimes in the various states are sufficiently short compared with the correlation time $[l(l+1)D_r]^{-1}$ in these states. In the latter case, however, the reorientation time in the unbound state is infinitely short, so that the diffusionlike behavior can never be attained for finite rotational steps.

Before we discuss further the rotational motion of the water molecule we describe the intermolecular propagator for the translational relative motion of different water molecules, $P(r_0, r, t)$, testing the conditions under which the solution of the translational diffusion equation might not apply. Again we have to treat the jump model and the two-state diffusion treatment in a rigorous manner. A propagator for the translational random-walk jump model has been given by Torrey[781] as

$$P(\mathbf{r}_0, \mathbf{r}, t) = (2\pi)^{-3} \int \exp\{-(2t/\tau_{b'})[1 - A(\boldsymbol{\rho})]\} \exp[-i\boldsymbol{\rho}(\mathbf{r}-\mathbf{r}_0)]\, d\boldsymbol{\rho} \qquad (119)$$

where $A(\boldsymbol{\rho}) = A(\varrho)$ is the three-dimensional Fourier transform of $P_1(\mathbf{r})$, the probability that one simple jump displaces the particle by $\mathbf{r}$:

$$A(\varrho) = \int P_1(\mathbf{r}) \exp(i\boldsymbol{\varrho} \cdot \mathbf{r})\, d\mathbf{r}$$

Equation (119) is the translational analog to the Ivanov expression previously mentioned. Torrey proposed the following expression for $P_1(\mathbf{r})$:

$$P_1(\mathbf{r}) = (1/4\pi D\tau_{b'}r) \exp[-r/(D\tau_{b'})^{1/2}]$$
$$= (1/\tfrac{2}{3}\pi\langle r^2\rangle r) \exp[-r(6/\langle r^2\rangle)^{1/2}] \tag{120}$$

where $\langle r^2\rangle = 6\tau_{b'}D$ is the mean square displacement of the molecule caused by one jump. From eqn. (120) we obtain

$$A(\varrho) = 1/(1 + D\tau_{b'}\varrho^2) = 1/(1 + \tfrac{1}{6}\langle r^2\rangle\varrho^2) \tag{121}$$

In the limiting case of vanishing jumping time and jumping length one finds the diffusion propagator from eqn. (119) which corresponds to eqn. (36a). An expression equivalent to eqn. (118) has not yet been derived. We do not give the explicit formulas of the propagator for the two-state diffusional model; it has the form

$$P(\mathbf{r}_0, \mathbf{r}, t) = (2\pi)^{-3} \int \{p_a''[\exp(-2t/\tau_{ta}')] + p_b'' \exp(-2t/\tau_{tb}')\}$$
$$\times \exp[-i\boldsymbol{\varrho}(\mathbf{r} - \mathbf{r}_0)]\, d\boldsymbol{\varrho} \tag{122}$$

where the expression in the curly brackets has to be constructed as the translational analog of eqn. (106) (see below).

The translational expressions are of minor importance for the following reason: All conclusions which can be derived from nuclear magnetic relaxation times of diamagnetic electrolyte solutions stem from an analysis of the intramolecular rotational part of the motion. This part implies that motion of only one intramolecular neighbor proton of the reference proton is involved. In contrast, the translational part implies the interaction with $\mathscr{N} - 1$ other water molecules, so that it is understandable that the information we could obtain from intermolecular relaxation using the translational propagator is necessarily of a lower quality.

2.4.2. Proton Relaxation Rate in Pure Water*

We now pass to the further evaluation of the rotational correlation function. Due to the orthogonality relations of the spherical harmonics,

* See also Volume 1, Chapter 6.

the correlation function according to eqn. (26) with $r_0 = r = \text{const} = b$ and the propagator of eqn. (105) is easily obtained as

$$g(t) = (1/4\pi b^6)P_2(t) \tag{123}$$

with the two limiting cases

$$g(t) = (1/4\pi b^6) \exp -t\{(1/\tau_{a2})p_0 + [(1 - p_0)/\tau_{b'2}]\} \tag{124}$$

for $\tau_a, \tau_{b'} \ll \tau_{a2}, \tau_{b'2}$; and

$$g(t) = (1/4\pi b^6)\{p_0[\exp(-t/\tau_{a2})] + (1 - p_0) \exp(-t/\tau_{b'2})\} \tag{125}$$

for $\tau_a, \tau_b \gg \tau_{a2}, \tau_{b'2}$; and with the propagator of eqn. (118),

$$g(t) = (1/4\pi b^6) \exp -t\{(1/\tau_{b'2}) + [(1 - C_2)/\tau_{b'}]\} \tag{126}$$

where b is the proton–proton distance in the water molecule ($b = 1.52$ Å).

Since all the time constants are very small compared with the period of the Larmor precession, we have the so-called extreme narrowing case and the Fourier transform of the time correlation function degenerates to a simple time integration

$$J(0) = 2 \int_0^\infty g(t)\, dt = (1/4\pi b^6) \int_{-\infty}^{+\infty} P_2(t)\, dt \tag{127}$$

The resulting intramolecular relaxation rate is easily calculated from eqns. (21), (21c), and (21d) as

$$(1/T_1)_{\text{intra}} = \tfrac{3}{2}\gamma_I^4 \hbar^2 \tau_c / b^6 \tag{128}$$

where τ_c is given as

$$\tau_c = \frac{\tau_{a2}\tau_{b'2}[(1/\tau_a) + (1/\tau_{b'})] + p_0\tau_{a2} + (1 - p_0)\tau_{b'2}}{(\tau_{a2}/\tau_a) + (\tau_{b'2}/\tau_{b'}) + 1} \tag{129}$$

and with

$$\frac{\tau_a}{\tau_{b'}} = \frac{1 - p_0}{p_0}; \qquad \tau_{b'} = \text{const}$$

$$\tau_c = \frac{\tau_{a2}\tau_{b'2} + (1 - p_1)^2\tau_{a2}\tau_{b'} + (1 - p_1)p_1\tau_{b'2}\tau_{b'}}{\tau_{b'2} + p_1(\tau_{a2} - \tau_{b'2}) + (1 - p_1)\tau_{b'}} \tag{130}$$

where p_1 is the probability for a water molecule to be in the bound state;

$p_1 = (1 - p_0)$. This τ_c is obtained from the time integral (127) according to eqn. (123) and rearrangement of terms.

For $\tau_{b'} \gg \tau_{a2}, \tau_{b'2}$ eqn. (129) reduces to

$$\tau_c = \tau_{a2}p_0 + \tau_{b'2}p_1 = \tau_{a2}(1 - p_1) + \tau_{b'2}p_1 \tag{131}$$

whereas the other limiting case, $\tau_{b'} \ll \tau_{a2}, \tau_{b'2}$, gives

$$\tau_c = \tau_{a2}\tau_{b'2}/(\tau_{b'2}p_0 + \tau_{a2}p_1)$$
$$= \tau_{a2}\tau_{b'2}/[\tau_{b'2}(1 - p_1) + \tau_{a2}p_1] \tag{132}$$

Finally, use of the time correlation function for nondiffusive jumps [eqn. (105)] yields

$$1/\tau_c = (1/\tau_{b'2}) + [(1 - C_2)/\tau_{b'}] \tag{133}$$

τ_c is the rotational or reorientational correlation time and we see that this correlation time may be expressed in a number of different forms depending on the structural and microdynamic details in liquid water. But, due to the fact that we have the extreme narrowing situation, i.e., we are performing "zero-frequency" measurements with water and diamagnetic electrolyte solutions, we have no experimental access to the structural particularities of τ_c. Worse than this, as already indicated, we do not even have an independent means of measuring τ_c. At best we measure the product of τ_c with an interaction factor e.g., b^{-6}. Still we are confident nowadays that $\tau_c = 2.5 \times 10^{-12}$ sec $\pm 10\%$ at 25°C.[283,347] The only information about the dynamic details of rotational motion of H_2O comes from the comparison of τ_c, the correlation time of the second-order spherical harmonics, with the reorientation time τ_r, the correlation time of the spherical harmonics of first order.

According to eqns. (44a) and (44b), we generally expect

$$\tau_r = 3\tau_c$$

if the diffusion equation is satisfied. Let us write an expression for τ_r which is analogous to eqn. (130) and which for sufficiently short $\tau_{b'}$ has a physical meaning:

$$\tau_r = \frac{\tau_{a1}\tau_{b'1} + (1 - p_1)^2\tau_{a1}\tau_{b'} + (1 - p_1)p_1\tau_{b'1}\tau_{b'}}{\tau_{b'1} + p_1(\tau_{a1} - \tau_{b'1}) + (1 - p_1)\tau_{b'}} \tag{130a}$$

with $\tau_{b'1} = 3\tau_{b'2}$, $\tau_{a1} = 3\tau_{a2}$. Equation (130a) shows that if the inequalities leading to eqn. (132) are fulfilled, then we have $\tau_r = 3\tau_c$; however, if $\tau_{b'} \gtrsim \tau_{a2}, \tau_{b'2}$, we have $\tau_{b'} \approx \tau_{a1}, \tau_{b'1}$ and as may be seen from eqns. (130) and (130a), $\tau_r/\tau_c < 3$.

Furthermore, from eqn. (117), $C_2 = C_1^3$; thus for our rotational jump formula (133):

$$\frac{\tau_r}{\tau_c} = \frac{(1/\tau_{b'2}) + [(1 - C_1^3)/\tau_{b'}]}{(1/3\tau_{b'2}) + [(1 - C_1)/\tau_{b'}]}$$

$$= 3\,\frac{\tau_{b'} + (1 - C_1^3)\tau_{b'2}}{\tau_{b'} + 3(1 - C_1)\tau_{b'2}} < 3$$

and $\tau_r/\tau_c \to 1$ if $C_1 \to 0$ and $\tau_{b'} \ll \tau_{b'2}$.

The dielectric relaxation time of water at 25°C is $\tau_d = 8.25 \times 10^{-12}$ sec (Chapter 8, Table I). This is the relaxation time for the (macroscopic) dielectric polarization. Unfortunately, the connection between the times τ_d and τ_r is not very well established; $1 \leq \tau_d/\tau_r < 2^{(118,211)}$ (see also Chapter 8, Section 1.2). Even with $\tau_d/\tau_r = 2$ we get $\tau_r = 4.1 \times 10^{-12}$ sec, i.e., $\tau_r \neq \tau_c$, which excludes the extreme limit of a jump model but on the other hand it is very likely that $\tau_r/\tau_c < 3$, so the strict diffusive model is not quite appropriate either. Probably a residence time in the bound state of about $\sim 10^{-12}$ sec is a good approximation.

Next we discuss the formula for the intermolecular relaxation rate in the case of extreme narrowing, which is obtained with the propagator as given in eqns. (119) and (120):

$$(1/T_1)_{\text{inter}} = (\pi\gamma_I^4\hbar^2\tilde{c}_I\tau_{b'}/a^3)[1 + (2a^2/5D\tau_{b'})] \tag{134}$$

where $\tilde{c}_I$ is the number of spins (protons/cm³) and a is the closest distance of approach between two protons on different water molecules. As $\tau_{b'} \to 0$, eqn. (134) reduces to the simpler form

$$(1/T_1)_{\text{inter}} = \tfrac{2}{5}\pi\gamma_I^4\hbar^2\tilde{c}_I/aD \tag{135}$$

which would of course be obtained directly if the propagator as resulting from the diffusion equation was used from the very beginning; in other words, eqn. (135) is the extreme narrowing case of the like-spin analog of eqn. (53). We note that unfortunately there is no full correspondence between the results given for the intramolecular and intermolecular relaxation rates. Such correspondence is achieved only when $\tau_{a2} = \tau_{b'2}$ [see eqns. (128) and (129)], i.e., the one-state strict microstep diffusion model applies, but the general formula, eqn. (129), with the special cases of eqns. (131)–

(133) requires another model than that underlying eqn. (134). The calculation of the intermolecular relaxation rate for models leading to eqns. (129) and (131)–(133) has not yet been performed. We shall see shortly that it is just the structural implication in the eqns. (129) and (131)–(133) which renders them so useful for the study of diamagnetic electrolyte solutions. Thus, in the following our interest will be entirely concentrated on the behavior of the intramolecular relaxation rate.

2.4.3. *Experimental Results for Proton Relaxation Rates in Diamagnetic Electrolyte Solutions*

In the preceding section we have derived formulas for the intramolecular proton relaxation rate which can also be applied to diamagnetic electrolyte solutions and we saw that the intramolecular relaxation rate is more appropriate for the study of these solutions. However, the intention to focus interest on the intramolecular relaxation rate is not of much value unless one is able to separate the two contributions to the relaxation rate experimentally. The primary experimental result is the total relaxation rate $1/T_1$,

$$(1/T_1) = (1/T_1)_{intra} + (1/T_1)_{inter} \qquad (136)$$

and these contributions are almost equal; e.g., at 25°C for water $1/T_1 = 0.28 \text{ sec}^{-1}$, $(1/T_1)_{intra} = 0.18 \text{ sec}^{-1}$, and $(1/T_1)_{inter} = 0.1 \text{ sec}^{-1}$.[283,347] One way to find $(1/T_1)_{intra}$ would be to calculate the intermolecular relaxation rate according to eqn. (134) or eqn. (135), which is possible if D is known and a can be reasonably estimated. However, it turns out experimentally that $1/T_1$ and $1/D$ are roughly proportional to one another in diamagnetic electrolyte solutions.[195] Thus with eqn. (135) [the difference between eqns. (134) and (135) is small]

$$1/T_1 = (1/T_1)_{intra} + \alpha(1/T_1)$$

or

$$(1/T_1)_{intra} = (1/T_1)(1 - \alpha) \qquad (137)$$

where α is approximately constant for solutions of many simple salts[195] (an exception was observed for $AlCl_3$).[359] The approximate validity of eqn. (137) gives us a simple means of studying the behavior of the intramolecular relaxation rate, that is, we may discuss the effects of dissolved salts on the total relaxation rate and we know that they are at the same time the effects on the intramolecular relaxation rate.

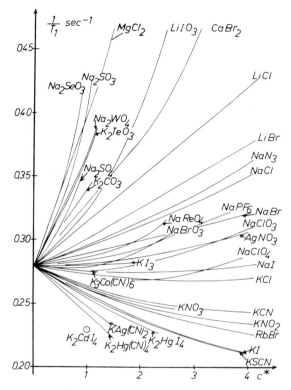

Fig. 7. Proton relaxation rates in aqueous solutions of a number of electrolytes as a function of the molal concentration c^*. Temperature, 25°C.[195,196]

The proton relaxation rate of water in diamagnetic electrolyte solutions generally depends on the concentration.[195,196,206,207,359,361,417,793] The concentration dependence is much weaker than that observed for paramagnetic electrolyte solutions. Figure 7 shows a number of typical examples. The relative variation of the proton relaxation rate is of the same order of magnitude as the relative variation of the viscosity η of the solution, and, as already mentioned, of the same order as the relative change of the water self-diffusion coefficient.[195] Thus in analogy to the Jones–Dole relation we may write for the relative variation of the proton relaxation rate

$$1/T_1 = (1/T_1)_0(1 + B'c^* + C'c^{*2} + \cdots) \tag{138}$$

$(1/T_1)_0$ is the proton relaxation rate of pure water, B' and C' are constants, and c^* is the molal salt concentration. According to our previous statements, we also have

$$(1/T_1)_{\text{intra}} = (1/T_1)^0_{\text{intra}}(1 + B''c^* + C''c^{*2} + \cdots) \qquad (139)$$

with

$$B' \approx B'', \qquad C' \approx C'' \qquad (140)$$

Our main interest is devoted to the B' (and B'') coefficients. Numerical values for B' coefficients are available in the literature[195,196] but this is not the case for C' (and C''). As already indicated, the viscosity B coefficients

$$\eta = \eta_0(1 + Bc^* + Cc^{*2} + \cdots) \qquad (141)$$

are of similar magnitude but there is one important difference: whereas the viscosity coefficients are usually known with high precision, this is not so for B' (and B''). Experimental development has not yet reached the point so as to make precision measurements of B' coefficients possible.

Let us consider $1/T_1$ results at 25°C. We may divide the salts into two classes: a class for which $B' > 0$ and one for which $B' < 0$; $C' > 0$ in all cases. Without any explanation at the moment we denote the former class as structure-forming salts (at 25°C) and the latter class as structure-breaking salts (at 25°C). The reason for these names will be given below. Now we apply eqn. (128): if $(1/T_1)_{\text{intra}}$ varies with concentration, this may be due to a concentration dependence of b, the intramolecular proton–proton distance, or a concentration dependence of τ_c. We know from nuclear magnetic resonance studies of solid crystal hydrates[51] that for the hydration complexes b is the same as in water vapor. Thus we can exclude the possibility that a change of b with concentration contributes substantially to the variation of $(1/T_1)_{\text{intra}}$. Other experimental results lead to the same conclusion (see below). We may formulate the important result that any change of the intramolecular relaxation rate in diamagnetic electrolyte solution reflects a variation of the mean rotational correlation time of the water molecule in these solutions.

For the further analysis we need one other experimental observation: In KCl solutions $B' \approx 0$[195] (actually, B' is still slightly negative at 25°C, but at somewhat higher temperatures $B' = 0$). This means that the ions K^+ and Cl^- at moderate concentrations have no influence on the mean correlation time τ_c of water. We go one step further: Referring to the fact that the transference numbers of Cl^- and K^+ are almost equal in H_2O, we also assume that the change in the water correlation time produced by the ions K^+ and Cl^- separately is negligible. We now consider as an introductory example a solution of NaCl. For NaCl one finds $B' > 0$, i.e., the mean water correlation time increases with increasing NaCl concentration. We now interpret this finding.

2.4.4. *Interpretation of Relaxation Data*

Let us follow the path of one individual water molecule through the solution. We distinguish two different environments between which the water exchanges: (1) The water molecule has a Na^+ ion as a nearest neighbor. (2) The water molecule has no Na^+ as nearest neighbor. This is again our geometrical classification of hydration water, or water of the hydration sphere, state b, and bulk water, state a. But we shall have to consider solutions with high ion concentration, so the second class of water molecules may be better called free water. We may also say that the water of the solution is divided in two subliquids, one subliquid being the hydration water of Na^+ (b), the other subliquid containing the remainder of the solvent (a). This notation has the advantage that it does not imply binding energies; in an energetic sense an individual hydration water molecule may be bound or unbound to any other particle depending on our choice of classifying energy $\varDelta E$; the same holds true for subliquid a.

As usual, we have to construct the appropriate propagator which describes the dynamic situation prevailing in the liquid in question. And here we may just apply our formulas (105)–(114) and (129)–(133). Previously we defined the two different states occurring in these equations as being bound or nonbound according to a classifying energy $\varDelta E$. But in eqn. (105) the way of defining states does not enter; it is merely necessary to specify two distinguishable states. So these two states may be our geometrically defined subliquids b and a, and we may simply characterize them by their different rotational correlation times. In subliquid a the correlation time is τ_c°; in the hydration spheres, i.e., in subliquid b, the correlation time is τ_c^+. The lifetime of the water in the hydration sphere is τ_b. Then we arrive at eqn. (129) for the mean correlation time in the solution. However, in an electrolyte solution there will be contact of hydration spheres as the concentration increases; consequently the residence time in subliquid b will increase. We write as a simple approximation

$$\tau_b^{-1} = (1 - x_h^+)/k_b, \qquad k_b = \text{const}$$

We have

$$\tau_a/\tau_b = (1 - x_h^+)/x_h^+, \qquad x_h^+ = c^+ n_h^+/55.5$$

assuming that a water molecule never has two Na^+ nearest neighbors. Here c^+ is the molal Na^+ concentration and n_h^+ is the hydration number of the cation, i.e., the mean number of water molecule nearest neighbors of the cation (Na^+ here). Then, replacing p_0 by $(1 - x_h^+)$, τ_{a2} by τ_c°, $\tau_{b'2}$ by τ_c^+, and $\tau_{b'}$ by τ_b in eqn. (129), we derive

$$\tau_c = \frac{(\tau_c^\circ \tau_c^+/k_b) + (1 - x_h^+)\tau_c^\circ + x_h^+ \tau_c^+}{(1/k_b)[\tau_c^\circ x_h^+ + \tau_c^+(1 - x_h^+)] + 1} \tag{130b}$$

But the general formulas (129) and (130b) are too clumsy for further use, and we ask which; if either, of the limiting formulas (131) or (132) is the correct one? Our definition of the hydration sphere is a geometrical one; thus we estimate a limiting value of τ_b by geometrical arguments and we utilize the relation $\tau_b = \langle r^2 \rangle / 6D$. We put $\langle r^2 \rangle = (2.8 \text{ Å})^2$ for the mean square displacement of the water molecule, 2.8 Å being the diameter of the H_2O molecule. $D \sim 2 \times 10^{-5} \text{ cm}^2 \text{ sec}^{-1}$, giving $\tau_b \approx 8 \times 10^{-12} \text{ sec}$. Taking account of possible binding effects, we get

$$\tau_b \gtrsim 8 \times 10^{-12} \text{ sec}$$

τ_a is of the same order, as shown above.

In pure water $\tau_c^\circ = 2.5 \times 10^{-12} \text{ sec}$ at 25°C; in the hydration sphere of Na^+, $\tau_c^+ \gtrsim \tau_c^\circ$. We see that eqn. (132), the "fast exchange" limit, is certainly inappropriate. On the other hand, from the results given in Section 2.2 we know that $\tau_c^+ \lesssim 3 \times 10^{-11} \text{ sec}$, whereas in Sections 2.2 and 2.3 we learned that τ_b may attain very high values. Our result is

$$\tau_b/\tau_c^+ > 1$$

For strongly hydrated ions this ratio may become very large, $> 10^3$. Numerical examination of eqn. (130b) shows[345] that even for $\tau_b/\tau_c^+ = 1$ an equation of type (131) is still a good approximation. Thus we write from eqns. (128), (130b), and (131)

$$\left(\frac{1}{T_1}\right)_{\text{intra}} = \frac{3}{2} \frac{\gamma_I^4 \hbar^2}{b^6} [\tau_c^\circ(1 - x_h^+) + \tau_c^+ x_h^+]$$

$$= \frac{3}{2} \frac{\gamma_I^4 \hbar^2}{b^6} \cdot \left[\tau_c^\circ\left(1 - \frac{c^* n_h^+}{55.5}\right) + \tau_c^+ \frac{n_h^+ c^*}{55.5}\right], \qquad \frac{c^* n_h^+}{55.5} < 1 \tag{142}$$

or if we generalize this equation to an electrolyte solution where cations and anions have correlation times different from that of free water,

$$\left(\frac{1}{T_1}\right)_{intra} = \frac{3}{2}\frac{\gamma_I{}^4\hbar^2}{b^6}\left[\tau_c{}^\circ\left(1 - \frac{c^*v^+n_h{}^+}{55.5} - \frac{c^*v^-n_h{}^-}{55.5}\right)\right.$$
$$\left. + \tau_c{}^+\frac{c^*v^+n_h{}^+}{55.5} + \tau_c{}^-\frac{c^*v^-n_h{}^-}{55.5}\right] \tag{143}$$

$$\frac{c^*}{55.5}(v^+n_h{}^+ + v^-n_h{}^-) \leq 1$$

where 1 mol of salt dissociates in v^+ mol of cations and v^- mol of anions. Here $n_h{}^\pm$ indicates the hydration (i.e., first coordination) number of the cation and anion, respectively, and $\tau_c{}^+$ and $\tau_c{}^-$ are the rotational correlation times of water in the cationic and anionic hydration spheres, respectively.

Equation (143) corresponds to[879]

$$1/T_1 = x_1(1/T_{11}) + x_2(1/T_{12}) + x_3(1/T_{13}) + \cdots$$

T_{1i} is the relaxation time in component i; the mole fraction of component i is x_i. Exchange times between the various components are very much shorter than the T_{1i}.

The behavior of eqn. (143) outside the composition range indicated above will be discussed later.

The application of the "jump model" leading to the correlation time according to eqn. (133) is to be carried out along the following lines: We write separate expressions corresponding to eqn. (133) for the three sub-liquids, i.e., for the free water[345]

$$1/\tau_c{}^\circ = (1/\tau_2{}^\circ) + [(1 - C_2{}^\circ)/\tau^\circ]$$

and

$$1/\tau_c{}^\pm = (1/\tau_2{}^\pm) + [(1 - C_2{}^\pm)/\tau^\pm]$$

for the cationic and anionic hydration water, $\tau_2{}^\circ$ and $\tau_2{}^\pm$ are the purely diffusional correlation times which correspond to the (energetically) bound state in the respective subliquid. During the times τ° and $\tau^\pm$ the molecule is in a bound state. In the unbound state the jumps occur. Here $C_2{}^\circ$ and $C_2{}^\pm$ are the rotational step parameters for jumps leaving the molecule within the subliquid in which it started. One needs further rotational step constants for transitions from one subliquid to the other, e.g., $C_2^{+\circ} = C_2^{\circ+}$, which describes steps involving transition from the cationic hydration sphere to the free water and vice versa. Now, it has been shown[345] for electrolyte solutions containing only one ion with a correlation time different from that of water that the validity of (142) improves as $C_2{}^\circ \to 0$, $C_2{}^+ \to 0$, $C_2^{+\circ} \to 0$, *irrespective* of the ratios $\tau_b/\tau_c{}^+$ and $\tau_b/\tau_c{}^\circ$. By analogy, we expect the same to be true for the more general situation, eqn. (143). On the other hand, strong variation in the rotational step parameters along the path of the water molecule causes drastic changes in the concentration dependence of eqns. (142) and (143).[42,345] Furthermore, we expect that the behavior is similar to that predicted by the rotational jump model, when motion within and between the various subliquids is characterized properly by "diffusional jumps," i.e., fluctuations between states of high and low energy, but reorientational motion can always be described as diffusive.

So far we have defined the water to be in the hydration sphere of a cation when one nearest neighbor is a Na^+ ion, say. But there may be configurations where a given water has a second Na^+ in rather close neighborhood or it has both Na^+ and Cl^- as nearest neighbors. These events may be considered in terms of an approach and final overlap of hydration spheres. Likewise, the "free" water (a) was defined to be of such a nature that Na^+ is not a nearest neighbor of the molecule considered. However, the water may be in the second coordination sphere of Na^+, or it may be very far away from Na^+, which are clearly different environments. These complications will be the more important the higher the concentration. For our model they have two consequences[345]: (a) $\tau_c{}^\circ$ and $\tau_c{}^+$ vary with the concentration and (b) the hydration numbers $n_h{}^\pm$ decrease with concentration, as an effect of overlap of hydration spheres. However, we shall retain the concept of constant hydration numbers over the appropriate concentration ranges (see below) and treat only the correlation times as concentration-dependent. Thus our model is a microdynamic model of nonoverlapping hydration spheres. The resulting effective concentration-dependent correlation times are in principle not strictly measurable quantities.[345] We write for the concentration dependence of the correlation times in the three subliquids

$$\tau_c{}^\circ = \tau_c{}^{\circ\circ}[1 + a_0^+ c^+ + a_0^- c^- + a_0^{++}(c^+)^2 + a_0^{--}(c^-)^2 + a_0^{+-}c^+c^- + \cdots]$$

$$\tau_c{}^+ = \tau_c{}^{+\circ}[1 + a_+^+ c^+ + a_+^- c^- + a_+^{++}(c^+)^2 + a_+^{--}(c^-)^2 + a_+^{+-}c^+c^- + \cdots] \quad (144)$$

$$\tau_c{}^- = \tau_c{}^{-\circ}[1 + a_-^+ c^+ + a_-^- c^- + a_-^{++}(c^+)^2 + a_-^{--}(c^-)^2 + a_-^{+-}c^+c^- + \cdots]$$

Let us confine our treatment to 1–1 electrolytes. Then, by insertion of eqn. (144) into eqn. (143) and retaining only members up to the second order in c^*, we obtain

$$\left(\frac{1}{T_1}\right)_{\text{intra}} = \frac{3\gamma_I^4\hbar^2}{2b^6} \left(\tau_c^{\circ\circ} + \left[\tau_c^{\circ\circ}(a_0^+ + a_0^-)\right.\right.$$

$$+ \frac{n_h^+}{55.5}(\tau_c^{+\circ} - \tau_c^{\circ\circ}) + \frac{n_h^-}{55.5}(\tau_c^{-\circ} - \tau_c^{\circ\circ})\left]c^*\right.$$

$$+ \left\{[\tau_c^{+\circ}(a_+^+ + a_+^-) - \tau_c^{\circ\circ}(a_0^+ + a_0^-)]\frac{n_h^+}{55.5}\right.$$

$$+ [\tau_c^{-\circ}(a_-^+ + a_-^-) - \tau_c^{\circ\circ}(a_0^+ + a_0^-)]\frac{n_h^-}{55.5}$$

$$\left.\left. + \tau_c^{\circ\circ}(a_0^{++} + a_0^{+-} + a_0^{--})\right\}(c^*)^2\right) \tag{145}$$

and taking into account that for pure water

$$(1/T_1)_{\text{intra}}^{\circ} = \tfrac{3}{2}(\gamma_I^4\hbar^2/b^6)\tau_c^{\circ\circ}$$

we find according to eqn. (139) that

$$B'' \equiv \left[\left(\frac{1}{T_1}\right)_{\text{intra}}^{\circ}\right]^{-1}\left[\frac{d(1/T_1)_{\text{intra}}}{dc^*}\right]_{c^*=0}$$

$$= a_0^+ + a_0^- + \frac{n_h^+}{55.5}\left(\frac{\tau_c^{+\circ}}{\tau_c^{\circ\circ}} - 1\right) + \frac{n_h^-}{55.5}\left(\frac{\tau_c^{-\circ}}{\tau_c^{\circ\circ}} - 1\right)$$

$$C'' = \left[\frac{\tau_c^{+\circ}}{\tau_c^{\circ\circ}}(a_+^+ + a_+^-) - (a_0^+ + a_0^-)\right]\frac{n_h^+}{55.5}$$

$$+ \left[\frac{\tau_c^{-\circ}}{\tau_c^{\circ\circ}}(a_-^+ + a_-^-) - (a_0^+ + a_0^-)\right] + (a_0^{++} + a_0^{+-} + a_0^{--}) \tag{146}$$

This is our final result and gives the interpretation of the concentration dependence (up to second order in c^*) of the intramolecular proton relaxation rate in diamagnetic electrolyte solutions in terms of a number of structural and microdynamic properties. We see from eqn. (146) that the B'' coefficients are additive with regard to ionic contributions (whereas the C'' coefficients are not):

$$B'' = B''^+ + B''^-$$

or since

$$B'' \approx B', \qquad B' \approx B'^- + B'^+$$

as previously mentioned, we set

$$B'(K^+) = B'(Cl^-)$$

which permits the determination of all ionic $B'^{\pm}$ coefficients if the experimental B values are known. In Table II a number of ionic $B'^{\pm}$ coefficients are collected. The rotational correlation time in the first hydration sphere of an ion relative to the rotational correlation time in pure water may be evaluated utilizing eqn. (146):

$$\tau_c^{\pm \circ}/\tau_c^{\circ \circ} = (B'^{\pm} - a_0^{\pm})(55.5/n_h^{\pm}) + 1 \qquad (147)$$

Usually $n_h^{\pm}$ and $a_0^{\pm}$ are unknown quantities. $\tau_c^{\pm \circ}$ is defined as the correlation time of a water molecule at $c^* \to 0$ having the respective ion as a nearest neighbor; $n_h^{\pm}$ is the mean number of water molecules being nearest neighbors of the ion at $c^* \to 0$. If no direct information on $n_h^{\pm}$ is available, simple geometry, i.e., the number of water molecules in a volume given by the relation

$$\Delta V = (4\pi/3)[(r_{ion} + 2r_{H_2O})^3 - r_{ion}^3]$$

will be the best guide. a_0 is also difficult to obtain. For structure-forming ions $a_0^{\pm} \approx 0.4B'^{\pm}$ seems to be the best approximation available. This estimate stems from a comparison of the water self-diffusion coefficient with the ionic self-diffusion coefficient and from the fact that for $c^* \to 0$ the hydration water self-diffusion coefficient and ionic self-diffusion coefficient are equal[354,359] (see below).

It remains for us to discuss the concentration dependence of $(1/T_1)_{intra}$ (or $1/T_1$) for higher concentrations:

$$(c^*/55.5)(n_h^+ + n_h^-) > 1$$

[see eqn. (143) with $v^+ = v^- = 1$]. The free water is used up when

$$(c^*/55.5)(n_h^+ + n_h^-) = 1$$

and for higher concentrations the hydration water of one of the ions, the anions say, is used up. Usually we assume that the hydration water is used up of that ion to which the binding is weaker, but the choice is arbitrary. In this concentration range eqn. (142) is valid. Finally, for

$$c^* n_h^+/55.5 > 1$$

we can say that all water present in the solution is hydration water of the cations, say, if binding to cations is stronger, even if the anions are situated in the cationic hydration water.

TABLE II. Ionic Coefficients B'^+ and B'^-[195,196]

Ion	0°C	25°C	50°C	80°C
Li^+	0.14	0.14	0.12	0.10
Na^+	0.05	0.06	0.06	0.05
K^+	−0.05	−0.01	0.00	0.01
Rb^+	−0.08	−0.04	−0.01	0.01
Cs^+	−0.09	−0.05	−0.01	0.01
H^+	0.04	0.06	—	—
Ag^+	0.05	0.06	—	—
Mg^{2+}	0.59	0.45	0.44	0.43
Ca^{2+}	0.29	0.27	0.25	0.27
Sr^{2+}	0.20	0.23	0.25	0.27
Ba^{2+}	0.14	0.18	0.19	0.18
Pb^{2+}	—	0.20	—	—
Al^{3+}	—	2.45[359]	—	—
Ga^{3+}	—	1.7[359]	—	—
F^-	0.16	0.14	0.16	0.15
Cl^-	−0.05	−0.01	0.00	0.01
Br^-	−0.08	−0.04	−0.02	−0.01
I^-	−0.15	−0.08	−0.06	−0.04
N_3^-	−0.08	0.00	—	—
I_3^-	—	0.02	—	—
HSO_4^-	—	0.02	—	—
IO_3^-	0.01	0.02	—	—
HCO_3^-	—	0.04	—	—
$HCOO^-$	0.0	0.05	—	—
CF_3COO^-	0.14	0.10	—	—
CD_3COO^-	—	0.18	—	—

TABLE II. (*Continued*)

Ion	0°C	25°C	50°C	80°C
$(CD_3)_4N^+$	—	0.18	—	—
OH^-	0.12	0.18	—	—
CCl_3COO^-	0.24	0.18	—	—
ReO_4^-	—	−0.03	—	—
CN^-	—	−0.04	—	—
NO_2^-	—	−0.05	—	—
NO_3^-	−0.10	−0.05	—	—
$AuBr_4^-$	—	−0.05	—	—
BrO_3^-	—	−0.06	−0.06	—
PF_6^-	—	−0.06	$−0.08_5$	—
BF_4^-	—	−0.07	—	—
SCN^-	—	−0.07	—	—
IO_4^-	—	$−0.07_5$	—	—
ClO_3^-	—	−0.08	—	—
ClO_4^-	−0.12	$−0.08_5$	—	—
$Ag(CN)_2^-$	—	−0.09	—	—
$Co(CN)_6^{3-}$	—	0.02	—	—
$S_4O_6^{2-}$	−0.06	0.05	—	—
SO_4^{2-}	0.15	0.12	—	—
WO_4^{2-}	0.20	0.17	—	—
SO_3^{2-}	—	0.22	—	—
CO_3^{2-}	0.26	0.25	—	—
SeO_3^{2-}	—	0.26	—	—
TeO_3^{2-}	0.40	0.31	—	—
HgI_4^{2-}	−0.11	−0.06	—	—
CdI_4^{2-}	—	−0.09	—	—
$Hg(CN)_4^{2-}$	−0.17	−0.11	—	—

If either of the concentrations

$$(c^*/55.5)(n_h{}^+ + n_h{}^-) = 1, \qquad c^*n_h{}^+/55.5 = 1$$

had real physical significance in the sense that water of discontinuous motional character is used up, then breaks in the slope of $(1/T_1)_{intra}$ should be observed at these concentrations.[345] Such breaks have occasionally been reported.[847,848] However, examples where independent confirmation from other laboratories exists are still lacking.

2.4.5. *Discussion of Correlation Times in Diamagnetic Electrolyte Solutions*

In Table III we have collected a number of results for $\tau_c^{\pm\circ}/\tau_c^{\circ\circ}$. All these numbers have been determined assuming $a_0{}^{\pm} = 0$. However, generally the magnitude of $\tau_c^{\pm\circ}/\tau_c^{\circ\circ}$ will be somewhat smaller if $\tau_c^{\pm\circ}/\tau_c^{\circ\circ} > 1$, and for $\tau_c^{\pm\circ}/\tau_c^{\circ\circ} < 1$ it will be greater when the a_0 corrections have been taken account of.

We may compare the numbers listed in Table III with the results presented in Table I. The latter correlation times concern the reorientation of the vector connecting the proton in the first hydration sphere with the center of the ion. The data given in Table III refer to the vector connecting two protons within the H_2O of the first hydration sphere. The correlation time of pure water $\tau_c^{\circ\circ} = 2.5 \times 10^{-12}$ sec. Thus for Mg^{2+}, according to Table III, we find $5.2 \times 2.5 \times 10^{-12}$ sec $= 1.3 \times 10^{-11}$ sec. But for Mn^{2+}, an ion very similar to Mg^{2+}, using the fact that Mn^{2+} is paramagnetic, $\tau_c = 3 \times 10^{-11}$ sec has been found (Table I). This difference between the two correlation times by about a factor of two must be due to a rotation of the hydration water molecule about the O–M axis, where M is the metal ion and O is the oxygen of the water[354] (see also Ref. 124). The correlation time for a given vector in the presence of internal rotation (in the case of extreme narrowing) is given by the Fourier transform (at $\omega = 0$) of eqn. (45):

$$\tau_c = \frac{\tau_c^*}{4}\left[(3\cos^2\theta - 1)^2 + 3(\sin^2 2\theta)\,\frac{1}{(\tau_c^*/\tau_i) + 1} \right.$$
$$\left. + 3(\sin^4\theta)\,\frac{1}{4(\tau_c^*/\tau_i) + 1} \right] \tag{148}$$

where θ is the angle between the vector whose reorientation is being studied and the O–M axis about which the rotation of the H_2O molecule occurs. τ_c^* is the rotational correlation time of this axis, i.e., the correlation time for the reorientation of the octahedral "skeleton" of the hydration complex, and τ_i is the reorientation time of the water about the O–M axis. The time which is measured in paramagnetic Mn^{2+} solutions is approximately

TABLE III. Ratio of the Correlation Time $\tau_c^{\pm}$ in the Hydration Sphere to the Correlation Time τ_c° in Pure Water for Various Hydration Numbers at 25°C[a]

Ion	$B'^{\pm}$	$n_h^{\pm} = 4$	6	8	10	12	14	16	20	24
Li^+	0.14	2.9	2.3	—	—	—	—	—	—	—
Na^+	0.06	1.8	1.6	1.4	—	—	—	—	—	—
K^+	−0.01	—	0.9	0.9	—	—	—	—	—	—
Mg^{2+}	0.45	—	5.2	—	—	—	—	—	—	—
Ba^{2+}	0.18	—	2.7	2.2	—	—	—	—	—	—
F^-	0.14	2.9	2.3	2.0	—	—	—	—	—	—
Br^-	−0.04	—	0.6	0.7	0.8	—	—	—	—	—
I^-	−0.08	—	—	0.4	0.6	—	—	—	—	—
TeO_3^{2-}	0.31	—	3.8	3.1	2.7	2.4	2.2	—	—	—
CO_3^{2-}	0.25	—	3.3	2.7	2.4	2.1	2.0	—	—	—
CCl_3COO^-	0.18	—	—	—	2.0	1.8	1.7	1.6	—	—
SO_4^{2-}	0.12	—	—	1.9	1.7	1.6	1.5	1.4	—	—
CN^-	−0.04	—	0.6	0.7	0.8	—	—	—	—	—
HgI_4^{2-}	−0.06	—	—	—	—	—	0.7	0.8	0.8	0.8
ClO_4^-	−0.08$_5$	—	—	0.4	0.5	0.6	0.7	0.7	—	—
$Ag(CN)_2^-$	−0.09	—	—	0.3	0.5	0.6	0.6	—	—	—
$Hg(CN)_4^{2-}$	−0.11	—	—	—	—	—	0.6	0.6	0.7	0.8

[a] All a coefficients are set equal to zero.

τ_c^*. Then with an angle $\theta = 90°$ for the intramolecular proton–proton vector one expects $\tau_c^*/\tau_i \approx 1$ in order to account for the ratio $\tau_c^*/\tau_c^+ = 2$. In the Al^{3+} hydration sphere rotation about the O–M axis is absent[359] (see below) but for all other hydration complexes so far studied (including probably Ga^{3+})[359] the water molecules rotate about the O–M axis on a time scale of about 10^{-11} sec. According to eqn. (148), $\tau_c^\pm/\tau_c^* \geq 1/4$, but for some of the structure-breaking (or weakly structure-forming) ions we find $\tau_c^\pm/\tau_c^* < 1/4$ if we put $\tau_c^* \approx 2 \times 10^{-11}$ sec. Obviously, the coupling of the water of the hydration aggregate, which is expressed via the condition $\tau_c^\pm/\tau_c^* \geq 1/4$, is removed for these ions, which means that the residence time of the water molecule in the first hydration sphere cannot be very long compared with the correlation time τ_c^*; rather it should be of the order of $\tau_c^\pm$ itself, i.e., $\sim 5 \times 10^{-12}$ sec, a figure which would result from purely translational diffusion of H_2O toward and away from the ion, as estimated above.

Another result which deserves special comment is that even anions with charge $z = 2$ may produce a $B' < 0$. So the structure-breaking effect is by no means limited to ions with unit charge, the important factor being the ratio z/r_{ion}, where r_{ion} is the ionic radius. A qualitative treatment discussing the various effects which determine the sign of B' has been given elsewhere.[196] Some exceptional cases where one expects a negative B' but finds a positive B' have been described.[196]

In all those cases where B' has been found to be negative, a $\tau_c^{\pm\circ}/\tau_c^{\circ\circ} < 1$ has been tabulated in Table III. By definition $\tau_c^{\pm\circ}$ describes the reorientational motion in the first hydration sphere of the ion, i.e., on the immediate surface of the ion. But eqn. (145) does not depend explicitly on the localization of the different regions in the solution, the only important fact being that the space is divided in distinguishable regions with different τ_c's. Thus one might argue[232] that for typical structure-breaking ions the water molecules with shorter reorientation times are those in the second coordination sphere, whereas those in the first layer have normal or even longer reorientation times than pure water. However, it should be kept in mind that the concentration range for which the mean water correlation time is shorter than that of pure water extends up to ~ 6 m. At such high concentration easier reorientation can no longer be a second-sphere effect, it must be due to the first coordination sphere. Thus, if an enhanced reorientation rate were a second-sphere effect at very low concentration, then at moderate concentrations some kind of rearrangement of the hydration shell should occur where slowly moving and ordered water molecules are replaced by less ordered and faster-moving ones. But all concentration dependences so far observed, be it of the water relaxation or of the ionic nuclear relaxation,

show a completely smooth and nonspecific behavior from $c^* \to 0$ up to
the highest concentrations, so that no experimental support seems to be
available for the hypothesis that at low concentration the structure-breaking
effect is just a second coordination sphere effect. We shall therefore adopt
the point of view that the high fluidity of water extends just to the surface
of the structure-breaking ions.[196,302,711,713]

As already mentioned, the numerical values of B' are similar to those
of the viscosity B coefficients. Thus, we expect the temperature dependence
of these two coefficients also to resemble one another. The temperature
dependence of B' may be described in the following way: We put

$$\tau_c^{\pm\circ} = \tilde{\tau}_0^{\pm} \exp(E_A^{\pm}/RT)$$
$$\tau_c^{\circ\circ} = \tilde{\tau}_0 \exp(E_A^{\circ}/RT)$$
(149)

Furthermore,

$$\Delta E_A^{\pm} = E_A^{\pm} - E_A^{\circ}$$
(150)

where $E_A^{\pm}$ is the activation energy for the reorientation of water in the
cationic or anionic hydration sphere, and E_A° is the corresponding activation energy in pure water.

It is convenient to rewrite eqn. (146) as

$$B' = a_0^+ + a_0^- + n_h^+ G^+ + n_h^- G^-$$
(151)

with

$$G^{\pm} = (1/55.5)\{g \exp(\Delta E_A^{\pm}/RT) - 1\}, \qquad g \gtrsim 1$$
(152)

Approximate experimental data for $E_A^{\pm}$ and E_A° are available[195] and are
collected in Table IV. It can be seen from these data that $\Delta E_A^{\pm} \gtrsim 0$ for
structure-forming ions ($B' > 0$ at 25°C) and $\Delta E_A^{\pm} \lesssim 0$ for structure-
breaking ions ($B' < 0$ at 25°C).

Thus as a rough estimate $dB'/dT < 0$ for the former and $dB'/dT > 0$
for the latter ions. But from the comparison of the experimental results
for dB'/dT and $dG^{\pm}/dT$ one finds[195]

$$dB'/dT > n_h^+(dG^+/dT) + n_h^-(dG^-/dT)$$

Consequently we must have

$$d(a_0^+ + a_0^-)/dT > 0$$
$$dn_h^{\pm}/dT > 0 \qquad \text{if} \quad G^{\pm} > 0$$
$$dn_h^{\pm}/dT < 0 \qquad \text{if} \quad G^{\pm} < 0$$

TABLE IV. Activation Energies for the Rotational Motion of Water Molecules in the Hydration Spheres of Various Ions[195]

Ion	$E_A{}^\circ$, $E_A{}^\pm$, kcal mol^{-1}, high-temperature range 35–80°C	$E_A{}^\circ$, $E_A{}^\pm$, kcal mol^{-1}, low-temperature range 0–25°C
H_2O	3.3	4.6
Li^+	4.3	4.7
Na^+	4.2	6.3
K^+	3.3	3.5
Rb^+	2.9	3.5
Cs^+	2.7	3.3
Mg^{2+}	5.2	7.8
Ca^{2+}	4.6	6.2
Sr^{2+}	3.7	5.1
Ba^{2+}	3.7	4.6
F^-	4.3	5.1
Cl^-	3.3	3.5
Br^-	2.9	3.3
I^-	2.7	2.9

The first inequality is probably the more important one, and it expresses the following fact: With increasing temperature the effect of the second (and higher) coordination sphere of an ion changes in the sense such as to make molecular motion slower, or, in other words, the breakdown of water structure further away from the ion, which, together with the first sphere effect, causes the slowing down of molecular motion in aqueous ionic solution is relatively weak at low temperatures,[196] and disappears with increasing temperature. It is due to this particular property of water that in certain temperature ranges one observes $dB'/dT > 0$ even in solutions where $B' > 0$. Corresponding behavior has also been observed for the viscosity B coefficients.[596,758]

2.4.6. *Rotational Correlation Times of Interparticle Vectors in the Hydration Sphere from Proton Relaxation Measurements*

Two proton relaxation studies have been performed with diamagnetic electrolyte solutions which are of a different type from those so far described.

The objective of these investigations was not to find the rotational correlation time of the intramolecular proton–proton vector in the hydration complex, but the correlation time for the reorientation of the vector connecting the proton in the first hydration sphere with the ion in the center of the complex. These studies are therefore the diamagnetic analogs of the investigations on paramagnetic ions like Mn^{2+} and Cu^{2+} described in Section 2.2.

In order to understand the principle of these measurements, we start from eqn. (136). The first example is a solution of LiCl in H_2O.[205,348] Apart from the proton–proton contribution, the intermolecular relaxation rate also contains a measurable proton–^7Li contribution since the magnetic moment of ^7Li is sufficiently large. Fabricand and Goldberg prepared solutions of ^7LiCl and ^6LiCl and measured the proton relaxation rate as a function of the concentration for both these LiCl solutions. Lithium-6 has a magnetic moment which is very small and thus its relaxation contribution is negligible. The difference of the two proton relaxation rates

$$(1/T_1)_{^7LiCl} - (1/T_1)_{^6LiCl} = 1/T_1'$$

is due to the desired proton–^7Li interaction.

As indicated, we have the diamagnetic analog to the proton relaxation caused by paramagnetic ions. Consequently, we can use eqn. (51) with $\omega_S \tau_c^{**} \to 0$,[348] which gives

$$\frac{1}{T_1'} = \frac{4}{3} \frac{\gamma_I^2 \gamma_S^2 \hbar^2 S(S+1) c^* n_h^+}{55.5 R_0^6} \left(\tau_c^{**} + \frac{\tau_t \tau_c^*}{\tau_c^* + \tau_b} \right) + \left(\frac{1}{T_1} \right)_{inter-H^7Li}$$

(153)

Here the new term with the correlation time $\tau_t \tau_c^*/(\tau_c^* + \tau_b)$[346] takes account of the fact that for sufficiently short lifetime τ_b the diffusive translational motion of ^7Li relative to the water protons is the process which determines decay of correlation; τ_t is the time after which the Li^+ ion has suffered a displacement of the order of the water diameter, i.e., $\tau_t \approx 5 \times 10^{-12}$ sec.[346] We have

$$\left(\frac{1}{T_1} \right)_{inter-H^7Li} = \frac{16}{45} \frac{\pi \gamma_I^2 \gamma_S^2 \hbar^2 S(S+1)}{aD} \tilde{c}_{Li}$$

(154)

according to eqn. (53) [apart from a factor 2/3, this formula is the same as eqn. (135)]. S is the spin of ^7Li ($S = 3/2$); γ_S is the gyromagnetic ratio of ^7Li.

The experimental result of Fabricand and Goldberg was

$$1/T_1' = 0.009 \pm 0.004 \text{ sec}^{-1}$$

For $(1/T_1)_{\text{inter-H}^7\text{Li}}$ we estimate with $a = 3$ Å

$$(1/T_1)_{\text{inter-H}^7\text{Li}} = 2 \times 10^{-3} \text{ sec}^{-1}$$

With $b = 2.7$ Å (see below) we find that

$$\tau_c^{**} = (2.6 \pm 1.4) \times 10^{-11} \text{ sec} \qquad \text{at } 25°\text{C}$$

Fabricand and Goldberg propose, without citing a reference, a smaller value for b, namely $b = 2.6$ Å; this would make τ_c^{**} smaller. We neglected the second term in the parentheses of eqn. (153). We note that our final numerical result differs from that given by Fabricand and Goldberg.

The other example to be discussed here concerns the hydration complex of Al^{3+}.[359] Again the intermolecular proton relaxation rate of eqn. (136) contains a contribution due to the proton–^{27}Al dipole–dipole interaction, i.e., the magnetic moment of ^{27}Al is large enough so as to produce a measurable relaxation effect. However, while no stable Al isotope without magnetic moment exists, the water protons can be replaced by deuterons, the magnetic moment of which is much smaller than that of ^{27}Al. The proton relaxation rate of a solution of $AlCl_3$ in D_2O_4 has been studied in the presence of a very small amount of H_2O. The few protons present in the solution also undergo weak relaxation effects by the deuterons and the protons present; both these contributions can be calculated. The observed relaxation rate is greater than the one thus calculated and the difference is caused by the dipole–dipole interaction with the Al^{3+} nuclei. If we again apply eqns. (153) and (154), we obtain the experimental result for the relaxation rate of a water proton in the first hydration sphere caused by the Al:

$$\left(\frac{1}{T_1}\right)_{\text{rot}}^{\text{HAl}} = \frac{4}{3} \frac{\gamma_I^2 \gamma_S^2 \hbar^2 S(S+1)}{R_0^6} \tau_c^{**} = 0.093 \text{ sec}^{-1}$$

(in D_2O), as limiting value for $c_{Al^{3+}} \to 0$. For ^{27}Al, $S = 5/2$ and with $R_0 = 2.58$ Å we find (after a slight isotope correction) for the hydration complex in H_2O

$$\tau_c^{**} = (5.3 \pm 1.3) \times 10^{-11} \text{ sec}$$

at 25°C. This result compares satisfactorily with the data obtained for paramagnetic Cr^{3+} as presented in Table I.

The reorientational correlation time of the intramolecular proton–proton vector in the first hydration sphere of Al^{3+} should be compared with this result. For such a comparison it is important to take account of the second coordination sphere effect on the proton relaxation rate of the solvent water, which we have expressed as $a_0 = a_0^+ + a_0^-$ in eqn. (146) (here we should write $a_0 = a_0^+ + 3a_0^-$). The appropriate corrections have been performed (for details see Ref. 359) and the following result was obtained:

$$\tau_c^{+\circ} = (4.4 \pm 0.6) \times 10^{-11} \text{ sec}$$

Since $\tau_b \gg \tau_c^*$ (see below) we obtain from

$$\tau_c^{**} = \tau_c^* \tau_b / (\tau_b + \tau_c^*)$$

$$\tau_c^{**} = \tau_c^*$$

and consequently we have the result

$$\tau_c^{+\circ} \approx \tau_c^*$$

i.e., the rotational correlation times of the two vectors, proton–Al and proton–proton, are equal within appropriate limits. The hydration complex of Al^{3+} behaves like a rigid molecular complex on the time scale of 10^{-10} sec and rotation of water about the O–M axis is absent.

2.4.7. Transverse Proton Relaxation Rates in Diamagnetic Electrolyte Solutions

So far when speaking about diamagnetic electrolyte solutions we have only mentioned the longitudinal relaxation time T_1. Most measurements on diamagnetic electrolyte solutions are indeed T_1 measurements. The time T_1 is of the order of seconds, and T_2 values of this length are very difficult to measure. On the other hand, measurements of T_2 would not be of much value because in most cases the condition of extreme narrowing is fulfilled. Thus $\tau_c \omega \ll 1$ and thus $T_2 = T_1$. Only in neutral water do we have $T_2 < T_1$, which is due to the presence of a small amount of $H_2^{17}O$ and to the relatively slow proton exchange of the water molecule ($\tau_b \approx 2 \times 10^{-3}$ sec at room temperature[275,564] at $pH = 7$). As soon as the pH deviates by about one unit in either direction from $pH = 7$ the proton exchange becomes faster and one finds $T_1 = T_2$. So, apart from this effect, one expects to find $T_1 = T_2$ in most diamagnetic electrolyte solutions. This has been checked in a number of cases[525] and the only solution where $T_2 < T_1$ was an $AlCl_3$ solution. The explanation for this effect is the slow exchange of the water protons in the hydration sphere of Al^{3+}. It may be shown[566] that the difference $(1/T_2) - (1/T_1)$ is given by the relation

$$(1/T_2) - (1/T_1) = (\Delta\omega_b{}^*)^2 x_h{}^+ (1 - x_h{}^+)\tau_b$$

This formula is equivalent to eqn. (100), to which it reduces if $x_h{}^+ \to$ small. However, $\Delta\omega_b{}^*$ has an entirely different origin. It is the chemical shift the proton suffers in the hydration sphere of Al^{3+} relative to pure water as a consequence of the strong ionic electric field effect. Since $\Delta\omega_b{}^*$ is proportional to the magnetic field, $(1/T_2) - (1/T_1)$ should also become greater for higher magnetic fields, and this has been verified by Lohmann.[525] Furthermore, the chemical shift of the proton in the $AlCl_3$ solution has been measured and the final result was $\tau_b \approx 1 \times 10^{-5}$ sec at 25°C, the proton residence time being somewhat dependent on pH and on the concentration. At about $-50°$C hydration water and free water give separate proton NMR absorption lines.[239,728] Extrapolation of exchange data obtained at this temperature to 25°C yields $\tau_b = 10^{-6}$ sec. With addition of paramagnetic ions separate ^{17}O signals are observable at room temperature.[121,403] the exchange time of water oxygen is much longer, $\tau_b > 0.01$ sec.[33,216]

2.5. Quadrupolar Relaxation of Water in Diamagnetic Electrolyte Solutions

The two other water molecules whose relaxation times may be studied in diamagnetic electrolyte solutions are $\underline{D}_2O$ and $H_2{}^{17}\underline{O}$. As already mentioned, the nuclei D and ^{17}O both have $I > 1/2$ and thus possess an electric quadrupole moment. This nuclear quadrupole moment interacts with the electric field gradient at the position of the nucleus produced by the surrounding electric charges and causes an additional nuclear magnetic relaxation mechanism. As for most other molecules, where nuclei relax by quadrupole interaction, in $\underline{D}_2O$ and $H_2{}^{17}\underline{O}$ the relaxation effect due to magnetic dipole–dipole interaction with other nuclei is negligible when compared with the quadrupolar part. The reason for the spin relaxation by quadrupole interaction is always the statistical fluctuation of the electric field gradient at the nuclear site caused by molecular motion. As shown in Section 2.1, the spectral density of the interaction energy at the magnetic resonance frequency and related frequencies determines the rate of relaxation. We are fairly confident that in pure water only intramolecular field gradients are important, i.e., the components of the field gradient tensor are fixed quantities in the molecular frame. In electrolyte solutions, in the vicinity of small, highly charged ions intermolecular contributions to the electric field gradient cannot be excluded *a priori*. However, since the experimental data so far available do not provide any striking indication in this direction, we shall assume that in electrolyte solutions as well electric field gradients are only of intramolecular origin (see also Ref. 124). Of course, in the laboratory system which defines the axis of quantization of the spin system the tensor components of the electric field gradient fluctuate if the molecule undergoes reorientation processes. It is reasonable to assume that at least in pure water the reorientational motion of the water molecule is isotropic, i.e., there is no molecular axis about which preferential tumbling occurs. In all our discussion of proton relaxation we made this tacit assumption. In this situation, just as for the intramolecular magnetic dipole–dipole interaction, the time correlation function which enters in the theory and of which the Fourier transform (the spectral density) has to be taken is the correlation function of the second-order spherical harmonics[1]:

$$g(t) = (1/4\pi) \int Y_2^{m*}(\Omega_0)P(\Omega_0, \Omega, t)Y_2^{m}(\Omega) \, d\Omega_0 \, d\Omega \qquad (155)$$

where $g(t)$ does not depend on m, and where Ω_0 and Ω are the directions of the principal axis of the electric field gradient tensor at times $t = 0$ and $t = t$, respectively, or since reorientation is assumed to be isotropic, the orientation of any intramolecular vector relative to the laboratory coordinate system. Again the propagator $P(\Omega_0, \Omega, t)$ gives the probability

$$dP = P(\Omega_0, \Omega, t) \, d\Omega$$

of finding the molecular orientation to be Ω in the interval $d\Omega$ at time t if it had the orientation Ω_0 at time zero. We see that, apart from a constant factor, eqn. (155) and the first term of eqn. (40), which is the rotational part of the correlation function for dipole–dipole interaction, are identical. But our treatment of the intramolecular proton relaxation rate in diamagnetic electrolyte solutions was also based on this same rotational time correlation function of the second-order spherical harmonics, so all results obtained for the intramolecular proton relaxation rate as a consequence of the nature of the time correlation function, *and in particular of the propagator $P(\Omega_0, \Omega, t)$*, are also valid for the quadrupolar relaxation rate of both nuclei ^2H and ^{17}O in water. Then from eqns. (21i) and (155) the formula (156) may be derived exactly along the same lines as presented for the dipole–dipole intramolecular relaxation rate:

$$\frac{1}{T_1} = \frac{1}{T_2} = \frac{3}{40} \frac{2I + 3}{I^2(2I - 1)} \left(1 + \frac{\tilde{\eta}^2}{3}\right)\left(\frac{eQ}{\hbar} \frac{\partial^2 V}{\partial z'^2}\right)^2 \tau_c \qquad (156)$$

which of course only applies for the situation of extreme narrowing. The z' axis is the direction of the maximum electric field gradient in the molecular frame; it is the direction of the D–O vector for D_2O and the x axis for $H_2{}^{17}O$ when the water molecule is placed in the x–y plane with the y axis bisecting the H—O—H angle.[124] We note that eqn. (156) gives the total relaxation rate, which is at the same time the intramolecular relaxation rate, as was pointed out previously. The contribution of the asymmetry parameter is usually small and is omitted in the following discussion.

From the quadrupole coupling constants eQ/h of ^{2}H and ^{17}O in water and the respective experimental relaxation rates[124,347,845] one finds τ_c to be in very good agreement with the value obtained from the proton relaxation rate: $\tau_c = 2.9 \times 10^{-12}$ sec for D_2O and $\tau_c = 2.4 \times 10^{-12}$ sec for $H_2{}^{17}O$. Knowledge of the quadrupole coupling constant in the liquid state and knowledge of the isotopic corrections present certain difficulties but we do not propose to go into details. Having established that τ_c has substantially the same value as derived from the relaxation of all water nuclei, this provides support for our treatment, from which it follows that the same "structural" expressions are valid for τ_c which were derived for the intramolecular proton relaxation rate, namely the formulas (129)–(133). But these were the formulas which led us to our results for diamagnetic electrolyte solutions, i.e., eqn. (143). Hence we may write for the relaxation rate of $\underline{D}_2O$ or $H_2{}^{17}\underline{O}$

$$\frac{1}{T_1} = \frac{3}{40} \frac{2I+3}{I^2(2I-1)} \left(\frac{eQ}{\hbar} \frac{\partial^2 V}{\partial z'^2}\right)^2 \left(1 + \frac{\bar{\eta}^2}{3}\right)$$

$$\times \left[\tau_c^0\left(1 - \frac{c^* v^+ n_h{}^+}{55.5} - \frac{c^* v^- n_h{}^-}{55.5}\right) + \frac{c^* v^+ n_h{}^+}{55.5}\tau_c^+\right.$$

$$\left. + \frac{c^* v^- n_h{}^-}{55.5}\tau_c^-\right], \qquad \frac{c^*}{55.5}(v^+ n_h{}^+ + v^- n_h{}^-) \leq 1 \qquad (157)$$

Here we have assumed that the quadrupole coupling constant is the same for all three environments: the cationic and anionic hydration spheres and the free water. Small variations of the intramolecular field gradient in the various environments may be possible. As the next consequence of eqn. (157), we expect that the coefficients B'' and C'' as given in eqn. (146) are the same for all three water nuclear relaxation rates in diamagnetic electrolyte solutions, i.e., the relative concentration dependence should be substantially the same for the proton, deuteron,[358,361,796,848] and ^{17}O relaxation rates.[218] One sees from Fig. 8, which allows a direct comparison of the relative concentration dependences of the proton and deuteron

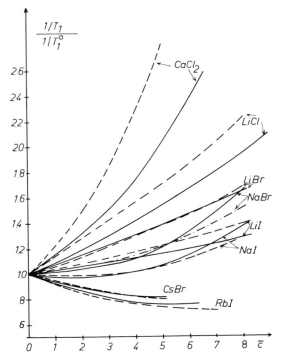

Fig. 8. Deuteron relaxation rates of a number of electrolyte solutions in D_2O as a function of the concentration $\bar{c}$ (aquamolality, moles of salt per 55.5 mol of water). Data are given as relative quantities relative to the relaxation rate of pure water. Temperature, 25°C. The dashed lines represent the corresponding relative proton relaxation rates.[358,361,796]

relaxation rates, that the predicted agreement does exist. It should be kept in mind that the experimental error of the data reported is at best $\pm 5\%$, so that searching for physically significant differences is not yet possible. We may mention one deviation of the relative proton relaxation rates from those of the deuterons which one would expect for comparatively strongly hydrated ions like Mg^{2+}, Ca^{2+}, or Li^+. This comes from the fact that we expect the identical relative concentration dependence for protons and deuterons only if the reorientational motion is isotropic. Now we concluded that in most cases the hydration water performs a rotation about the O–M axis (or corresponding axis for the F^- ion). With regard to the water molecule this is an anisotropic rotational motion; formula (148) is to be applied in eqns. (143) and (157). Here $\tau_c{}^*$ has, apart from isotope effects, the same value for protons and deuterons, but the angle θ is now $104.5/2°$ $= 52.2°$, whereas for protons it is $90°$. However, systematic experiments to study this effect have not yet been performed, to the author's knowledge.

Figure 9 shows ^{17}O relaxation rates[218] in a number of aqueous electrolyte solutions. The qualitative behavior is again similar to that of the proton and deuteron relaxation rate, but the decrease of the relaxation rate in the presence of structure-breaking ions is less pronounced. The reason for this pecularity seems to be known now. As may be seen from the figure, the experimental data presented are linewidth measurements. According to Section 2.1.1 a linewidth measurement is a measurement of the transverse relaxation rate. We have mentioned that in neutral water for

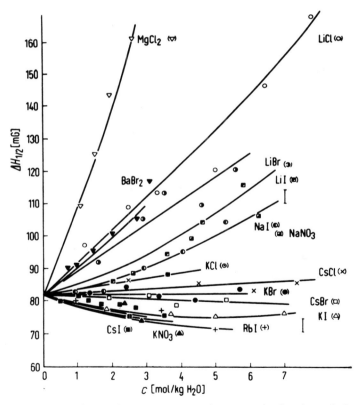

Fig. 9. Linewidths of ^{17}O resonance of water as a function of the concentration for a number of electrolyte solutions. The linewidth is proportional to the relaxation rates $1/T_2 \approx 1/T_1$. Temperature, 25°C.[218]

the proton $T_2 < T_1$, which is a consequence of the proton exchange of water. Since for the proton the fact that $T_2 < T_1$ is caused by the spin–spin coupling between the protons and the ^{17}O nucleus, one also expects and observes $T_2 < T_1$ for the ^{17}O relaxation.[564] For the measurements represented in Fig. 9 a small amount of NaOH was added to the solution which changed the pH sufficiently so as to produce fast proton exchange, making $T_2 = T_1$ in pure water. But at the time these experiments were performed it was not yet known that in the presence of ions (probably preferentially structure-breaking ions) the proton exchange in the water molecule becomes slower.[355] Thus we are not sure that the added small amount of NaOH resulted in linewidth measurements that really give T_1 measurements. More careful repetition of these, as well as of the proton and deuteron, measurements is desirable.

2.6. Relaxation Times of Ionic Nuclei Relaxing by Magnetic Dipole–Dipole Interaction

Since most of the nuclei with a magnetic moment have $I > 1/2$, we expect that in many cases the nuclei which reside in the solute particle, i.e., in the ions in our systems, undergo magnetic relaxation via the quadrupole interaction. Thus all nuclei in halide anions with the exception of F^- and all alkali metal nuclei relax by quadrupole interaction (partly with more than one isotopic species). But there is one favorable peculiarity: The ion ^7Li relaxes only to about 50% by quadrupole interaction, the remaining contribution being due to magnetic dipole–dipole interaction. The magnetic moments of the alkaline earth metal nuclei are zero or very small, so that measurements of relaxation rates would be very difficult; still, they have $I > 1/2$. Furthermore, all Group III metals have nuclei which relax by quadrupole interaction.

The interpretation of the ionic nuclear relaxation rate caused by quadrupole interaction requires a knowledge of the electric field gradient at the center of the spherical ion due to the surrounding particles. We are able to predict fairly well the ionic quadrupolar relaxation rate of halide and alkali metal ionic nuclei in the limit of high dilution.[343,352,356,360] The concentration dependence in many cases seems also to be understood in its general principles.[353,356,360] However, our knowledge of the theory is not yet good enough to permit the determination of relevant and meaningful microdynamic data from experimental quadrupolar relaxation rates, in addition to and in comparison with those presented on the preceding pages. Therefore a discussion of ionic nuclear relaxation rates will not be given for those cases where quadrupole interaction provides the main relaxation mechanism.

Even if $I = 1/2$, one cannot be sure that the ionic nucleus relaxes by magnetic dipole–dipole interaction. Very often when the nucleus resides in a molecular ion which lacks spherical symmetry other relaxation mechanisms occur which may compete with or entirely dominate over the magnetic dipole–dipole interaction. These are the relaxations by anisotropic chemical shift and by spin–rotation interaction. The so-called scalar interaction of second kind may also be effective in certain cases. Deviation from the predicted order of magnitude, temperature dependence, and H_2O–D_2O dependence indicates that one of these mechanisms dominates. Examples are the ^{15}N ($I = 1/2$) relaxations in the anions $^{15}\underline{N}O_3^-$, $^{15}\underline{N}O_2^-$, $S\underline{C}^{15}N^-$, which occur by anisotropic chemical shift and/or by spin–rotation interaction.[658] The ions $^{15}\underline{N}H_4^+$ and $(CH_3)_3{}^{15}\underline{N}H^+$ relax by intramolecular dipole–dipole interaction.[658] First estimates of the rotational correlation times are $\tau_c \approx 10^{-12}$ sec for the former and $\tau_c \approx 10^{-11}$ sec for the latter ion.[658] In both cases the concentration was very high, $\sim 20\ m$. Even in enriched samples the ^{15}N magnetic resonance signal is very weak, and since measurements at lower concentrations are desirable, progress in experimental techniques is awaited which will yield more interesting information.

Even if better experimental data were available for ions such as $^{15}NO_3^-$, e.g., at lower concentration, their interpretation would present difficulties because we have almost no knowledge of the (anisotropic) chemical shift tensor and of the spin-rotation interaction tensor. We see from these considerations that the only ions which may be treated with some success are those where a nucleus relaxes by magnetic dipole–dipole interaction. Here we have interaction of the reference nucleus with one or more magnetic point dipole(s); in this case the nature of interaction is simple and well known, but a problem arises since we are not always sure about the geometry, i.e., the distances of the point dipoles and the number of point dipoles at the respective sites.

F^-: The relaxation rate of the F^- ion in aqueous solution is due to the interaction of the reference ^{19}F nucleus in the center of the F^- ion with all other nuclear magnetic point dipoles in the solutions. In the limit of very low F^- concentrations all these magnetic point dipoles are water protons. We distinguish two classes of water protons: (1) those protons which are members of the first hydration sphere of the reference F^- ion selected at random and (2) all other water protons. The complex formed from the

$n/2 = n_h^-$ water molecules plus the ion is considered as a molecule, the lifetime of one water molecule being τ_b. If we consider the vector connecting the ^{19}F with a proton of the first hydration sphere, then of course the propagator which describes the motion of this vector of constant length is purely rotational in nature during the mean lifetime τ_b. Let R_0 again be the distance between the F nucleus and the first hydration proton; then we obtain the same result as eqn. (128), namely that the relaxation rate is

$$\gamma_S^2 \gamma_I^2 \hbar^2 \tau_c / R_0^6$$

A factor $\frac{3}{2}\gamma_I^2$ is replaced by γ_S^2, as may be shown from the theory; this is a consequence of the fact that "unlike spins" having different gyromagnetic ratios γ_I and γ_S are interacting here.[1] Further implications of the fact that the two spins are unlike and which gain importance for the ^{1}H–^{19}F spin pair, namely that relaxation is described here by a system of two coupled differential equations instead of the simple Bloch equation, need only be mentioned but not treated quantitatively; the numerical consequences are not essential.[443] But the ^{19}F interacts with n protons in the first hydration sphere, and we consider all these n contributions to be independent ("uncorrelated"), which again neglects a complication which we only mention without taking account of its small numerical corrections. Thus

$$(1/T_1)_{\text{intra}'} = \gamma_S^2 \gamma_I^2 \hbar^2 n \tau_c'' / R_0^6 \tag{158}$$

The subscript intra' indicates that we have an "intramolecular" relaxation rate as regards the hydration complex. With τ_c^* the rotational correlation time of the vector F–H toward the first hydration sphere, τ_c'' is given by

$$\tau_c'' = \tau_c^*(\tau_b + \tau_t)/(\tau_b + \tau_c^*) \tag{159}$$

and takes account of the lifetime τ_b of water and the diffusion time τ_t of the water when leaving the complex. This expression has already been introduced in eqn. (153).

Equation (158) gives the contribution of the first hydration water to the relaxation rate of F$^-$; all other protons contribute to the usual translational intermolecular relaxation rate:

$$(1/T_1)_{\text{inter}} = (4\pi/15)\gamma_S^2 \gamma_I^2 \hbar^2 \tilde{c}_I / a D_{\text{FH}} \tag{160}$$

where a is the distance of closest approach between the fluoride nucleus and the water outside the first hydration sphere; $D_{\text{FH}} = \frac{1}{2}(D + D_{\text{F}^-})$, where D and D_{F^-} are the self-diffusion coefficients of water and F^-, respectively; $\tilde{c}_I$ is the proton concentration in spins per cm³. We have

$$1/T_1 = (1/T_1)_{\text{intra}'} + (1/T_1)_{\text{inter}} \tag{161}$$

Calculation of $(1/T_1)_{\text{inter}} = 0.05$ sec⁻¹ at 25°C, together with the experimental result $1/T_1 = 0.49$ sec⁻¹, yields $(1/T_2)_{\text{intra}'} = 0.44$ sec⁻¹. In order to calculate τ_c'', R_0 and n must be known. Two possible hydration numbers should be considered: $n_h^- = 6$ or 4; furthermore the first hydration waters may be symmetrically bound to the F^-, the O–F axis bisecting the H—O—H angle (model I), or they may be arranged so that one of the protons is on the axis connecting F and O (model II). For model I one finds from the geometry of water and F^- $R_0 = 2.28$ Å and $n = 2n_h^-$; for model II the corresponding values are $R_0 = 1.80$ Å and $n = n_h^-$, since one proton of H_2O points away from F^- and the interaction can be neglected ($1/R^6$ dependence of relaxation contribution). One finds the following results[354]:

Model I

$$n_h^- = 4, \quad \tau'' = 1.5 \times 10^{-11} \text{ sec}$$
$$n_h^- = 6, \quad \tau'' = 1.0 \times 10^{-11} \text{ sec}$$

Model II

$$n_h^- = 4, \quad \tau'' = 0.7 \times 10^{-11} \text{ sec}$$
$$n_h^- = 6, \quad \tau'' = 0.5 \times 10^{-11} \text{ sec}$$

The uncertainty of these results is estimated to be $\pm 25\%$.

Let us assume for the moment that $\tau_b \gg \tau_c^*$, so that these results are τ_c^* values, i.e., they concern reorientational motion of the F–H vector. The time constants for model I are about the same as those found for a transition metal ion hydration complex, $\tau_c = 2\text{–}3 \times 10^{-11}$ sec; of course the lower charge of F^- reduces the stiffness of the complex. However, the results for model II can only be understood if there is considerable lateral freedom in the hydration complex so that we have more or less uncorrelated motion of $F–H_2O$ pairs around F^-, where F^- is of course the common pair partner for all water molecules.

The proton–proton vector of the water in the first hydration sphere has correlation times (at 25°C).[354]

$$n_h^- = 4 \quad \tau_c^{-\circ} = 5.1 \pm 0.3 \times 10^{-12} \text{ sec}$$
$$n_h^- = 6 \quad \tau_c^{-\circ} = 4.2 \pm 0.3 \times 10^{-12} \text{ sec}$$

These times are calculated with appropriate corrections for the concentration dependence of the correlation times τ_c^- and τ_c°, i.e., of the a_0 and a_- coefficients in eqns. (146) and (147). We see that for model I, $\tau_c^{-\circ} \approx \frac{1}{2}\tau_c^*$, and we conclude that the water molecule rotates about the F–O axis, which means that the correlation time for the O–F axis is slightly longer than τ_c^*. In model II lateral motion of the O–F vector and rotation of H_2O about this axis may occur. In no case is the hydration complex of F^- a rigid aggregate during a time of the order of 10^{-10} sec. We quote the correlation times of two fluorine-containing ions which have been reported (both at 25°C).[354] For HF_2^-, $\tau_c = 3.2 \times 10^{-12}$ sec, for BF_4^-, $\tau_c = 2.4 \times 10^{-12}$ sec (for HF, $\tau_c = 1.5 \times 10^{-12}$ sec). Slightly above room temperature, relaxation by spin–rotation interaction becomes important for BF_4^-.[354]

Li^+: At 25°C the ^7Li relaxation rate in water as $c \to 0$ has been found to be[359] (see also Refs. 147 and 847)

$$(1/T_1)_{H_2O}^{Li} = 0.055 \text{ sec}^{-1}$$

In D_2O the corresponding limiting value is

$$(1/T_1)_{D_2O}^{Li} = 0.027 \text{ sec}^{-1}$$

Apart from some small corrections[359] the difference between these two relaxation rates is caused by the magnetic dipole–dipole interaction between the ^7Li nucleus and the water protons:

$$(1/T_1)_{dH} \approx (1/T_1)_{H_2O}^{Li} - (1/T_1)_{D_2O}^{Li} \tag{162}$$

For this dipolar part the same formulas as eqns. (158)–(161), apply. Again $(1/T_1)_{inter}$ can be calculated according to eqn. (160) and equals 8×10^{-3} sec^{-1}; then by eqn. (161) $(1/T_1)_{intra'} = 2.7 \times 10^{-2}$ sec^{-1}. We assume $n_h^+ = 4$, i.e., $n = 8$, and with $R_0 = 2.7$ Å we find[359]

$$\tau_c'' \approx \tau_c^* = (1.5 \pm 0.3) \times 10^{-11} \text{ sec}$$

From the proton relaxation rates after application of the a_0 correction in eqn. (147) the intramolecular proton–proton correlation time in the first hydration sphere with $n_h^+ = 4$ is found to be

$$\tau_c^{+\circ} = 4.5 \times 10^{-12} \text{ sec}$$

Again we have $\tau_c^{+\circ} \approx \frac{1}{2}\tau_c^*$ and conclude that water molecules rotate about the O–M axis with a time constant $\tau_i \approx 10^{-11}$ sec.

It is interesting that the quantity $(1/T_1)_{dH}$ defined in eqn. (162) decreases as the concentration of the salts LiCl, LiBr, and LiI increases. This should be due partly to a decrease of τ_b in eqn. (159). The concentric forces which originate from the Li^+ ion and which build up an ordered hydration complex around Li^+ gradually vanish as the number of Li ions increases, i.e., we have transition from an electrolyte solution with hydrated ions in the strict sense to a melt of the crystal hydrate. This picture may also be inferred from the behavior of the static dielectric constant (see Chapter 8), from the self-diffusion coefficients (see below), and from the disappearance of symmetry effects in the quadrupolar relaxation rate of ^7Li with increasing concentration.

2.7. Self-Diffusion Coefficients in Electrolyte Solutions

It is well known that nuclear magnetic spin echo studies permit the measurement of self-diffusion coefficients in liquids. In electrolyte solutions the self-diffusion coefficients which have been obtained by the NMR method are mainly those of the solvent water.[193,195,354,358,541,785] The reason for this is again that so many ionic nuclei relax by quadrupole interaction. As mentioned above, quadrupole interaction is often a very strong relaxation mechanism, i.e., both relaxation times, longitudinal and transverse, are short, often of the order of milliseconds, and a short transverse relaxation time precludes the application of the corresponding nuclear magnetic signal for diffusion measurements. Pulsed gradient methods being developed at several laboratories may help in some cases. However, another difficulty is provided by the rather weak signals produced from many ionic nuclei, in particular at lower concentration, where results are of greatest interest. For now even the H_2O self-diffusion coefficients in ionic solutions are of poor quality, self-diffusion measurements by NMR pulse techniques still present appreciable problems, and data reported from different laboratories, though in qualitative agreement as regards general trends, show severe deviations from one another in a stricter quantitative comparison. The general trend of the self-diffusion coefficient of water in electrolyte solutions has already been described by $1/T_1 \propto 1/D$, which holds for all three nuclear relaxation rates. If it were only for this experimental and qualitative behavior of D, the present section would not be needed. However, it is our objective in this section to demonstrate the connection of self-diffusion data with the general framework underlying our discussion of nuclear magnetic relaxation rates in diamagnetic electrolyte solutions.

As before, our starting point is the propagator $P(\mathbf{r}_0, \mathbf{r}, t)$. We may now choose $\mathbf{r}_0$ to be the origin of our coordinate system, which is the laboratory system. We are no longer interested in the motion of one particle relative to another one; rather, we study the diffusive motion in the solution relative to an arbitrary origin fixed in the solution. Thus we write $P(\mathbf{r}, t)$ and

$$dP = P(\mathbf{r}, t)\, d\mathbf{r}$$

is the probability of finding a particle at $\mathbf{r}$ in $d\mathbf{r}$ at time t when it was at the origin $\mathbf{r} = 0$ at time zero. It is useful to start again with the electrolyte solution which is composed of only two subliquids, as we did on p. 363, so our electrolyte solution is NaCl, LiCl, ... or KF, KI, ..., composed of the two subliquids a and b (b is the subliquid where molecular motion differs from that of pure water at $c \to 0$). Let us consider once again the rotational propagator of eqn. (105). We recall that the limiting case which leads to eqn. (113) is not realized in electrolyte solutions. At best

$$\tau_a, \tau_b < \tau_{a1}, \tau_{b1} \tag{163}$$

However, for all other τ_{al} and τ_{bl}

$$\tau_a, \tau_b \gtrsim \tau_{al}, \tau_{bl}, \qquad l > 1$$

and in this event for "long" observation times the diffusion behavior is only just approximated, since

$$P_1(t) = \exp\{-[(1/\tau_{a1})p_0 + (1 - p_0)(1/\tau_{b1})]t\} = \exp(-2t\bar{D}_r)$$
$$P_l(t) \approx 0 \qquad \text{for} \quad l > 1 \tag{164}$$
$$\bar{D}_r = p_0 D_r{}^\circ + (1 - p_0)D_r{}^b; \qquad 2D_r{}^\circ = 1/\tau_{a1}, \qquad 2D_r{}^b = 1/\tau_{b1}$$

Usually, when hydration is stronger, even inequality (163) is not fulfilled; one never observes diffusionlike behavior with *one* mean rotational diffusion coefficient; the propagator now corresponds to the superposition of two separate diffusion processes. The reason for this situation is of course that the "observation times" for rotational diffusion are so short compared with the lifetime of the molecule in the different subliquids. These observation times are always of the order of magnitude of the correlation times, i.e., $\tau_{a1}, \tau_{a2}, \tau_{b1}, \tau_{b2}, \ldots$, because for times $t \gg \tau_{al}, \tau_{bl}$, $l = 1, 2, \ldots$, we see from eqn. (164) that $P_l(t) = 0$ for $l > 0$ and from eqn. (105) that $P(\Omega_0, \Omega, t) = 1/4\pi$. We have a random distribution of orientations. It should, however, be noted that the behaviors of the two propagators, one for slow and one for fast exchange, approach one another in the limiting case $(1 - p_0) = p_1 \to 1$. Then, according to eqns. (113) and (114),

$$P_l(t) = e^{-t/\tau_{bl}}, \qquad l > 0$$

in both situations.

Let us return to the translational diffusion. The corresponding propagator is now [see eqn. (122)]

$$P(\mathbf{r}, t) = (2\pi)^{-3} \int \{p_a'' [\exp(-t/\tau_{ta}')] + p_b'' \exp(-t/\tau_{tb}')\} \exp(-i\boldsymbol{\rho} \cdot \mathbf{r})\, d\boldsymbol{\rho}$$

$$= (2\pi)^{-3} \int P_\varrho(r) \exp(-i\boldsymbol{\rho} \cdot \mathbf{r})\, d\boldsymbol{\rho} \tag{165}$$

The quantities $1/\tau_{ta}'$, $1/\tau_{tb}'$, p_a'', and p_b'' are exactly the same as given in eqns. (107)–(110) with $1/\tau_{al}$, $1/\tau_{bl}$, and $1/\tau_{b'}$ being replaced by $D_a\varrho^2$, $D_b\varrho^2$, and $1/\tau_b$, respectively. Now we translate inequality (163) and eqn. (164) literally: If we have

$$\tau_a, \tau_b \ll 1/D_a\varrho^2, 1/D_b\varrho^2 \tag{166}$$

then

$$P_\varrho(t) = \exp(-\bar{D}t\varrho^2) = \exp\{-[D_a p_0 + D_b(1 - p_0)]\varrho^2 t\} \tag{167}$$

[see eqn. (36a)]. But here the observation time is of the order of $(D\varrho^2)^{-1}$, $D \approx 10^{-5}$ cm^2 sec^{-1}, $0 < \varrho < \infty$; the lower limit of ϱ is given by the macroscopic size of the system, i.e., $\varrho \gtrsim 1$ cm^{-1}. Thus, for the translational diffusion the observation time may be made very long. For NMR measurements of self-diffusion the observation time is of the order of 10^{-3} sec, much shorter than for tracer methods. We see that the inequality (166) is nearly always fulfilled.

Thus for the translational diffusion process generally we have the "fast exchange" limit relative to the observation time; it is in fact very difficult to attain the slow exchange limit. Neutron scattering is the only available technique so far, although pulsed gradient NMR experiments should be useful when the respective techniques are sufficiently developed.

As a consequence of eqn. (167), we may now write the water self-diffusion coefficient in a (two-subliquid) electrolyte solution, applying again the special nomenclature for electrolyte solutions,

$$D = D_0[1 - (n_h {}^\pm c^*/55.5)] + D^\pm(n_h {}^\pm c^*/55.5) \tag{168}$$

Here D_0 is the self-diffusion coefficient of the free water (subliquid a),

$D^{\pm}$ is the self-diffusion coefficient of the water that is a member of the cationic or anionic hydration sphere. Generalized to a 1–1 electrolyte, the reciprocal self-diffusion coefficient is

$$1/D = 1/[D_0(1 - x_h^+ - x_h^-) + D^+ x_h^+ + D^- x_h^-] \qquad (169)$$

whereas the rotational correlation time (for second-order spherical harmonics) is

$$\tau_c = \tfrac{1}{6}[(1/D_r^\circ)(1 - x_h^+ - x_h^-) + (1/D_r^+)x_h^+ + (1/D_r^-)x_h^-]$$
$$x_h^{\pm} = n_h^{\pm} c^*/55.5 \qquad (170)$$

Here D_r° and $D_r^{\pm}$ are the rotational diffusion coefficients of free water and of water in the cationic or anionic hydration sphere, respectively. Thus we see that $1/D$ represents the reciprocal of the mean translational self-diffusion coefficient in the solution; the correlation time τ_c measures the mean of the reciprocal rotational diffusion coefficients in the solution. When, e.g., $x_h^+ \to 1$, D^{-1} and τ_c give the reciprocal translational and rotational diffusion coefficients of the hydration water, respectively.

It is plausible that in the hydration spheres the translational and rotational diffusion coefficients undergo about the same changes relative to pure water, which explains the approximate proportionality between $(1/T_1)_{\text{intra}}$ and $(1/T_1)_{\text{inter}}$ $[(1/T_1)_{\text{inter}} \infty 1/D]$ and thus between $1/T_1$ and $1/D$, as described above.

It is clear that eqns. (168) and (169) are only of a formal character, as was explained for the rotational correlation time. If, for instance, we write

$$D = D_0\left(1 - \frac{n_h^+ c^*}{55.5} - \frac{n_h^- c^*}{55.5}\right) + D^+ \frac{n_h^+ c^*}{55.5} + D^- \frac{n_h^- c^*}{55.5} \qquad (171)$$

then we must keep in mind that all self-diffusion coefficients D_0, D^+, D^- are concentration-dependent:

$$D_0 = D_0^\circ[1 - (d_0^+ c^+ + d_0^- c^-) + \cdots]$$
$$D^+ = D^{+\circ}[1 - (d_+^+ c^+ + d_+^- c^-) + \cdots]$$
$$D^- = D^{-\circ}[1 - (d_-^+ c^+ + d_-^- c^-) + \cdots]$$

where d's are constants corresponding to the a's in eqn. (144). The $D^{\pm\circ}$ can again be determined with

$$D^{+\circ}(\text{K}^+) = D^{-\circ}(\text{Cl}^-) \qquad (172)$$

which is a consequence of equal transference numbers of both these ions; of course, for evaluation of the $D^\pm$ estimates of the d's should be available.

At this point we have to discuss the question of whether the self-diffusion coefficient of hydration water $D^\pm$ and the self-diffusion coefficient of the ion have to be equal. In a way this corresponds to the comparison of the rotational correlation time of the intramolecular proton–proton vector in the first hydration sphere with the correlation time of the vector connecting such a proton with the ion in the center of the hydration complex. Usually for even fairly strongly hydrated ions we may safely assume that $D^\pm = D_{ion}$, where D_{ion} is the ionic self-diffusion coefficient. Experimental data for D_{ion}, particularly over a wide concentration range, are scarce. The study of water and ionic self-diffusion in KF and LiCl solutions for which

$$D = D_0°[1 + d_0{}^\pm c^* + O(c^{*2})][1 - (c^* n_h{}^\pm/55.5)] + D^{\pm°}[1 + d_\pm{}^\pm c^* + O(c^{*2})]$$
$$\times (c^* n_h{}^\pm/55.5) \tag{173}$$

has shown[354,359] that one can indeed put $D^\pm = D_{ion}$ for F^- and Li^+, and one obtains reasonable d_0 and $d_\pm$ values, that is, their signs and magnitudes are comparable with the gross concentration changes of other quantities. It is characteristic for these systems that the identical diffusion coefficient for hydration water and ion does not conflict with eqn. (171) for low or moderate concentrations. At higher concentrations water normally begins to diffuse faster than the corresponding ion. Thus, finally, when all water is formally hydration water, $D > D_{ion}$. For $CaCl_2$ comparison of diffusion coefficients is also possible and leads to the same result.[350] However, as already mentioned, the number of examples for which diffusion measurements of water and the ion over a wide concentration range are available is limited.

What lifetime of the water molecule in the hydration sphere is implicit in the statement $D_{ion} = D^\pm$? The velocity correlation function $\overline{v_x(0)v_x(t)}$ (e.g., of the x component) of a particle has decayed to zero usually after a time τ_v of the order of 10^{-13} sec. The self-diffusion coefficient is[188,678]

$$D = \int_0^\infty \overline{v_x(0)v_x(t)}\, dt \tag{174}$$

Thus, if two particles are connected to one another for a time very much longer than the correlation time of the velocity correlation function, then, since they must have essentially the same velocity correlation function, they have the same self-diffusion coefficient. With $\tau_v \approx 10^{-13}$ sec the time of attachment to produce the same D is estimated to be $\sim 10^{-11}$ sec.

TABLE V. Ratio of Water Self-Diffusion Coefficient $D^{\pm}$ in the Hydration Sphere to the Self-Diffusion Coefficient D_0 in Pure Water for Various Temperatures and Hydration Numbers[195]a

Ion	0°C $\frac{1}{D_0}\left(\frac{dD}{dc^*}\right)^{\pm}_0$	$n_h{\pm} =$ 6	8	10	25°C $\frac{1}{D_0}\left(\frac{dD}{dc^*}\right)^{\pm}_0$	$n_h{\pm} =$ 6	8	10	80°C $\frac{1}{D_0}\left(\frac{dD}{dc^*}\right)^{\pm}_0$	$n_h{\pm} =$ 6	8	10	D_{Ion} at 25°C 10^{-5} cm² sec⁻¹
Li⁺	-0.12	—	—	—	-0.12	—	—	—	-0.20	—	—	—	1.03
Na⁺	-0.10	—	—	—	-0.08	—	—	—	-0.12	—	—	—	1.33
K⁺	0.04	1.4	1.2	1.2	0.01	1.1	1.1	1.1	-0.02	0.8	0.8	0.9	1.96
Rb⁺	0.05	1.4	1.4	1.2	0.01	1.1	1.1	1.1	-0.02	0.8	0.8	0.9	2.07
Cs⁺	0.09	2.0	1.7	1.4	0.03	1.2	1.2	1.1	-0.02	0.8	0.8	0.9	2.05
Mg²⁺	—	—	—	—	-0.28	—	—	—	—	—	—	—	0.70
Ca²⁺	—	—	—	—	-0.21	—	—	—	—	—	—	—	0.79
Sr²⁺	—	—	—	—	-0.18	—	—	—	—	—	—	—	0.77
Ba²⁺	—	—	—	—	-0.14	—	—	—	—	—	—	—	0.84
F⁻	-0.14	—	—	—	-0.12	—	—	—	-0.10	—	—	—	1.47
Cl⁻	0.04	1.4	1.2	1.2	0.01	1.1	1.1	1.1	-0.02	0.8	0.8	0.9	2.02
Br⁻	0.05	1.4	1.4	1.2	0.04	1.4	1.2	1.2	-0.01	0.9	0.9	0.9	2.08
I⁻	0.07	1.7	1.4	1.4	0.05	1.4	1.4	1.2	0.00	1.0	1.0	1.0	2.02
H₂O	$D_0 = 1.06 \times 10^{-5}$ cm² sec⁻¹				$D_0 = 2.31 \times 10^{-5}$ cm² sec⁻¹				$D_0 = 7.5 \times 10^{-5}$ cm² sec⁻¹				

a All d coefficients are set equal to zero. $(dD/dc^*)_{c^* \to 0} = \nu^+(dD/dc^*)_0^+ + \nu^-(dD/dc^*)_0^-$.

Finally, we raise the question: Is the self-diffusion coefficient of water in the hydration sphere of structure-breaking ions equal to that of the ion defining these hydration spheres? For instance, D_{ion} of the anion I^- decreases with increasing KI concentration; however, D of water increases.[350] Thus, common self-diffusion at low concentrations would imply that here fluidity increase is confined to second spheres, and that at higher concentrations drastic rearrangement would have to occur. As already explained, there is no experimental evidence available for such rearrangement and we conclude that diffusion of water and ions is uncorrelated in hydration spheres of structure-breaking ions, the lifetime of water in these spheres being $< 10^{-11}$ sec. In Table V a number of $D^{\pm}/D_0$ results are listed. They are obtained from the data presented in Ref. 195 according to eqns. (171) and (172); for their evaluation all d coefficients [eqn. (173)] have been set equal to zero. Where the resulting $D^{\pm}$ was smaller than D_{ion}, no values are given in Table V; here $D^{\pm} = D_{ion}$ and the d coefficients must be nonzero.

2.8. Structural Interpretation of Microdynamic Data

Figure 10 represents two examples of the microdynamic structure of an electrolyte solution, namely aqueous NaI and $MnCl_2$. The various regions are characterized by three dynamic quantities: τ_c, D, and τ_b. In our model in Fig. 10 all dynamic pecularities are localized in the first hydration spheres of the ions (all a and d coefficients are zero). We have shown how these results can be obtained from the experimental data.

As we have indicated in the introduction, we wish to give a brief sketch of a transcription of the microdynamic structure to a set of parameters which are related to the "true" structure of the electrolyte solution. The "true" structure of a liquid is given by a set of molecular distribution functions which give the probability (density) of finding a certain number of molecules or other solute particles (ions) in a given configuration relative to a selected reference particle. The simplest form of such a molecular distribution function is a pair distribution function which merely concerns the configuration of one other molecule (or ion) relative to the reference particle. It is easily recognized that the probability density $p(\mathbf{r})$ occurring in eqn. (26) represents a pair distribution function for a water molecule–ion pair. Such a pair distribution function may be written as[371]

$$p(\mathbf{r}_0) = \text{const} \times \exp[-E(\mathbf{r}_0)/RT] \tag{175}$$

where $E(\mathbf{r}_0)$ is the potential of mean force acting between the two particles.

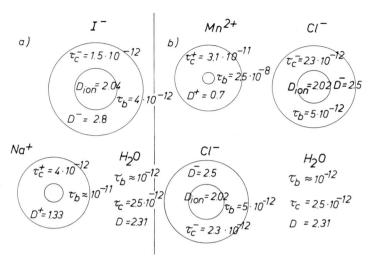

Fig. 10. "Microdynamic structure" for two-electrolyte solutions: (a)
(a) NaI, (b) MnCl₂. τ_c, τ_b are given in seconds, D is given in 10^{-5} cm²
sec⁻¹. Temperature, 25°C.

On the other hand, we may use the elementary kinetic formula for the lifetime of pair attachment τ_b:

$$\tau_b = \tau_0 e^{E_A/RT} \tag{176}$$

where E_A is the activation energy for the dynamic process of separation of
the ion–water molecule pair. Now for a number of ions we found from the
Fourier transform of the time correlation function of second-order spherical
harmonics that the propagator $P(\mathbf{r}_0, \mathbf{r}, t)$ is of rotational form, leaving
$r_0 = R_0 = \text{const}$. We found a rotational correlation time $\tau_c^* \approx 3 \times 10^{-11}$
sec $\ll \tau_b$. Then from eqn. (176) with $\tau_0 \approx 10^{-14}$ sec we derive $E_A \gtrsim 4.5$
kcal mol⁻¹. The E_A in eqn. (176) and $(E(\mathbf{r}_0))_{\text{max}}$, the maximum depth of
$E(\mathbf{r}_0)$, should not be very different quantities; thus $(E(\mathbf{r}_0))_{\text{max}} = E_A$, and
we have the result that the depth of $E(\mathbf{r}_0)$ determining $p(\mathbf{r}_0)$ is $\gtrsim 4.5$ kcal
mol⁻¹ when we find the propagator for the relative motion of H_2O and the
ion to be of rotational form. But we saw that this conclusion can only be
drawn directly if the interacting ion carries an unpaired electron. For
diamagnetic ions we have no information about the Fourier transform of
$g(t)$ at higher frequencies. However, if we find $\tau_c^* = 2\text{–}3 \times 10^{-11}$ sec, then
we may still argue that $\tau_b \gg \tau_c^*$ and thus $(E(\mathbf{r}_0))_{\text{max}} \gtrsim 4.5$ kcal mol⁻¹.
If, moreover, one finds $\tau_c^+ = \tau_c^* = 2\text{–}3 \times 10^{-11}$ sec, then it can be concluded
that even for the rotation of the water molecule about the axis O–M the

potential depth, which determines the orientation-dependent pair distribution function $p(\phi)$ for two water molecules in the first hydration sphere,

$$p(\phi) = \text{const} \times e^{-E(\phi)/RT}$$

we have $(E(\phi))_{\text{max}} \gtrsim 4.5$ kcal mol^{-1}. Here $p(\phi)$ is the probability (density) of finding a water partner in the first hydration sphere with an angle ϕ about the O–M axis (as measured from the reference molecule). It has been found that $(E(\phi))_{\text{max}} > 4.5$ kcal mol^{-1} for Al^{3+}[359] (see p. 378); however, $(E(\phi))_{\text{max}} \lesssim 4.5$ kcal mol^{-1} for F$^-$ and Li$^+$,[354,359] i.e., the rotational potential is flatter and the orientational distribution function more isotropic. According to the discussion in Section 2.4.5, in the hydration sphere of most other ions the distribution $p(\phi)$ will not have very deep energy minima.

Other structural information may be obtained in the following way: Consider a water molecule in liquid water. H_2O is a small molecule and from the comparison with other small molecules like NH_3, C_2H_6, etc. we expect a correlation time $\tau_c \approx 5 \times 10^{-13}$ sec at 25°C, but one actually finds $\tau_c = 2.5 \times 10^{-12}$ sec. Now apply eqn. (81),

$$\tau_c = 4\pi R^3 \eta^* / 3kT \qquad (177)$$

where we have replaced the viscosity η by a "microviscosity" η^* (see below). R is the radius of the particle in question. Since $\tau_c(H_2O) > \tau_c(NH_3)$, the product $R^3\eta^*$ in eqn. (177) must be larger for water than for ammonia. The reason for this is the strong hydrogen bonding of H_2O. We recall that we have defined the bound state of a water molecule by the energy limit $\Delta E = -kT$ (see p. 352). For the unbound state ($E > -kT$) the microviscosity η^* is about equal to that of ammonia, denoted as η_0^*, and the large τ_c for water is due to a larger effective R in eqn. (177), which we denote as R^*. Then we rewrite eqn. (177) as[349–351]

$$\frac{1}{\tau_c} = \frac{3kT}{4\pi\eta_0^*(R^*)^3} = \frac{3kT}{4\pi\eta_0^*}\left(\frac{p_0}{R_0^3} + \frac{p_1}{R_1^3} + \frac{p_2}{R_2^3} + \cdots\right) \qquad (178)$$

where R_0 is the radius of the unbound water molecule, R_1 is the radius of an aggregate formed from two water molecules (neglecting anisotropy effects), R_2 is the radius of an aggregate formed from three water molecules, etc.; p_i, $i = 0, 1, 2, \ldots$, are the corresponding probabilities for the occurrence of aggregates of $i + 1$ particles. Obviously eqn. (178) is a generalized form of the fast exchange formula eqn. (132) for more than two states.

This generalization may be used if rotational diffusion coefficients for the various aggregates are defined. In analogy to eqn. (174), this is so when the lifetime of the aggregate is sufficiently longer than $\tau_\omega^{(i)}$, the time after which correlation in the angular velocity, e.g., of the ith aggregate, $\omega_x^{(i)}$, has decayed to zero (see p. 393); thus

$$D_{ri} = (1/6\tau_c^{(i)}) = \int_0^\infty \overline{\omega_x^{(i)}(0)\omega_x^{(i)}(t)}\, dt$$

where $(\tau_c^{(i)})^{-1} \times 3kT/4\pi\eta_0^*R_i^3$. But our definition of binding by the energy $\Delta E = -kT$ is chosen in a way which approximately satisfies this requirement.[351]

Now the $p_i = p_i(w_1)$ in eqn. (178) may be expressed in terms of pair probabilites w_1 which give the probability that a pair of water molecules is bound to another for a sufficiently long time, i.e., with an energy $< -kT$. Since $\tau_{c0} = 4\pi\eta_0^*R_0^3/3kT = 5\times10^{-13}$ sec and $\tau_c = 2.5\times10^{-12}$ sec, w_1 may be calculated from an appropriate equation based on eqn. (178); and, using the fact that $w_1 \to 0$ as the depth of the pair energy $E \to -kT$, the actual depth E_{max} can be estimated when $w_1 \neq 0$ is given.[351] The result for pure water which was derived in this way is $E_{max} \gtrsim 2.3$ kcal mol^{-1},[351] which agrees with our general ideas of H-bonding in pure water.

In electrolyte solutions the geometry of the hydration sphere requires a different construction of the $p_i(w_1)$ from w_1,[349,351] Also the binding to the ion has to be taken into account. Furthermore, the quantity η_0^* undergoes alterations (the "background effect") because packing of water is modified

TABLE VI. Orientational Potential Depth for a Water Molecule in the Hydration Sphere of a Number of Ions

Ion	E_{max}, kcal mol^{-1}
Pure water	≥ 2.3
I$^-$	≥ 1.25
Br$^-$	≥ 1.7
Cl$^-$	≥ 2.2
F$^-$	≥ 4.5
K$^+$ ($= $ Cl$^-$)	≥ 2.2
Na$^+$	≥ 4.5
Li$^+$	≥ 4.5

in the electrolyte solutions; we have electrostrictive compression. By this effect the repulsive or hard-core contribution to η_0^* will also increase. The latter situation becomes particularly important for strongly hydrated ions with high charge and small radius.

In Sections 2.4.3–2.4.5 we found increases of τ_c in the hydration sphere of ions with small radius, part of this increase of τ_c being caused by an increase of η_0^* in eqn. (178) but part of the increase is due to an increase of pair probabilities w_1 [i.e., of the p_i, $i > 0$, in eqn. (178)] related to an increase of the interparticle potential depth. This provides justification of the term "structure-former." Conversely, when τ_c is smaller in the hydration sphere than in pure water, the w_1 become smaller [i.e., the p_i for $i > 0$ in eqn. (178)], which means that the interparticle potential depths become flatter, and so we talk of "structure-breakers." Some numerical examples are given in Table VI. The above discussion has been limited to rotational motion of water molecules, but data for translational motion can be derived along the same lines.

CHAPTER 8

Dielectric Properties

Reinhard Pottel

Drittes Physikalisches Institut
Universität Göttingen
Göttingen, Germany

1. BASIC THEORY

1.1. Types of Dielectric Polarization and Its Decay

Suppose a macroscopic uniform electric field **E** with constant strength (due to free extraneous charges) is present in some aqueous solution of simple ions during the time interval $-\infty < t < 0$. Moreover, suppose that the field is switched off instantaneously at the time $t = 0$ and remains zero during the time interval $0 \geq t > +\infty$ (also that no macroscopic field due to any bound or free charges exists in the liquid). At times $t < 0$ the electric field maintains a constant total dielectric polarization (dipole moment density) $\mathbf{P}_{tot}$ (≤ 0) within the liquid. This polarization consists of several contributions P_i of different kinds which, after the field has been switched off, decay with different decay rates (Fig. 1). The origin, the approximate decay time $\mathscr{E}_i$ (at 25°C), and the approximate magnitude P_i (per unit electric field strength) of the polarization of the various types are as follows.

(a) The distortion of the electronic structure of the water molecules and ions; $\mathscr{E}_e < 10^{-14}$ sec; $4\pi P_e/E = \varepsilon_e - 1$; electronic permittivity $\varepsilon_e \approx 2$.

(b) The small reversible shifts ($\ll$atomic diameter) of the atoms and the small reversible rotations ($\ll 2\pi$) of the molecules away from such temporary equilibrium positions around which oscillations are possible;

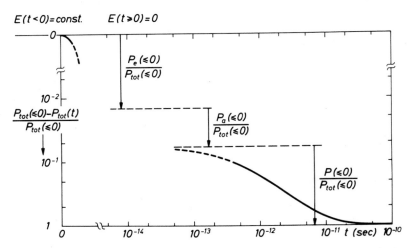

Fig. 1. Decay of the total dielectric polarization $\mathbf{P}_{tot}(t)$ of an aqueous ionic solution after a formerly static electric field $\mathbf{E}$ has been switched off at the time $t = 0$. (The decay of P_e and P_a depends on the realizable decay of practical applied electric fields.) Temperature about 25°C.

$\mathscr{C}_a < 10^{-12}$ sec; $4\pi P_a/E = \varepsilon_\infty - \varepsilon_c$; "infinite-frequency" permittivity $4 < \varepsilon_\infty < 6$.

(c) The irreversible rotations of the polar water molecules; $\mathscr{C} \approx 10^{-11}$ sec; $4\pi P/E = \varepsilon_0 - \varepsilon_\infty$; "zero-frequency" or "static permittivity $80 > \varepsilon_0 > 60$.

(d) The irreversible rotations of ion pairs; 10^{-12} sec $< \mathscr{C}_{IP} < 10^{-10}$ sec; $4\pi P_{IP}/E \approx (\varepsilon_0 - \varepsilon_\infty)(c_{IP}/c_w)(\mu_{IP}/\mu_w)^2$, depending on the concentration c_{IP} and dipole moment μ_{IP} of the ion pairs relative to those of water, c_w, μ_w.

(e) The distortion of the spatial ion distribution; $\mathscr{C}_{II} > \mathscr{C}$; $4\pi P_{II}/E \ll \varepsilon_0 - \varepsilon_\infty$ for up to about 1 mol liter^{-1} monovalent ions.

(f) The spatial asymmetry, due to relaxation, of the orientational polarization of the water around a moving ion generated by its field; $\mathscr{C}_{Iw} < \mathscr{C}$; $4\pi P_{Iw}/E \ll \varepsilon_0 - \varepsilon_\infty$ for up to about 1 mol liter^{-1} monovalent ions.

We confine our considerations to solutions of simple monovalent ions with concentrations mainly up to about 1 mol liter^{-1}. For these solutions numerical estimates (see Section 6) show the polarization contributions (d)–(f) to be negligibly small in comparison to the total polarization.

We focus attention on the decay of the polarization only for times longer than about 10^{-12} sec. With respect to this time scale a decay within shorter times is considered here to happen instantaneously.

Within the above limitations there only remains the orientation polarization P of the water as polarization contribution with finite decay rate. Its response to switching off at $t = 0$ a formerly constant applied electric field is described by (Fig. 2)

$$P(t) = P(0)\Psi(t) \tag{1}$$

At $t = 0$ with $E = 0$ the liquid may be thought to be in a state corresponding to a very infrequently occurring, large *spontaneous* thermal fluctuation of the zero-field *equilibrium* polarization. The decay function of thermal polarization fluctuations at equilibrium is the time autocorrelation function of the polarization noise $\tilde{\mathbf{P}}(t)$. After switching off the electric field the decay of the nonequilibrium polarization ratio $P(t)/P(0)$ also takes place according to

$$\Psi(t) = \langle \tilde{\mathbf{P}}(0) \cdot \tilde{\mathbf{P}}(t) \rangle / \langle \tilde{\mathbf{P}}(0) \cdot \tilde{\mathbf{P}}(0) \rangle \tag{2}$$

where the average has to be taken over an ensemble of equal systems in thermal equilibrium without an applied electric field.

The zero-field spontaneous fluctuations $\tilde{\mathbf{P}}(t)$ originate in fluctuations of the permanent electric dipole moments $\boldsymbol{\mu}(t)$ of the water molecules due to their random rotational motions. The correlation between subsequent values of that component of the molecular electric dipole moment vector which is observed disappears with time according to

$$\psi(t) = \langle \boldsymbol{\mu}(0) \cdot \boldsymbol{\mu}(t) \rangle / \langle \boldsymbol{\mu}(0) \cdot \boldsymbol{\mu}(0) \rangle \tag{3}$$

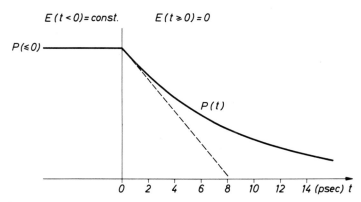

Fig. 2. Decay of the dielectric orientational polarization $\mathbf{P}(t)$ of the water in an aqueous ion solution after a formerly static electric field $\mathbf{E}$ has been switched off at the time $t = 0$. Temperature about 25°C.

where again the average has to be taken at thermal equilibrium in the absence of a macroscopic electric field.

The aim of dielectric relaxation spectroscopy is to obtain an appropriately defined decay time of $\psi(t)$, the "reorientation time" τ of the water molecule, from the polarization decay function $\Psi(t)$ derived from measurements. Because of technical reasons the rapidly decreasing polarization decay function $\Psi(t)$ of water and of aqueous solutions cannot yet be determined directly from time-domain measurements of the polarization response to a step-function electric field. Therefore an alternating electric field $E(t) = \hat{E}e^{i\omega t}$ is applied to the liquid (see Section 2) by means of which frequency-domain measurements of the complex permittivity $\varepsilon(\omega) = \varepsilon'(\omega) -i\varepsilon''(\omega)$, defined by

$$4\pi P(t)/E(t) = \varepsilon(\omega) - \varepsilon_\infty \tag{4}$$

are performed. Its relation to $\Psi(t)$ is established from a synthesis of $E(t)$ by superposing step functions[370]:

$$[\varepsilon(\omega) - \varepsilon_\infty]/(\varepsilon_0 - \varepsilon_\infty) = \int_0^\infty [-d\Psi(t)/dt]e^{-i\omega t}\,dt \equiv \mathscr{L}(-\dot{\Psi}) \tag{5}$$

The $\varepsilon''(\omega)$ here does *not* contain the conductivity (σ) contribution $4\pi\sigma/\omega$ (see Section 3).

In the measurements to be reported a macroscopic dielectric relaxation time $\mathscr{E}$ is derived from the frequency ω_r at which the imaginary part of the permittivity reaches its maximum value:

$$[d\varepsilon''(\omega)/d\omega]_{\omega=\omega_r} = 0 \Rightarrow \mathscr{E} = 1/\omega_r \tag{6}$$

With respect to a comparison with the molecular orientational correlation time as derived from the proton magnetic resonance relaxation (see Section 1.4), the molecular dielectric relaxation time is appropriately defined by

$$\tau = \int_0^\infty \psi(t)\,dt \tag{7}$$

There is no generally valid theory relating $\psi(t)$ and $\Psi(t)$, and thus τ and $\mathscr{E}$, to each other. Up to now a tractable relation is only available for the Onsager model of a pure polar liquid.[589]

1.2. Dielectric Relaxation in a Model of Pure Water

In Fröhlich's version of the Onsager model of a pure polar liquid (Fig. 3) a polar molecule is supposed to be a sphere with a nonpolarizable

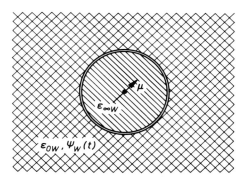

Fig. 3. Onsager model (Fröhlich's version) of a water molecule in pure liquid water: Dielectric sphere (permittivity $\varepsilon_{\infty w}$) with a permanent electric point dipole (moment μ) embedded in a continuous dielectric medium [permittivity ε_{0w}, polarization decay function $\Psi_w(t)$ of bulk water].

permanent point dipole with moment μ at its center. The sphere consists of a homogeneous dielectric medium of permittivity $\varepsilon_{\infty w}$ by which the distortional polarizability of the molecule is simulated [permanent electric dipole moment of the molecule in the gas phase $3\mu/(\varepsilon_{\infty w} + 2)$]. All other molecules of the liquid, including the nearest neighbors around the arbitrary central molecule, are treated as a homogeneous dielectric medium characterized macroscopically by the static permittivity ε_{0w} and the polarization decay function $\Psi_w(t)$ of the liquid.

With respect to the rather open structure of liquid water (see Volume 1, Chapter 14), the Onsager model seems to be a poor approximation to the real liquid. But if one accepts the statement of Litovitz[738,883] that the rate of change of the environment of a water molecule is much faster (about ten times) than the rate of change of its orientation, then a water molecule during its reorientation time "sees" a nearly constant average environment which may probably be appropriately represented by the homogeneous dielectric medium in the Onsager model. This view is compatible with the fact that the static permittivity ε_{0w} of pure water calculated on the basis of this model agrees very well with the measured ε_{0w} values in the temperature range 0–60°C.[369]

According to these arguments, the Nee and Zwanzig theory[589] of dielectric relaxation in polar liquids, based on the Onsager model, is used here, at first for pure water. If, in agreement with the statement by Litovitz, the water molecule is supposed to reorient in small-angle steps ("rotational Brownian motion"), and if, because of $\varepsilon_{0w} \gg \varepsilon_{\infty w}$ for water, terms $(\varepsilon_{\infty w}/\varepsilon_{0w})^{n>1}$ are neglected in comparison to one, the following results can be derived from this theory.

"Dielectric friction" due to the relaxation of the dipole reaction field causes the molecular dielectric decay function $\psi_w(t)$ to be nonexponential. The macroscopic dielectric decay function $\Psi_w(t)$ very nearly shows exponential behavior except at times appreciable shorter than the decay time $\mathscr{C}_w$.

Up to frequencies not too much beyond the relaxation frequency ω_r ($0 \leq \omega \lesssim 2.5\omega_r$, say), the Laplace transforms [see eqn. (5)] $\mathscr{L}(-\dot{\Psi}_w)$ and $\mathscr{L}(-\dot{\psi}_w)$ of the time derivative of the decay functions $\Psi_w(t)$ and $\psi_w(t)$ are approximately related according to

$$\mathscr{L}(-\dot{\Psi}_w) \approx \frac{\mathscr{L}(-\dot{\psi}_w) - (\varepsilon_{\infty w}/2\varepsilon_{0w})}{1 - (\varepsilon_{\infty w}/2\varepsilon_{0w})} \tag{8}$$

if terms $|\varepsilon_{\infty w}/[\varepsilon_w(\omega) - \varepsilon_{\infty w}]|^{n>1}$ are neglected in comparison to one. Within the same frequency range and to the same degree of approximation the theory yields

$$\mathscr{L}(-\dot{\psi}_w) \approx \frac{1}{1 + i\omega\tau_w[1 - i\omega(\varepsilon_{\infty w}/\varepsilon_{0w})\tau_w]} \tag{9}$$

with τ_w as defined in eqn. (7). On combining the eqns. (9), (8), and (5), one arrives at

$$\frac{\varepsilon_w(\omega) - \hat{\varepsilon}_{\infty w}}{\varepsilon_{0w} - \hat{\varepsilon}_{\infty w}} \approx \frac{1}{1 + i\omega\mathscr{C}_w}, \qquad \omega\mathscr{C}_w \lesssim 2.5 \tag{10}$$

with the extrapolated "infinite"-frequency permittivity $\hat{\varepsilon}_{\infty w} \approx 1.5\varepsilon_{\infty w}$. The resulting eqn. (10) is in agreement with results from measurements (see Volume 1, Chapter 7).

The macroscopic dielectric relaxation time $\mathscr{C}_w$ defined by eqn. (6) is related to the molecular dielectric relaxation time τ_w by

$$\mathscr{C}_w \approx [1 + (\varepsilon_{\infty w}/\varepsilon_{0w})]\tau_w \; (\approx \tau_w) \tag{11}$$

The reason for the macroscopic relaxation time $\mathscr{C}_w$ being nearly equal to the molecular relaxation time τ_w is that with $\varepsilon_{0w} \gg \varepsilon_{\infty w}$ the Onsager cavity field (due to the polarization by the applied electric field E) is relaxing much more rapidly than the macroscopic orientational polarization P_w.

If one imagines all neighbors around an arbitrary central water molecule within an appropriately large sphere to have no permanent electric dipole moments, then the dielectric friction (due to electrostatic interactions between the reference dipole and its neighbor dipoles) disappears, and the τ_w of the central molecule reduces to the fictitious molecular relaxation time τ_{0w} which is due only to short-range interaction. The following relation holds between τ_w, including dielectric friction, and τ_{0w} for pure water:

$$\tau_w \approx 2[1 - 2(\varepsilon_{\infty w}/\varepsilon_{0w})]\tau_{0w} \ (\approx 2\tau_{0w}) \tag{12}$$

1.3. Dielectric Relaxation in a Model of Aqueous Ionic Solutions

Up to now a generally valid theory describing the dielectric relaxation in solutions of nondipolar solutes in polar solvents is not available. So in treating aqueous solutions of simple (nondipolar) ions, one can only try to make a crude extension of the model and of the theory used for pure water (see Section 1.2) in order to estimate how the dielectric relaxation of the water molecules is influenced by the ions.

In the model used here for an ionic solution only the molecular decay function $\psi(t)$ of those water molecules *immediately* adjacent to the nondipolar solute particles ("hydration water") is assumed to be influenced by the solute particles, while the $\psi(t)$ of the other ("free") water molecules is assumed to behave as in pure water. The electric dipole–dipole interaction (Onsager cavity field and reaction field) between the latter molecules can then be treated just as in the case of pure water (Fig. 3), but the homogeneous dielectric medium surrounding the molecular cavity has to be characterized by the macroscopic data ε_0 and $\Psi(t)$ of the solution (Fig. 4).

The complete description of the local field within each molecular cavity must also include a depolarizing field, due to the electric polarization charges around the nondipolar solute particles. This additional field has been chosen to take the form of the Lorentz field which here has the form

$$E_1 = -\frac{v(\varepsilon_0 - \varepsilon_e)/(2\varepsilon_0 + \varepsilon_e)}{1 + v(\varepsilon_0 - \varepsilon_e)/(2\varepsilon_0 + \varepsilon_e)} E \left(\approx -\frac{\frac{1}{2}v}{1 + \frac{1}{2}v} E \right) \tag{13}$$

where v denotes that fraction of the total volume of the solution which is occupied by the solute particles, regarded as dielectric spheres with permittivity $\varepsilon_e \approx 2$. A solute particle in this model may consist not only of an ion but also of those adjacent water molecules which have lost their orientational polarizability by the influence of the ion (see Section 4). The depolarizing field E_1 relaxes just as does the usual Onsager cavity field.

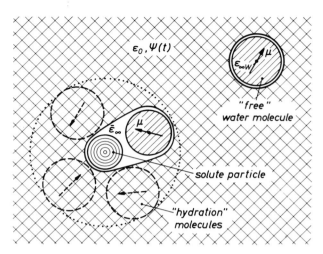

Fig. 4. Modified Onsager cavity for a water ("hydration")
molecule adjacent to a solute particle in an aqueous ionic
solution. This cavity and that of a "free" water molecule are
surrounded by a continuous dielectric medium [permittivity
ε_0, polarization decay function $\Psi(t)$ of the solution].

The dipole–dipole interaction of a water molecule immediately adjacent
to a solute particle with the surrounding medium is estimated by considering
this molecule *together* with the solute particle to be located within an ellipsoidal cavity (Fig. 4). Just as in the case of pure water, the cavity field
in this cavity is relaxing much more rapidly than the macroscopic orienta-
tional polarization P, because $\varepsilon_0 \gg \bar{\varepsilon}_\infty$ also holds in the solution ($\bar{\varepsilon}_\infty$ is
the average background permittivity in the water–solute cavity, $\bar{\varepsilon}_\infty < \varepsilon_{\infty w}$).
Thus the contribution of the relaxation of the hydration water to the decay
of P is nearly independent of the cavity field.

The molecular relaxation time τ_h of the hydration water relative to τ_w
of the other water molecules is changed for two reasons. The reaction field
in the ellipsoidal cavity and consequently the dielectric friction are smaller,
so that instead of eqn. (12) we have to use the relation

$$2\tau_{0h} > \tau_h \gtrsim \tau_{0h}[1 + (V_w/V_h)(\varepsilon_{0w}/\varepsilon_0)(\tau_{0w}/\tau_{0h})] \tag{14}$$

where V_w and V_h are the volumes of a cavity containing a free water molecule or a hydration molecule and a solute particle, respectively. In addition,
the relaxation time τ_{0h} due to short-range interaction may be affected by
the solute particle.

A crude calculation shows that in the frequency interval $0 \leq \omega \lesssim 2.5\omega_r$, $\mathscr{L}(-\dot{\Psi})$ is related to $\mathscr{L}(-\dot{\psi}_i)$ of the cationic $(i = +)$ and anionic $(i = -)$ hydration water and of the free water $(i = w)$ according to

$$\mathscr{L}(-\dot{\Psi}) \approx \sum_{i=+,-,w} (c_i^*/c_w^*)\mathscr{L}(-\dot{\psi}_i) \tag{15}$$

if terms $|\,(\varepsilon_\infty/2)/[\varepsilon(\omega) - \varepsilon_\infty]\,|^{n\geq1}$ are neglected in comparison to one. The c_i^* denote the molal concentrations of the different kinds of water in the solution; $c_w^* = 55.5$ mol kg^{-1} in pure water. To the same degree of approximation $\mathscr{L}(-\dot{\psi}_i)$ should not deviate too much from

$$\mathscr{L}(-\dot{\psi}_i) \approx 1/(1 + i\omega\tau_i) \tag{16}$$

at frequencies ω not too much exceeding τ_i^{-1}.

From eqns. (5), (15), and (16) one obtains $(d\varepsilon''/d\omega)_{\omega_r} = 0$, which determines the principal relaxation time $\mathscr{C}(c^*) = 1/\omega_r(c^*)$, and on differentiating $(d\varepsilon''/d\omega)_{\omega_r} = 0$ with respect to the solute molal concentration c^*, for $c^* \to 0$, one gets the limiting relative molal change of $\mathscr{C}$

$$\frac{1}{\mathscr{C}_w}\left(\frac{d\mathscr{C}}{dc^*}\right)_{c^*\to0} = \frac{2}{c^*}\sum_{h=+,-}\frac{(\tau_h/\tau_w) - (\tau_w/\tau_h)}{[(\tau_h/\tau_w) + (\tau_w/\tau_h)]^2}\frac{c_h^*}{c_w^*} \equiv \sum_{h=+,-}B_d^h \tag{17}$$

The dilution of the water by the nonpolar solute particles and the depolarizing field, eqn. (13), causes the static permittivity ε_0 of the solution to be reduced compared to ε_{0w} of pure water according to

$$\varepsilon_0 = \varepsilon_{0w}\frac{[1 + (\varepsilon_e/2\varepsilon_0)] - v\{1 - (\varepsilon_e/2\varepsilon_{0w})[3 - (\varepsilon_{0w}/\varepsilon_0)]\}}{1 + (\varepsilon_e/2\varepsilon_0) + \frac{1}{2}v[1 - (\varepsilon_e/\varepsilon_0)]}$$

$$\approx \varepsilon_{0w}\frac{1 - v}{1 + \frac{1}{2}v} \tag{18}$$

This formula has been tested experimentally with aqueous tetraalkylammonium bromide solutions for v up to about 0.3 and has been found reliable to within less than 2%.[644]

1.4. Comparison of the Molecular Rotational Correlation Times as Derived from Dielectric Relaxation and Proton Magnetic Resonance Relaxation

The molecular rotational correlation time

$$\tau = \int_0^\infty \psi(t)\,dt \tag{7}$$

derived from the measured dielectric relaxation time $\mathscr{C}$ [eqn. (6)] is determined by the decay of the time autocorrelation function [eqn. (3)]

$$\psi(t) = \langle \cos \vartheta(t) \rangle \tag{19}$$

of the electric dipole moment vector $\mu(t)$ executing directional fluctuations $\vartheta(t)$ (Fig. 5).

The molecular rotational correlation time

$$\tau_m = \tfrac{1}{2} \int_{-\infty}^{+\infty} \chi(t)\, dt \tag{20}$$

derived from the measured intramolecular proton magnetic relaxation rate $1/T_1$ is determined by the decay of the time autocorrelation functions

$$
\begin{aligned}
\chi(t) &= \frac{\langle \sin \theta(0) \cos \theta(0)\, e^{i\Phi(0)} \sin \theta(t) \cos \theta(t)\, e^{-i\Phi(t)} \rangle}{\langle \sin^2 \theta(0) \cdot \cos^2 \theta(0) \rangle} \\
&= \frac{\langle \sin^2 \theta(0)\, e^{2i\Phi(0)} \sin^2 \theta(t)\, e^{-2i\Phi(t)} \rangle}{\langle \sin^4 \theta(0) \rangle}
\end{aligned} \tag{21}
$$

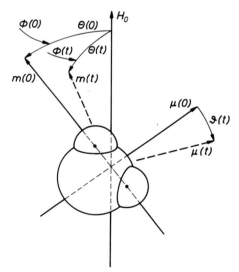

Fig. 5. Orientational fluctuations of a water molecule: directional fluctuations $\vartheta(t)$ of the electric dipole moment vector μ effective in dielectric relaxation, and directional fluctuations $\Phi(t)$ and $\theta(t)$ (with respect to the static magnetic field H_0) of the proton–proton connecting vector m effective in proton magnetic relaxation.

These are calculated from the fluctuating magnetic fields set up by one proton magnetic dipole at the position of the other proton due to variations of the orientation $\theta(t)$, $\Phi(t)$ of the proton–proton connecting vector **m** (Fig. 5).

The dielectric friction[589] acts anisotropically on the rotational motion $\dot{\theta}$, $\dot{\Phi}$ effective in proton magnetic relaxation. In order to avoid this friction anisotropy in the calculation, an average dielectric friction coefficient has been used. After calculating $\mathscr{L}(-\dot{\chi})$ from the diffusion equation for rotational Brownian motion with frequency-dependent friction[589] one gets for pure water

$$\tau_{mw} \approx \frac{\tau_{0w}}{3} + \frac{5}{8}\left(1 - 4\,\frac{\varepsilon_{\infty w}}{\varepsilon_{0w}}\right)\frac{\tau_{0w}}{3} \qquad (22)$$

and with eqn. (12)

$$\frac{\tau_w}{\tau_{mw}} \approx 3\,\frac{16}{13}\left(1 - \frac{6}{13}\,\frac{\varepsilon_{\infty w}}{\varepsilon_{0w}}\right) \qquad (23)$$

and with eqn. (11)

$$\frac{\mathscr{C}_w}{\tau_{mw}} \approx 3\,\frac{16}{13}\left(1 + \frac{7}{13}\,\frac{\varepsilon_{\infty w}}{\varepsilon_{0w}}\right) \approx 3.7 \qquad (24)$$

The experimental values of the latter ratio have been quoted as 3.3–3.4 at 25°C (Chapter 7, Section 2.4.2) and as 3.7 over the range 0–75°C.[479]

With the reasoning leading to eqn. (14) one finds for the correlation times τ_h, τ_{hm} of the hydration water in a solution

$$\frac{\tau_w}{\tau_{mw}} > \frac{\tau_h}{\tau_{mh}} \gtrsim \frac{13}{16}\,\frac{1 + (V_w/V_h)(\varepsilon_{0w}/\varepsilon_0)(\tau_{0w}/\tau_{0h})}{1 + (5/8)(V_w/V_h)(\varepsilon_{0w}/\varepsilon_0)(\tau_{0w}/\tau_{0h})}\,\frac{\tau_w}{\tau_{mw}} > 0.81\,\frac{\tau_w}{\tau_{mw}} \qquad (25)$$

With respect to a comparison (Section 5) of the measured macroscopic quantities $(1/\mathscr{C}_w)(d\mathscr{C}/dc^*)$ and $[1/(1/T_1)_w][d(1/T_1)/dc^*]$, it is important to note that the latter has the dependence

$$\left(\frac{1}{1/T_1}\right)_w\left(\frac{d(1/T_1)}{dc^*}\right) = \frac{1}{c^*}\sum_{h=+,-}\frac{\tau_{mh} - \tau_{mw}}{\tau_{mw}} \cdot \frac{c_h^*}{c_w^*} \equiv \sum_{h=+,-} B_m^{\ h} \qquad (26)$$

on τ_{mh}/τ_{mw} (Chapter 7, Section 2.4.4), which is different from the dependence of the former on τ_h/τ_w, eqn. (17)!

2. EXPERIMENTAL METHODS

In Volume 1, Chapter 7, Section 7 Hasted has already presented an extensive survey of electromagnetic microwave measurement techniques used for determining the complex permittivity $\varepsilon_w(\omega) = \varepsilon_w'(\omega) - i\varepsilon_w''(\omega)$ of pure water. These could also be applied to aqueous ionic solutions, but compared to water, the imaginary part of the permittivity of an ionic solution additionally contains the conductivity (σ) contribution $4\pi\sigma/\omega$ which may be of appreciable magnitude, especially at lower frequencies ($\omega < \omega_r$). The electromagnetic wave reflection at the liquid–air interface and the wave attenuation within the liquid are therefore larger for an ionic solution than for pure water. Measurements of the reflection coefficient of a single liquid–air interface or of a liquid layer backed by different terminations are thus even less suitable than for pure water.

The most reliable and accurate methods are cavity resonator measurements and interferometric transmission measurements.

A convenient form of a cavity resonator is a circular cylinder oscillating in the TM_{010} mode (Fig. 6). The liquid is contained in a tube made of a low-loss dielectric material (fused silica). The tube is located along the axis of the cylinder or parallel to the axis.[187,460,461] The resonance frequency ν_{res} and the width $\Delta\nu$ of the resonance curve (power versus frequency) have to be measured with (index 1) and without (index 0) liquid in the tube. In the case of an axially located narrow tube, $r_2/r_1 < 1/50$, with a thin wall, $(r_2 - r_3)/r_2 \ll 1$, the real and imaginary parts of the permittivity of the liquid can be calculated from the measured quantities ν_{res} and $\Delta\nu$ with the simple approximate formulas

$$\varepsilon' \approx 1 + 0.538\left(\frac{r_1}{r_3}\right)^2 \frac{(\nu_{res})_0 - (\nu_{res})_1}{(\nu_{res})_0} \tag{27}$$

$$\varepsilon_{tot}'' \approx 0.269\left(\frac{r_1}{r_3}\right)^2\left[\left(\frac{\Delta\nu}{\nu_{res}}\right)_1 - \left(\frac{\Delta\nu}{\nu_{res}}\right)_0\right] \tag{28}$$

In the case of wider or otherwise located tubes much more complicated formulas require computer calculation.

The interferometric measurement of amplitude and phase of a wave transmitted through a liquid sample becomes especially convenient if reflections at the points of entering and leaving the sample may be disregarded. The conditions, easily achieved with lossy liquids, are to have a propagating wave within the sample and to change the path length of the wave through the sample. Figure 7 shows an appropriate apparatus which can be constructed with coaxial lines or waveguides.[81,82,642] Within the

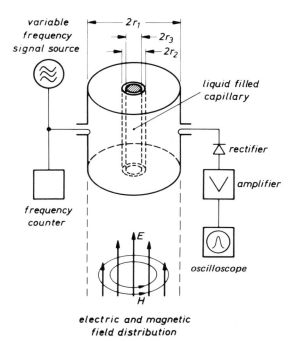

Fig. 6. Experimental arrangement for complex permittivity measurements on aqueous ionic solutions in a TM_{010} mode cavity resonator.

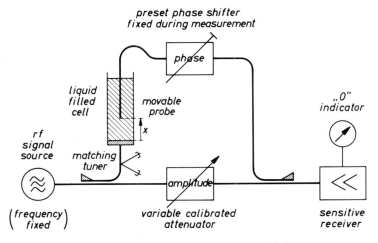

Fig. 7. Experimental arrangement for complex permittivity measurements on aqueous ionic solutions by an interferometric transmission method.[642]

liquid-filled line a receiving probe is shifted along the direction x of the propagating wave, the electric field strength of which varies as $E = \hat{E} \times \exp[i\omega t - (\alpha + 2\pi i/\lambda)x]$. The probe position and, by a variable, calibrated attenuator, the amplitude of a reference signal with fixed phase are adjusted toward zero interference of the probe signal with the reference signal. The differences between subsequent readings of such probe positions and attenuator settings immediately yield the wavelength λ and the attenuation exponent $\alpha\lambda$, respectively, of the wave within the liquid. From the measured λ and $\alpha\lambda$ the real and imaginary parts of the permittivity follow according to

$$\varepsilon' = \left(\frac{\lambda_0}{\lambda}\right)^2\left[1 - \left(\frac{\alpha\lambda}{2\pi}\right)^2\right] + \left(\frac{\lambda_0}{\lambda_c}\right)^2 \tag{29}$$

$$\varepsilon''_{\text{tot}} = \left(\frac{\lambda_0}{\lambda}\right)^2\left(\frac{\alpha\lambda}{\pi}\right) \tag{30}$$

with λ_0 the free-space wavelength and λ_c the cutoff wavelength of the (empty) line used as sample container.

In order to investigate dielectric relaxation in aqueous ionic solutions, besides the high-frequency complex permittivity the dc conductivity σ has to be measured. This is usually performed by using a liquid-filled two-electrode cell the impedance of which is determined with an impedance bridge at frequencies in the kHz range.[808]

A direct measurement of the static permittivity ε_0 (in the kHz frequency range) would also be highly desirable, but up to now no reliable experimental method is known which will not be disturbed by the high conductivity of aqueous ionic solutions.

3. CHARACTERISTIC QUANTITIES DERIVED FROM COMPLEX PERMITTIVITY MEASUREMENTS

Since the electromagnetic microwave measurement techniques, as used today, were developed (in the 1940's) many measurements of the complex permittivity at GHz frequencies have been made on aqueous ionic solutions. Results are reported of solutions of mononuclear and multinuclear ions of different valences at concentrations between about 0.1 M and saturation, at temperatures between about 0 and 60°C, and at frequencies between about 0.1 and 40 GHz.[268,271,304,319,321–323,461,642,643,645,667–669,770,827]

The present treatise is mainly confined to solutions of monovalent ions with concentrations up to about 1 M because of the following reasons: With polyvalent ions the presence of ion pairs is to be expected,[642,770] but most dielectric data available on solutions of polyvalent ions are insufficient for separating accurately enough the effects of ion-pair and water orientational polarization which is required in order to clearly recognize the ionic influence on the dielectric relaxation of the solvent water. For solutions with higher ionic concentrations models are lacking as basis for the evaluation and interpretation of the measured dielectric data.

In most publications measured permittivity values are reported at three or two frequencies or even at one frequency only. Obviously the conclusions drawn from such data with respect to dielectric relaxation times and static permittivities do not entirely suffice for getting a consistent total picture of the influence of ions on the solvent water. It therefore seems to be reasonable to consider a selection of data which are based on measurements of three or more frequencies.

From the values of the imaginary part, $\varepsilon''_{tot}(\nu)$, of the permittivity measured at frequencies ν $(= \omega/2\pi)$ has been subtracted the conductivity contribution $2\sigma/\nu$, due to the ionic drift, with the *dc* conductivity used for σ. So the dielectric loss number

$$\varepsilon''(\nu) = \varepsilon''_{tot}(\nu) - [2\sigma(\nu \approx 0)/\nu] \tag{31}$$

to be discussed with respect to dielectric relaxation may include a frequency-dependent part of the conductivity [which could be due to the polarization mechanisms (e) and (f) mentioned in Section 1.1. See also Section 6.1].

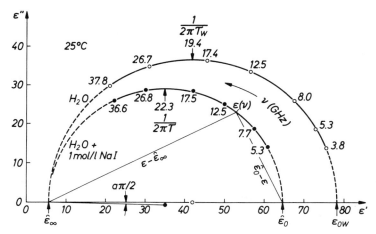

Fig. 8. Imaginary part $\varepsilon''(\nu)$ versus real part $\varepsilon'(\nu)$ of the complex permittivity ε (without conductivity contribution) of an aqueous 1 M NaI solution and of pure water from measurements at various frequencies ν. Temperature 25°C.[271]

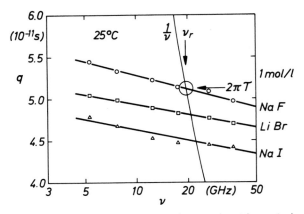

Fig. 9. Ratio q of the chord lengths $|\hat{\varepsilon}_0 - \varepsilon|$ and $|\varepsilon - \hat{\varepsilon}_\infty|$ (Fig. 8) divided by the frequency ν [eqn. (32)] versus the frequency ν on a log scale, from measurements on three 1 M alkali halide solutions. The additional curve $1/\nu$ versus ν (on a log scale) serves to determine precisely the relaxation frequency ν_r and the relaxation time $\mathscr{C}$, respectively. Temperature 25°C.[271]

If measured $\varepsilon''(\nu)$ and $\varepsilon'(\nu)$ values are plotted in an ε'', ε' plane, circular arcs can be fitted to the $\varepsilon(\varepsilon''(\nu), \varepsilon'(\nu))$ points for each of the 1–1 electrolyte solutions investigated. The centers of the circles lie slightly below the ε' axis. An example[271] is shown in Fig. 8. Extended down to the ε' axis, the circular arc intersects the ε' axis at the points $(0, \hat{\varepsilon}_\infty)$ and $(0, \hat{\varepsilon}_0)$. The angle between the ε' axis and the straight line connecting the center of the circle and the point $(0, \hat{\varepsilon}_\infty)$ is denoted by $a\pi/2$. The type of the frequency distribution along the $\varepsilon(\varepsilon''(\nu), \varepsilon'(\nu))$ circular arc plot can be seen from a plot of

$$q \equiv (1/\nu) | \hat{\varepsilon}_0 - \varepsilon |/| \varepsilon - \hat{\varepsilon}_\infty | \tag{32}$$

versus $\log \nu$. The characteristic of this plot is its linearity with negative slope, as demonstrated in Fig. 9.[271]

From the characteristics of both the $\varepsilon''(\varepsilon')$ plot (Fig. 8) and the $q(\log \nu)$ plot (Fig. 9) it follows that

$$\varepsilon(\nu) = \varepsilon'(\nu) - i\varepsilon''(\nu) \approx \hat{\varepsilon}_\infty + \frac{\hat{\varepsilon}_0 - \hat{\varepsilon}_\infty}{1 + (i)^{1-a}(2\pi\nu\mathscr{C})^{1-b}} \tag{33}$$

with $1 > b > a > 0$; this provides an excellent description of the measured $\varepsilon(\nu)$ values within the limits of experimental errors ($\lesssim 1\%$ for ε', $\lesssim 2\%$

for $\varepsilon''^{(271)}$). The parameters $\hat{\varepsilon}_\infty$, $\hat{\varepsilon}_0$, and a are explained in Fig. 8. The principal relaxation time $\mathscr{C}$ is determined by the frequency ν_r at which $d\varepsilon''/d\nu = 0$ [eqn. (6)] and $q = 1/\nu$ [eqn. (32), Fig. 9]. The parameter b is given by meaning

$$b = 1 + \left[\frac{1}{\varepsilon''} \frac{d\varepsilon'}{d(2\pi\nu\mathscr{C})} \right]_{2\pi\nu\mathscr{C}=1} \cos\frac{a\pi}{2} \tag{34}$$

and is related to the $q(\log \nu)$ plot by

$$b \approx -\frac{\log e}{2\pi\mathscr{C}} \frac{\Delta q}{\Delta(\log \nu)} \tag{35}$$

The parameters $\hat{\varepsilon}_\infty$, $\hat{\varepsilon}_0$, and a, and $\mathscr{C}$ and b may be determined graphically from the $\varepsilon''(\varepsilon')$ or $q(\log \nu)$ plots, respectively (special method for $\hat{\varepsilon}_0$ and $a^{(271)}$) or by computer regression analysis (both methods have been applied by the author's co-workers K. Giese and D. Adolph).

In the description by eqn. (33) with $1 > b > a > 0$ the dielectric relaxation of aqueous ionic solutions clearly differs from that of pure water and aqueous solutions of nonelectrolytes. The $\varepsilon_w(\nu)$ of pure water with $b_w \approx 0.01$ and $a < 0.01$ (25°C) closely approximates the Debye function [eqn. (10)] following from eqn. (33) with $b = a = 0$, while $\varepsilon(\nu)$ of aqueous solutions of organic nonelectrolytes[644] satisfies the Cole–Cole function, eqn. (33) with $1 > b = a > 0$.

The function $\varepsilon(\nu)$, eqn. (33), empirically defined only within a frequency interval of about $2.5 > \omega\mathscr{C} > 0.2$, does not satisfy the generally valid Kramers–Kronig relations between $\varepsilon'(\nu)$ and $\varepsilon''(\nu)$[246] in the actual case of $b > a \geq 0$. Consequently it cannot apply beyond the former frequency interval, so that the terminal points $(\hat{\varepsilon}_\infty, 0)$ and $(\hat{\varepsilon}_0, 0)$ of the circular arc plot of eqn. (33) (Fig. 8) do not represent the correct asymptotic values of the permittivity at "infinite" frequency, $\varepsilon(\infty) \equiv \varepsilon_\infty$, and at zero frequency, $\varepsilon(0) \equiv \varepsilon_0$, respectively. With the help of eqns. (5), (15), and (16) the relative deviation $|\hat{\varepsilon}_0 - \varepsilon_0|/\varepsilon_0$ may be estimated as not exceeding the experimental error.

The values of the characteristic quantities $\mathscr{C}$, $\hat{\varepsilon}_0$, $\hat{\varepsilon}_\infty$, b, and a derived from complex permittivity measurements on various $1\,M$ solutions of monovalent ions are collected in Tables I and II. From measurements performed on solutions at various concentrations it has been found that $\mathscr{C}$ varies linearly with the molal concentrations up to about $1\,m$, and that $\hat{\varepsilon}_0$ decreases linearly at low molar concentrations (Figs. 10 and 11).

TABLE I. Characteristic Quantities Derived from Complex Permittivity Measurements on 1 M Solutions at 25°C[271,645][a]

Solute	$100(\mathscr{E} - \mathscr{E}_w)/\mathscr{E}_w c^*,$ $(\mathrm{mol/kg\ H_2O})^{-1}$	$\hat{\varepsilon}_0$	$100b$	$100a$
LiCl	−4	64.9	3.2	1
LiBr	−7	64.4	3.5	1
LiI	−10	63.4	3.7	1
NaF	−1	68.3	4.7	1
NaCl	−7	66.7	4.8	2
NaBr	−9	65.9	4.4	2
aNI	−13	64.8	4.1	1
KF	−2	69.0	5	1
KCl	−8	68.1	4.5	3
KBr	−10	67.7	4.1	3
KI	−12	67.1	4.3	2
RbF	−2	69.6	5.0	1
RbCl	−6	69.1	4.9	3
RbBr	−10	68.5	4.9	2
RbI	−12	67.8	4.9	2
CsF	+1	71.0	3.3	2
CsCl	−6	70.6	3	2
CsBr	−7	70.2	4	2
CsI	−9	69.8	4	2
$(CH_3)_4NBr$	+14	66.6	3.6	1
$(C_2H_5)_4NBr$	+36	61.0	5.5	3
$(C_3H_7)_4NBr$	+65	55.0	7.2	6
$(C_4H_9)_4NBr$	+86	50.6	8.1	7
Pure H_2O	$\mathscr{E}_w = 8.25$ psec	78.3	1	0.4

[a] $\hat{\varepsilon}_\infty = 5.4 \pm 0.6$ common to water and all solutions.

TABLE II. Characteristic Quantities Derived from Complex Permittivity Measurements on 1 M Solutions at Various Temperature[a]

Solute	t, °C	$100(\mathscr{C} - \mathscr{C}_w)/\mathscr{C}_w c^*$, $(mol/kg\ H_2O)^{-1}$	$\hat{\varepsilon}_0$	$100b$	$100a$
LiCl	0	-7	72.0	2	1
	25	-4	64.9	3.2	1
	50	0	58.2	3.3	2
LiBr	5	-9	70.0	2.9	0
	25	-7	64.4	3.5	1
NaCl	0	-17	73.5	2.1	1
	15	-9	69.2	3.6	1
	25	-7	66.7	4.8	2
	35	-6	63.5	5.1	3
	50	-4	59.6	4.7	3
CsF	25	$+1$	71.0	3.3	2
	50	$+1$	63.0	4.5	3
CsI	5	-14	76.8	3.2	2
	25	-9	69.8	4	2
$(CH_3)_4NBr$	5	$+12$	73.5	3.4	2
	25	$+14$	66.6	3.6	1
Pure H_2O	0	$\mathscr{C}_w = 17.6$ psec	88.2	2	1
	5	14.8	85.8	1	0.5
	15	10.7	81.9	—	—
	25	8.25	78.3	1	0.4
	35	6.5	74.8	0	0
	50	4.8	70.0	0	0

[a] Based on Refs. 271 and 645 as well as unpublished results of D. Adolph.
$\hat{\varepsilon}_\infty = 5.5 \pm 0.6$ common to water and all solutions.

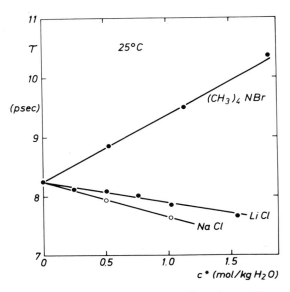

Fig. 10. Dielectric relaxation time $\mathscr{E}$ of three different salt solutions versus the salt molality c^*. Temperature 25°C. (Based on data from Refs. 271 and 645 as well as unpublished data of D. Adolph.)

4. INFORMATION FROM THE "STATIC" PERMITTIVITY

The values of the static permittivity $\hat{\varepsilon}_0$ of the electrolyte solutions in Tables I and II are smaller than the value for pure water. In order to obtain information from the difference values $\hat{\varepsilon}_{0w} - \hat{\varepsilon}_0$, the model of a solution as described in Section 1.3 is accepted. Here that means in particular the solution is considered to consist only of water molecules with the orientational polarizability of pure water and of nonpolar solute particles which may include water molecules without orientational polarizability.

Using eqn. (18), $\hat{\varepsilon}_0$ allows the determination of the volume $(1 - v)V$ of that water in the solution (with total volume V) which has an orientational polarizability identical to that of pure water. The number of moles of water contained in this volume, $(1 - v)V/V_M^{H_2O}$ ($V_M^{H_2O} = 18$ cm³ mol⁻¹), is compared with the total water content of the solution.

In *all* tetraalkylammonium bromide solutions ($0.11 \leq v \leq 0.28$)[644] these two quantities of water differ by less than 1%, which is within the limits of experimental error. This indicates that eqn. (18) is a fairly correct description of the interdependence of the static permittivity $\hat{\varepsilon}_0$ and the volume fraction v of particles without orientational polarizability in the solution. It also indicates that the orientational polarizability of the hydration water of the tetraalkylammonium ions *and* of the bromide ion are equal to that of pure water.

In the $(1\ M)$ alkali halide solutions the quantity of orientationally polarizable water is up to 10% less than the total quantity of water. We define z_f as the number of moles of water without orientational polarizability per mole of electrolyte; these are almost independent of the halide anion (F^- included!). Regarding additionally the former statement about the bromide ion, this "dielectrically saturated" water has to be completely attributed to the cations. The z_f values, together with estimated coordination numbers z_+ of the alkali ions,[271] are presented in the Table III. Water molecules with reduced but nonzero orientational polarizability may also contribute to the z_f values. A comparison between z_f and z_+ shows that the inner hydration layer of only the small cations Li^+ and Na^+ is com-

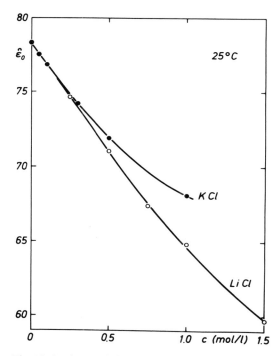

Fig. 11. Static permittivity $\hat{\varepsilon}_0$ of two alkali halide solutions versus the salt molarity c. Temperature 25°C. (KCl up to 0.5 mol liter^{-1} [827]; LiCl, unpublished data of D. Gottlob.)

TABLE III. Numbers z_f of Water Molecules without Orientational Polarizability and Total Numbers z_+ of Water Molecules (Coordination Numbers) in the Inner Hydration Layer of the Alkali Ions[271]

t, °C	z_f (± 0.5)				
	Li$^+$	Na$^+$	K$^+$	Rb$^+$	Cs$^+$
0	6.2	5.7	—	—	—
25	5.6	4.6	3.2	2.3	1.6
50	5.6	4.6	—	—	1.6
z_+	4.9	6.8	7.7	8.2	9.3

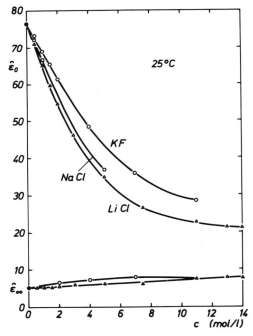

Fig. 12. Static permittivity $\hat{\varepsilon}_0$ and "infinite-frequency" permittivity $\hat{\varepsilon}_\infty$ of three alkali halide solutions versus the salt molarity c. Temperature 25°C. (Unpublished data of D. Gottlob.)

pletely or predominantly saturated dielectrically by the strong ionic electric field.

The dielectric saturation of water molecules immediately adjacent to small cations is not preserved at very high electrolyte concentrations c. This may be concluded from the decreasing slope of the $\hat{\varepsilon}_0(c)$ curves toward the saturation concentration, as shown in the Fig. 12 for LiCl, NaCl, and KF solutions.

5. INFORMATION OBTAINABLE FROM THE DIELECTRIC RELAXATION TIME WITH THE HELP OF THE PROTON MAGNETIC RELAXATION RATE

The values of the dielectric relaxation time $\mathscr{C}$ of most of the alkali halide solutions are smaller and those of the tetraalkylammonium bromide solutions are larger than the value of pure water (see Tables I and II). In order to get information from the values of the relative molal difference $B_d \equiv (1/\mathscr{C}_w)(d\mathscr{C}/dc^*)$, the model of a solution as described in Section 1.3 is employed. In the present context this means that the water in the solution is considered to consist of the "free," the cationic, and the anionic "hydration" molecules with the dielectric molecular rotational correlation times τ_w, τ_+, and τ_-, respectively [eqn. (7)], the dielectric relaxation of which is described by eqns. (15) and (16). These equations are compatible with the empirical eqn. (33) within the frequency interval of the measurements, if the concentration of hydration water is much smaller than that of free water.

According to eqns. (15)–(17), the relative molal shift of the dielectric relaxation time is composed additively of contributions from the cationic and anionic hydration water, as

$$\frac{1}{\mathscr{C}_w}\left(\frac{d\mathscr{C}}{dc^*}\right)_{c^*\to 0} \equiv B_d = B_d{}^+ + B_d{}^- \tag{36}$$

The $B_d{}^+$ and $B_d{}^-$ coefficients cannot be separated without additional appropriate information. This is available from the relative molal shift of the intramolecular proton magnetic relaxation rate $1/T_1$ [eqn. (26)]:

$$\frac{1}{(1/T_1)_w}\left(\frac{d(1/T_1)}{dc^*}\right)_{c^*\to 0} \equiv B_m = B_m{}^+ + B_m{}^- \tag{37}$$

where $B_m{}^+$ and $B_m{}^-$ are known separately (see Chapter 7, Sections 2.4.3 and 2.4.4).

For purposes of a qualitative comparison the measured B_d and B_m values of alkali halide solutions (from Table I in this Chapter and Table II in Chapter 7) are represented in Fig. 13 by the length of the bars; the distance between them indicates the number z_f of dielectrically saturated water molecules per cation (see Section 4). Only in the case of solutions with the largest cations (smallest z_f) and largest anions do we find B_d and B_m values of equal sign and similar magnitude.

For a quantitative comparison of B_d and B_m the different dependences on the molecular rotational correlation time ratio τ_h/τ_w [eqn. (17)] or τ_{mh}/τ_{mw} [eqn. (26)], respectively, and a possible difference between these

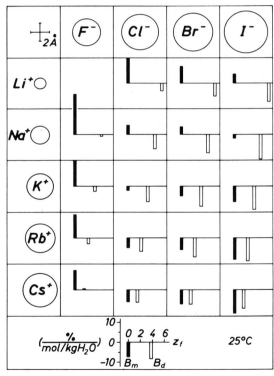

Fig. 13. Measured relative molal shifts of the dielectric relaxation time, B_d, and of the intramolecular proton magnetic relaxation rate, B_m, of aqueous alkali halide solutions (Table I and Chapter 7, Table II). Temperature 25°C. The distance between the bars indicates the number z_f of dielectrically saturated water molecules per cation. The circles indicate the ionic diameters in crystals.

ratios have to be taken into account. With the relation (25), τ_{mh}/τ_{mw} can be estimated not to deviate significantly from τ_h/τ_w. With the help of eqns. (26) and (17), taking $\tau_{mh}/\tau_{mw} \approx \tau_h/\tau_w$, the B_m^- and B_m^+ values of the alkali halide solutions (Table II in Chapter 7) have been converted to B_d^- and B_d^+ values, denoted by $B_{(m)d}^-$ and $B_{(m)d}^+$.[271] From the B_d values together with these data two important conclusions can be drawn.

The differences $B_d - B_{(m)d}^-$ are roughly independent of the anions, whereas $B_{(m)d}^-$ varies between about $+0.05$ and -0.07 (mol/kg)$^{-1}$ from F$^-$ to I$^-$ (Table IV). So we have

$$B_{(m)d}^- \approx B_d^- \tag{38}$$

This means the dielectric and proton magnetic relaxation data contain the same information about the influence of the halide anions on the adjacent hydration molecules (see Chapter 7, Section 2.4.5 and Table III).

Because of eqn. (38) B_d^+ is given by

$$B_d^+ \approx B_d - B_{(m)d}^- \tag{39}$$

As Table V shows, the B_d^+ values of the small cations Li$^+$ and Na$^+$ essentially differ from the respective $B_{(m)d}^+$ values, whereas those of the large Cs$^+$ ion agree with one another. The dielectric and proton magnetic relaxation data must therefore contain different information about the influence of small cations (with large z_f) on the adjacent hydration molecules (see Chapter 7, Section 2.4.5 and Table III, and this chapter, Section 6).

In the case of the tetraalkylammonium bromide solutions the B_d values, as taken from Table I, nearly agree with the B_m values (Chapter 7, Table II), but not with $B_{(m)d}$! This finding is not compatible with eqns. (17) and (26). Inspection of the function $\mathscr{C}(c^*)$ [eqns. (5), (6), (15), and (16)], however, shows that one cannot expect to get the slope B_d for infinite

TABLE IV. Relative Molal Change, Due to the Halide Ions, of the Dielectric Relaxation Time, $B_d^- \approx B_{(m)d}^-$, As Converted from the Relative Molal Change of the Proton Magnetic Relaxation Rate B_m^- at 25°C, and Coordination Numbers z_- Used in the Conversion[271]

	F$^-$	Cl$^-$	Br$^-$	I$^-$
B_d^-, 10^{-2} (mol/kg H$_2$O)$^{-1}$	$+5$	-1	-3	-7
z_-		5.1	6.1	6.7	7.8

TABLE V. Relative Molal Change, Due to the Alkali Ions, of the Dielectric Relaxation Time, $B_d{}^+$, the Corresponding Values $B_{(m)d}^+$ as Converted from the Relative Molal Change of the Proton Magnetic Relaxation Rate, $B_m{}^+$, and Coordination Numbers z_+ Used in the Conversion[271]a

t, °C	Li$^+$		Na$^+$		K$^+$		Rb$^+$		Cs$^+$	
	$B_d{}^+$	$B_{(m)d}^+$	$B_d{}^+$	$B_{(m)d}^+$	$B_d{}^+$	$B_{(m)d}^+$	$B_d{}^+$	$B_{(m)d}^+$	$B_d{}^+$	$B_{(m)d}^+$
0	−2	+4	−12	+3	—	−6	—	−7	—	−8
25	−3	+4	−6	+4	−7	−1	−6	−3	−4	−4
50	0	+3	−4	+5	—	0	—	−1	−3	−1
z_+	4.9		6.8		7.7		8.2		9.3	

a $B_d{}^+$ and $B_{(m)d}^+$ are given in 10^{-2} (mol/kg H$_2$O)$^{-1}$.

dilution $(c^* \to 0)$ from $(\mathscr{E} - \mathscr{E}_w)/(c^*\mathscr{E}_w)$ of 1 M solutions (Table I) of ions as large as the tetraalkylammonium ions, the hydration water of which amounts to an appreciable fraction of the total water in the solution. Bearing this in mind, the evaluation of the finding $B_m \approx [(\mathscr{E} - \mathscr{E}_w)/(c^*\mathscr{E}_w)]_{c=1M}$ leads to the conclusion that the influence of the tetraalkylammonium ions $(\tau_+ > \tau_w)$ must extend beyond the inner hydration layer.[644]

6. THE INFLUENCE OF SMALL CATIONS ON THE DIELECTRIC RELAXATION TIME

The comparison between the cationic contributions to the relative molal change of the dielectric relaxation time, $B_d{}^+$, and of the proton magnetic relaxation rate, $B_m{}^+$, of solutions with the small cations Li$^+$ or Na$^+$ (Table V) yields the surprising result $B_d{}^+ < 0$, $B_{(m)d}^+ > 0$. That this is due to fundamental physical effects is drastically demonstrated in the divergent course of $\mathscr{E}/\mathscr{E}_w$ and $(1/T_1)/(1/T_1)_w$ at higher electrolyte concentrations, as shown in Fig. 14.

One reason for the contrast $B_d{}^+ < 0$, $B_{(m)d}^+ > 0$ obviously has to be sought in the fact that there is a reorientation of water molecules in the inner hydration layer of Li$^+$ and Na$^+$ which causes proton magnetic but not dielectric relaxation, because this reorientation (around the radially oriented electric dipole axis) does not contribute to the dielectric orientational polarization (see Section 4). However, the fact that $\mathscr{E}$ is smaller than

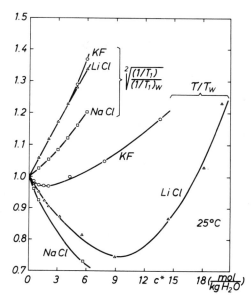

Fig. 14. Solution/water ratios of the dielectric relaxation time $\mathscr{C}$ and of the intramolecular proton magnetic relaxation rate $1/T_1$ of three alkali halide solutions versus the salt molality c^*. Temperature 25°C. ($1/T_1$ from Ref. 195; see also Chapter 7; $\mathscr{C}$ unpublished data of D. Gottlob.)

$\mathscr{C}_w$ in LiCl solutions with concentrations up to 17 mol (kg H_2O)$^{-1}$ (about 1 mol of LiCl per 3 mol of H_2O!, Fig. 14) cannot be explained in these terms. Therefore several mechanisms will be discussed below which reduce the relaxation time $\mathscr{C}$ but not the proton magnetic relaxation rate $1/T_1$ of electrolyte solutions in comparison to pure water.

6.1. Electrostatic Interaction Mechanisms Independent of the Water Structure

6.1.1. Ion Pairs

Lithium and sodium halide ion pairs might have a smaller reorientation time τ_{IP} than the water molecule (τ_w). Their contribution to the total orientational polarization could thus reduce the principal dielectric relaxation time $\mathscr{C}$ of a solution, but to account for $(\mathscr{C} - \mathscr{C}_w)/\mathscr{C}_w \approx -3\%$ for a 1 M LiCl solution, the concentration of Li$^+$Cl$^-$ ion pairs would have to be higher than about 0.1 mol liter^{-1} and this is hardly the case.

6.1.2. Ion Drift

The electric field of an ion polarizes the surrounding water. The water at the water–ion "interface" thus carries a polarization charge (bound charge) with sign opposite to that of the ion charge. If the ion is drifting under the influence of an applied electric field E, the distribution of the water polarization charge becomes asymmetric with respect to the ion, due to the orientational relaxation of the water. The Coulomb interaction between the asymmetrically distributed water polarization charge and the ionic charge influences the drift motion of the ion.[881] The ions thereby make a polarization contribution of the approximate magnitude

$$\frac{4\pi P_{\mathrm{Iw}}}{E} \approx 2\left(\frac{4\pi}{3}\right)^2 \frac{\varepsilon_{0\mathrm{w}} - \varepsilon_{\infty\mathrm{w}}}{\varepsilon_{0\mathrm{w}}\varepsilon_{\infty\mathrm{w}}} \left(\frac{\varepsilon_{\infty\mathrm{w}}}{\varepsilon_{0\mathrm{w}}} \mathscr{C}_{\mathrm{w}}\right)^2 \frac{\sigma^2}{v} \tag{40}$$

relaxing with the relaxation time

$$\mathscr{C}_{\mathrm{Iw}} \approx \left[1 - \frac{8\pi}{9} \frac{\varepsilon_{0\mathrm{w}} - \varepsilon_{\infty\mathrm{w}}}{\varepsilon_{0\mathrm{w}}\varepsilon_{\infty\mathrm{w}}} \left(\frac{\varepsilon_{\infty\mathrm{w}}}{\varepsilon_{0\mathrm{w}}} \mathscr{C}_{\mathrm{w}}\right) \frac{\sigma}{v}\right]\left(\frac{\varepsilon_{\infty\mathrm{w}}}{\varepsilon_{0\mathrm{w}}} \mathscr{C}_{\mathrm{w}}\right) \quad (\ll\mathscr{C}_{\mathrm{w}}) \tag{41}$$

where σ is the dc conductivity and v the solute volume fraction (Sections 1.3 and 4). From the relations (40) and (41) the contribution of the ion–water polarization mechanism, $(-B_d)_{\mathrm{IW}}$, to the B_d of LiCl and NaCl solutions is estimated to be of the order of magnitude 10^{-4} (mol/kg H_2O)$^{-1}$, which is much too small in comparison to the experimental $-B_{d^+}$ values 3×10^{-2} and 6×10^{-2} (mol/kg H_2O)$^{-1}$, respectively.

6.2. Mechanisms Depending on the Water Structure

6.2.1. Structural Mismatch

The structural order of water molecules adjacent to a small cation definitely differs from that of undisturbed water far away from the ion. As Frank and Wen[232] pointed out, there must consequently exist a structural mismatch region of water molecules with reduced reorientation time between the inner hydration sphere and the unperturbed water. As the inner hydration molecules of small cations do not contribute to the dielectric orientational polarization (see Section 4), the relative molal change of the dielectric relaxation time due to these cations, B_{d^+}, should only involve the effect of the more mobile water in the mismatch region, but no effect of the less mobile cationic inner hydration molecules. Then $B_{d^+} < 0$ is to be expected. In contrast, the cationic relative molal change of the proton magnetic relaxation rate, B_m^+, does involve an effect of the cationic inner hydration molecules which may outweigh the effect of the mismatch water, so that $B_m^+ > 0$ results.

In the Frank–Wen picture the experimental data $B_d^+ \approx -0.03$ (mol/kg H_2O)$^{-1}$ and $B_m^+ \approx 0.13$ (mol/kg H_2O)$^{-1}$ (at 25°C) due to the Li$^+$ ion lead to the following conclusion: The $z_f \approx 5$ inner hydration molecules without orientational polarizability have a reorientation time $\tau_+^{(1)} \approx 2.8\tau_w$, while the estimated 15 water molecules in the second hydration layer have a reorientation time $\tau_+^{(2)} \approx 0.9\tau_w$.

The Frank–Wen picture is no longer valid for solutions containing more than about 1 mol liter^{-1} salt, because there is not enough space for so many different structural regions to exist. In an 11 m LiCl solution only the (inner) "cationic" hydration water is left, so that one should expect $\mathscr{E} > \mathscr{E}_w$. However, the measurements yield $\mathscr{E} < \mathscr{E}_w$ up to 17 m (Fig. 14), so that, at least in highly concentrated LiCl and NaCl solutions, the experimental result $\mathscr{E} < \mathscr{E}_w$ must have an origin other than a structural mismatch.

6.2.2. Hydration Water Exchange

In solutions of small cations with orientationally nonpolarizable inner hydration spheres the decay of the orientational polarization may be accelerated, as compared to pure water, by an exchange of water molecules between the solvent regions with orientationally polarizable water and these hydration regions. Polarized water molecules entering the cationic inner hydration spheres lose their preferred orientation, and water molecules leaving these hydration spheres have no preferred orientation.

For a simplified treatment of the contribution of the hydration water exchange to the dielectric relaxation the solvent is assumed merely to consist of orientationally polarizable water (molality $c_w^* - z_f c^*$) with the dielectric relaxation time τ_w and of cationic hydration water (molality $z_f c^*$) without orientational polarizability. The dielectric relaxation time of such a solution is

$$\mathscr{E} \approx \frac{\tau_w}{1 + \{z_f(c^*/c_w^*)/[1 - (z_f c^*/c_w^*)]\}(\tau_w/\tau_f)} \quad (<\mathscr{E}_w!) \qquad (42)$$

with c^* the electrolyte molality, c_w^* the total water molality, z_f the number of cation hydration molecules (see Section 4), and τ_f the mean residence time of a water molecule within a cationic hydration sphere. Equation (42) has been derived with the simplifying assumptions that the molecular electric dipole orientational correlation time τ_w [eqn. (7)] of the orientationally polarizable water molecules in the solution is essentially longer than the mean time between translational jumps, and that τ_f is longer than τ_w. A more general treatment of the dielectric relaxation due to hydration water exchange has been performed by Giese[270] yielding the following main results.

With the water exchange mechanism the sign and magnitude of the experimental $B_d{}^+$ values of LiCl and NaCl solutions can be accounted for. The residence time τ_f of a water molecule in the inner hydration sphere of Li$^+$ and Na$^+$ is nearly equal to or shorter than the time τ_+ (derived from $B_m{}^+$) after which a water molecule reorientates in contact with the cation ($2.5\tau_w$ for Li$^+$, $1.5\tau_w$ for Na$^+$).

The discussion of the temperature dependence of $\mathscr{E}$ (compare Tables I and II) by Giese[270] shows that an increase of the temperature (e.g., $25 \to 50°C$) changes $\mathscr{E}$ as does an increase of the ionic radius.

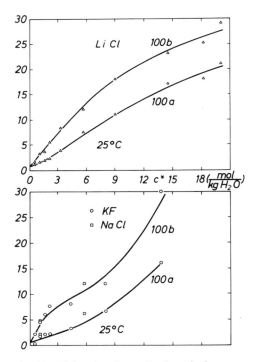

Fig. 15: Dielectric absorption-broadening parameters a and b [eqns. (33)–(35)] of three alkali halide solutions versus the salt molality c^*. Temperature 25°C. (Unpublished data of D. Gottlob.)

The water exchange mechanism also makes possible an understanding of the experimental finding $\mathscr{E} < \mathscr{E}_w$ in the highly concentrated LiCl and NaCl solutions (Fig. 14). In principle relation (42) remains applicable here, but one probably has to assume the quantities τ_w, τ_f, and z_f all to be functions of the electrolyte concentration c^* with probably $\tau_w(c^*) > \tau_w(0)$ and $z_f(c^*) < z_f(0)$. Despite this, formula (42) may imply $\mathscr{E} < \mathscr{E}_w$.

In the present context the strong broadening of the dielectric absorption curves $\varepsilon''(\nu)$ of some highly concentrated alkali halide solutions in comparison to that of pure water should be mentioned. This experimental result is expressed by the large measured values of the a and b parameters in eqn. (33) plotted versus the salt molality c^* in Fig. 15. The broadening is roughly proportional to $(1 + \frac{1}{4}a\pi)/(1 - b)$. There is still no means of extracting useful information from these data.

7. SUMMARY

Dielectric relaxation spectroscopy of aqueous solutions of simple ions yields two types of information about dynamic properties of the hydration water.

One type agrees with the information derived from the proton magnetic relaxation. This concerns the increased rotational mobility of the hydration water of anions and cations of medium size (e.g., Br^-, Cs^+) (negative hydration) and the decreased rotational mobility of the hydration water of very large ions [e.g., $(CH_3)_4N^+$] (hydrophobic or apolar hydration).

The other type of information concerning the dynamics of the hydration molecules of *small* cations (e.g., Li^+) deviates from the information obtained from the proton magnetic relaxation. While the latter mechanism indicates a decreased rotational mobility (positive hydration), dielectric relaxation is insensitive to that reorientation but indicates, in turn, the exchange of hydration molecules with other water molecules.

References

1. A. Abragam, "The Principles of Nuclear Magnetism," Clarendon, Oxford (1961).
2. T. Ackermann, *Z. Elektrochem.* **62**, 411 (1958).
3. J. N. Agar, *Ann. Rev. Phys. Chem.* **15**, 469 (1964).
4. J. C. Ahluwalia and J. W. Cobble, *J. Am. Chem. Soc.* **86**, 5381 (1964).
5. J. C. Ahluwalia, F. J. Millero, R. N. Goldberg, and L. G. Hepler, *J. Phys. Chem.* **70**, 319 (1966).
6. J. W. Akitt, A. K. Covington, J. G. Freeman, and T. H. Lilley, *Trans. Faraday Soc.* **65**, 2701 (1969).
7. D. M. Alexander and D. J. T. Hill, *Aust. J. Chem.* **22**, 347 (1969).
8. N. L. Alpert, W. E. Keiser, and H. A. Szymanski, "IR Theory and Practice of Infrared Spectroscopy," 2nd ed., Plenum, New York (1970).
9. A. P. Altshuller, *J. Chem. Phys.* **24**, 642 (1956).
10. A. P. Altshuller, *J. Chem. Phys.* **28**, 1254 (1958).
11. M. Amdur, J. Padova, and A. Schwartz, *J. Chem. Eng. Data* **15**, 417 (1970).
12. R. L. Amey, *J. Phys. Chem.* **72**, 3358 (1968).
13. H. L. Anderson and L. A. Petree, *J. Phys. Chem.* **74**, 1455 (1970).
14. H. L. Anderson, R. D. Wilson, and D. E. Smith, *J. Phys. Chem.* **75**, 1125 (1971).
15. J. E. Anderson, *J. Chem. Phys.* **47**, 4879 (1967).
16. J. E. Anderson and P. A. Fryer, *J. Chem. Phys.* **50**, 3784 (1969).
17. J. E. Anderson and W. H. Gerritz, *J. Chem. Phys.* **53**, 2584 (1970).
18. K. P. Anderson, D. A. Newell, and R. M. Izatt, *Inorg. Chem.* **5**, 62 (1966).
19. E. M. Arnett, *in* "Physico-Chemical Processes in Aqueous Solvents," (F. Franks, ed.), p. 105, Heinemann, London (1967).
20. E. M. Arnett, W. G. Bentrude, J. J. Burke, and P. M. Duggleby, *J. Am. Chem. Soc.* **87**, 1541 (1965).
21. E. M. Arnett and J. J. Campion, *J. Am. Chem. Soc.* **92**, 7097 (1970).
22. E. M. Arnett, W. B. Kover, and J. V. Carter, *J. Am. Chem. Soc.* **91**, 4028 (1969).
23. E. M. Arnett and D. R. McKelvey, *J. Am. Chem. Soc.* **88**, 2598 (1966).
24. E. M. Arnett and D. R. McKelvey, *J. Am. Chem. Soc.* **88**, 5031 (1966).
25. E. M. Arnett and D. R. McKelvey, *in* "Solute–Solvent Interactions," (J. F. Coetzee and C. D. Ritchie, eds.), Dekker, New York (1969).
26. S. Arrhenius, *Z. Phys. Chem.* **1**, 631 (1887).
27. M. Arshadi and P. Kebarle, *J. Phys. Chem.* **74**, 1483 (1970).
28. S. J. Ashcroft and C. T. Mortimer, "Thermochemistry of Transition Metal Complexes," Academic, New York (1970).
29. G. Atkinson and Y. Mori, *J. Phys. Chem.* **71**, 3523 (1967).
30. L. Avedikian and N. Dollet, *Bull. Soc. Chim. France* 4551 (1967).

31. A. M. Azzam, *Z. Elektrochem.* **58**, 889 (1954).
32. R. M. Badger and S. H. Bauer, *J. Chem. Phys.* **5**, 839 (1937).
33. H. W. Baldwin and H. Taube, *J. Chem. Phys.* **33**, 206 (1960).
34. S. K. Banerjee, K. K. Kundu, and M. N. Das, *J. Chem. Soc. A* **1967**, 166.
35. R. B. Barnes, R. C. Gore, U. Liddel, and V. Z. Williams, "Infrared Spectroscopy," Reinhold, New York (1944).
36. S. J. Bass, W. I. Nathan, R. M. Meighan, and R. H. Cole, *J. Phys. Chem.* **68**, 509 (1964).
37. R. G. Bates, "Determination of pH: Theory and Practice," Wiley, New York (1964).
38. R. G. Bates, *in* "Hydrogen-Bonded Solvent Systems," (A. K. Covington and P. Jones, eds.), Taylor and Francis, London (1968).
39. E. Bauer and M. Magat, *J. Phys. Radium* **9**, 319 (1938).
40. E. C. Baughan, *J. Chem. Soc.* **1940**, 1403.
41. J. G. Bayly, V. B. Kartha, and W. H. Stevens, *Infrared Phys.* **3**, 211 (1963).
42. D. Beckert and H. Pfeifer, *Ann. Physik* 7 Flg., **16**, 262 (1965).
43. P. Bender and W. J. Biermann, *J. Am. Chem. Soc.* **74**, 322 (1952).
44. A. Ben-Naim, *Trans. Faraday Soc.* **66**, 2749 (1970).
45. A. Ben-Naim and F. H. Stillinger, *in* "Water and Aqueous Solutions," (R. H. Horne, ed.), Wiley–Interscience, New York (1971).
46. J. D. Bernal and R. H. Fowler, *J. Chem. Phys.* **1**, 515 (1933).
47. J. D. Bernal and R. H. Fowler, *J. Chem. Phys.* **1**, 538 (1933).
48. J. D. Bernal and G. Tamm, *Nature* **135**, 229 (1935).
49. R. A. Bernheim, T. H. Brown, H. S. Gutowsky, and D. E. Woessner, *J. Chem. Phys.* **30**, 950 (1959).
50. S. L. Bertha and G. R. Choppin, *Inorg. Chem.* **8**, 613 (1969).
51. I. Berthold and Al. Weiss, *Z. Naturforsch.* **22a**, 1433, 1440 (1967).
52. G. L. Bertrand, F. J. Millero, C. H. Wu, and L. G. Hepler, *J. Phys. Chem.* **70**, 699 (1966).
53. K. E. Bett and J. B. Cappi, *Nature* **207**, 620 (1965).
54. O. N. Bhatnagar and C. M. Criss, *J. Phys. Chem.* **73**, 174 (1969).
55. D. R. Bidinosti and W. J. Biermann, *Can. J. Chem.* **34**, 1591 (1956).
56. W. J. Biermann and N. Weber, *J. Am. Chem. Soc.* **76**, 4289 (1954).
57. M. J. Blandamer, D. E. Clarke, N. J. Hidden, and M. C. R. Symons, *Trans. Faraday Soc.* **64**, 2691 (1968).
58. M. J. Blandamer and M. C. R. Symons, *J. Phys. Chem.* **67**, 1304 (1963).
59. N. Bloembergen, *J. Chem. Phys.* **27**, 572 (1957).
60. N. Bloembergen and L. O. Morgan, *J. Chem. Phys.* **34**, 842 (1961).
61. J. O'M. Bockris and A. K. N. Reddy, "Modern Electrochemistry," Vol. 1, Plenum, New York (1970).
62. A. Bodanszky and W. Kauzmann, *J. Phys. Chem.* **66**, 177 (1962).
63. W. Bol, "A Study of the Hydration of Divalent Ions in Aqueous Solution with the Method of Isomorphous Replacement," 4th Nat. Congr. Italian Assoc. Crystallography, Caglian, October 1970.
64. P. D. Bolton, *J. Chem. Ed.* **47**, 638 (1970).
65. P. D. Bolton and L. G. Hepler, *Quart. Rev.* **25**, 521 (1971).
66. O. D. Bonner and G. B. Woolsey, *J. Chem. Phys.* **67**, 2420 (1968).
67. F. Booth, *J. Chem. Phys.* **19**, 391, 1327, 1615 (1951).

68. M. Born, *Z. Physik* **1**, 45 (1920).
69. M. Born, *Z. Physik* **1**, 221 (1920).
70. R. H. Boyd, *J. Chem. Phys.* **35**, 1281 (1961).
71. R. H. Boyd, *J. Chem. Phys.* **39**, 2376 (1963).
72. R. H. Boyd and P. S. Wang, Abstract of Papers, 155th Nat. Mtg. Am. Chem. Soc., San Francisco, Calif., April 1968.
73. G. W. Brady, *J. Chem. Phys.* **28**, 464 (1958).
74. L. Brewer, *in* "The Chemistry and Metallurgy of Miscellaneous Materials: Thermodynamics" (L. L. Quill, ed.), McGraw-Hill, New York (1950).
75. G. Brink and M. Falk, *Can. J. Chem.* **48**, 2096 (1970).
76. G. Brink and M. Falk, *Can. J. Chem.* **48**, 3019 (1970).
77. H. Brintzinger and R. E. Hester, *Inorg. Chem.* **5**, 980 (1966).
78. T. L. Broadwater and R. L. Kay, *J. Phys. Chem.* **74**, 3802 (1970).
79. T. H. Brown, R. A. Bernheim, and H. S. Gutowsky, *J. Chem. Phys.* **33**, 1593 (1960).
80. S. B. Brummer and A. B. Gancy, Office of Saline Water, Research and Development Progress Report No. 643, U.S. Dept. of the Interior, 1971.
81. T. J. Buchanan, *Proc. IEE* **99**(III), 61 (1952).
82. T. J. Buchanan and E. H. Grant, *Brit. J. Appl. Phys.* **6**, 64 (1955).
83. A. D. Buckingham, *Disc. Faraday Soc.* **24**, 151 (1957).
84. A. D. Buckingham, *Proc. Roy. Soc. (London)* **A248**, 169 (1958).
85. A. D. Buckingham, *Trans. Faraday Soc.* **56**, 753 (1960).
86. F. Buckley and A. A. Maryott, *Nat. Bur. Std. Arc.* **1958**, 589.
87. K. Buijs and G. R. Choppin, *J. Chem. Phys.* **39**, 2035 (1963).
88. K. Buijs and G. R. Choppin, *J. Chem. Phys.* **40**, 3120 (1964).
89. K. W. Bunzl, *J. Phys. Chem.* **71**, 1358 (1967).
90. R. E. Burton and J. Daly, *Trans. Faraday Soc.* **66**, 1281 (1970).
91. R. E. Burton and J. Daly, *Trans. Faraday Soc.* **67**, 1219 (1971).
92. W. R. Busing and D. F. Hornig, *J. Phys. Chem.* **65**, 284 (1961).
93. R. Carnsey, R. J. Boe, R. Mahoney, and T. A. Litovitz, *J. Chem. Phys.* **50**, 5222 (1969).
94. H. Y. Carr and E. M. Purcell, *Phys. Rev.* **94**, 630 (1954).
95. B. Case, N. S. Hush, R. Parsons, and M. E. Peover, *J. Electroanalyt. Chem.* **10**, 360 (1965).
96. M. Ceccaldi, C. Thro, and E. Roth, Comm. Energie Atom. Report No. 649, 42 (1957).
97. A. C. Chapman and L. E. Trilwell, *Spectrochim. Acta* **20**, 937 (1964).
98. J. Chédin, *Ann. Chim.* **8**, 243 (1937).
99. H. Chen and D. E. Irish, *J. Phys. Chem.* **75**, 2672 (1971).
100. A. M. Chmelnick and D. Fiat, *J. Chem. Phys.* **47**, 3986 (1967).
101. A. M. Chmelnick and D. Fiat, *J. Chem. Phys.* **51**, 4238 (1969).
102. G. R. Choppin and K. Buijs, *J. Chem. Phys.* **39**, 2042 (1963).
103. J. J. Christensen and R. M. Izatt, *in* "Physical Methods in Advanced Inorganic Chemistry" (H. A. O. Hill and P. Day, eds.), Interscience, New York (1968).
104. J. J. Christensen, R. M. Izatt, L. D. Hansen, and J. A. Partridge, *J. Phys. Chem.* **70**, 2003 (1966).
105. J. J. Christensen, H. D. Johnston, and R. M. Izatt, *Rev. Sci. Instr.* **39**, 1365 (1968).
106. J. J. Christensen, J. H. Rytting, and R. M. Izatt, *J. Chem. Soc. A* **1969**, 861.
107. J. J. Christensen, J. H. Rytting, and R. M. Izatt, *J. Chem. Soc. B* **1970**, 1646.

108. J. J. Christensen, M. D. Slade, D. E. Smith, R. M. Izatt, and J. Tsang, *J. Am. Chem. Soc.* **92**, 4164 (1970).

109. J. J. Christensen, D. P. Wrathall, and R. M. Izatt, *Anal. Chem.* **40**, 175 (1968).

110. J. J. Christensen, D. P. Wrathall, J. O. Oscarson, and R. M. Izatt, *Anal. Chem.* **40**, 1713 (1968).

111. E. C. W. Clarke and D. N. Glew, *Trans. Faraday Soc.* **62**, 539 (1966).

112. J. H. R. Clarke and L. A. Woodward, *Trans. Faraday Soc.* **62**, 2226 (1966).

113. J. W. Cobble, *J. Chem. Phys.* **21**, 1443, 1451 (1953).

114. J. W. Cobble, *J. Chem. Phys.* **21**, 1446 (1953).

115. R. S. Codrington and N. Bloembergen, *J. Chem. Phys.* **29**, 600 (1958).

116. J. F. Coetzee and G. P. Cunningham, *J. Am. Chem. Soc.* **86**, 3403 (1964).

117. J. F. Coetzee and G. P. Cunningham, *J. Am. Chem. Soc.* **87**, 2529 (1965).

118. R. H. Cole, *in* "Magnetic and Dielectric Resonance and Relaxation" (J. Smidt, ed.), p. 96, North-Holland, Amsterdam (1963).

119. H. Coll, R. V. Nauman, and P. W. West, *J. Am. Chem. Soc.* **81**, 1284 (1959).

120. M. J. Colles, G. E. Walrafen, and K. W. Wecht, *Chem. Phys. Letters* **4**, 621 (1970).

121. R. E. Connick and D. Fiat, *J. Chem. Phys.* **39**, 1349 (1963).

122. B. E. Connick and R. E. Poulsen, *J. Chem. Phys.* **30**, 759 (1959).

123. R. E. Connick and R. E. Powell, *J. Chem. Phys.* **21**, 2206 (1953).

123a. R. E. Connick and E. D. Stover, *J. Phys. Chem.* **65**, 2074 (1961).

124. R. E. Connick and K. Wüthrich, *J. Chem. Phys.* **51**, 4506 (1969).

125. B. E. Conway, *Ann. Rev. Phys. Chem.* **17**, 481 (1966).

126. B. E. Conway, J. E. Desnoyers, and R. E. Verrall, *Trans. Faraday Soc.* **62**, 2738 (1966).

127. B. E. Conway and L. H. Laliberté, *in* "Hydrogen-Bonded Solvent Systems" (A. K. Covington and P. Jones, eds.), Taylor and Francis, London (1968).

128. B. E. Conway and L. H. Laliberté, *J. Phys. Chem.* **72**, 4317 (1968).

129. B. E. Conway and L. H. Laliberté, *Trans. Faraday Soc.* **66**, 3032 (1970).

130. B. E. Conway and R. E. Verrall, *J. Phys. Chem.* **70**, 3952 (1966).

131. B. E. Conway, R. E. Verrall, and J. E. Desnoyers, *Z. Physik. Chem.* **230**, 157 (1965).

132. B. E. Conway, R. E. Verrall, and J. E. Desnoyers, *Trans. Faraday Soc.* **62**, 2738 (1966).

133. R. P. J. Cooney and J. R. Hall, *Austr. J. Chem.* **22**, 337 (1969).

134. J. M. Corkill, J. F. Goodman, and J. R. Tate, *Trans. Faraday Soc.* **65**, 1742 (1969).

135. F. A. Cotton and G. Wilkinson, "Advanced Inorganic Chemistry," p. 43, Interscience, New York (1962).

136. A. M. Couture and K. J. Laidler, *Can. J. Chem.* **34**, 1209 (1956).

137. A. M. Couture and K. J. Laidler, *Can. J. Chem.* **35**, 202 (1957).

138. A. M. Couture and K. J. Laidler, *Can. J. Chem.* **35**, 207 (1957).

139. A. K. Covington, J. G. Freeman, and T. H. Lilley, *Trans. Faraday Soc.* **65**, 3136 (1969).

140. A. K. Covington, J. G. Freeman, and T. H. Lilley, *J. Phys. Chem.* **74**, 3773 (1970).

141. A. K. Covington and T. H. Lilley, Chem. Soc. Specialist Periodical Reports "Electrochemistry," Vol. 1 (1970).

142. A. K. Covington, T. H. Lilley, and R. A. Robinson, *J. Phys. Chem.* **72**, 2759 (1968).

143. A. K. Covington, R. A. Robinson, and R. G. Bates, *J. Phys. Chem.* **70**, 3820 (1966).

144. A. K. Covington, M. J. Tait, and W. F. K. Wynne-Jones, *Proc. Roy. Soc.* **A286**, 235 (1965).

145. M. C. Cox, D. H. Everett, D. A. Landsman, and R. J. Munn, *J. Chem. Soc. B* **1968**, 1373.
146. W. M. Cox and J. H. Wolfenden, *Proc. Roy. Soc. A* **145**, 486 (1934).
147. R. A. Craig and R. E. Richards, *Trans. Faraday Soc.* **59**, 1972 (1963).
148. C. M. Criss and J. W. Cobble, *J. Am. Chem. Soc.* **86**, 5385 (1964).
149. C. M. Criss and J. W. Cobble, *J. Am. Chem. Soc.* **86**, 5390 (1964).
150. C. M. Criss, R. P. Held, and E. Luksha, *J. Phys. Chem.* **72**, 2970 (1968).
151. G. P. Cunningham, D. F. Evans, and R. L. Kay, *J. Phys. Chem.* **70**, 3998 (1966).
152. G. P. Cunningham, G. A. Vidulich, and R. L. Kay, *Chem. Eng. Data* **12**, 336 (1967).
153. J. Daly and R. E. Burton, *Trans. Faraday Soc.* **66**, 2408 (1970).
154. A. D'Aprano, *J. Phys. Chem.* **75**, 3290 (1971).
155. D. H. Davies and G. C. Benson, *Can. J. Chem.* **43**, 3100 (1965).
156. T. Davies, S. S. Singer, and L. A. K. Stavely, *J. Chem. Soc.* **1954**, 2304.
157. A. R. Davis and D. E. Irish, *Inorg. Chem.* **7**, 1699 (1968).
158. A. R. Davis, J. W. Macklin, and R. A. Plane, *J. Chem. Phys.* **50**, 1478 (1969).
159. A. R. Davis and R. A. Plane, *Inorg. Chem.* **7**, 2565 (1968).
160. A. J. de Bethune, *J. Chem. Phys.* **29**, 616 (1958).
161. P. Debye and L. Pauling, *J. Am. Chem. Soc.* **47**, 2129 (1925).
162. J. Del Bene and J. A. Pople, *Chem. Phys. Letters* **4**, 426 (1969).
163. J. Del Bene and J. A. Pople, *J. Chem. Phys.* **52**, 4858 (1970).
164. C. L. de Ligny and M. Alfenaar, *Rec. Trav. Chim.* **84**, 81 (1965).
165. C. L. de Ligny and M. Alfenaar, *Rec. Trav. Chim.* **86**, 929 (1967).
166. C. L. de Ligny, M. Alfenaar, and N. G. van der Veen, *Rec. Trav. Chim.* **87**, 585 (1968).
167. M. de Paz, S. Ehrenson, and L. Friedman, *J. Chem. Phys.* **52**, 3362 (1970).
168. M. de Paz, A. G. Giardini, and L. Friedman, *J. Chem. Phys.* **52**, 687 (1970).
169. M. de Paz, J. J. Leventhal, and L. Friedman, *J. Chem. Phys.* **49**, 5543 (1968).
170. M. de Paz, J. J. Leventhal, and L. Friedman, *J. Chem. Phys.* **51**, 3748 (1969).
171. J. Depireux and D. Williams, *Nature* **191**, 699 (1962).
172. J. E. Desnoyers, M. Arel, G. Perron, and C. Jolicoeur, *J. Phys. Chem.* **23**, 3346 (1969).
173. J. E. Desnoyers, R. Francescon, P. Picker, and C. Jolicoeur, *Can. J. Chem.* **49**, 3460 (1971).
174. J. E. Desnoyers and C. Jolicoeur, *in* "Modern Aspects of Electrochemistry" (J. O'M. Bockris and B. E. Conway, eds.), No. 5, Plenum, New York (1969).
175. J. E. Desnoyers, G. E. Pelletier, and C. Jolicoeur, *Can. J. Chem.* **43**, 3232 (1965).
176. C. Deverell, *in* "Progress in Nuclear Magnetic Resonance Spectroscopy" (J. W. Emsley, J. Freeney, and L. H. Sutcliffe, eds.) Vol. 4, p. 235, Pergamon, Oxford (1969).
177. G. H. F. Diercksen, *Chem. Phys. Letters* **4**, 373 (1969).
178. J. S. Dohnanyi, *Phys. Rev.* **125**, 1824 (1962).
179. D. A. Draegert, N. W. B. Stone, B. Carnutte, and D. Williams, *J. Opt. Soc. Am.* **56**, 64 (1966).
180. D. A. Draegert and D. Williams, *J. Chem. Phys.* **48**, 401 (1968).
181. S. I. Drakin, L. V. Lantukhova, and M. Kh. Karapet'yants, *Russian J. Phys. Chem.* **40**, 240 (1966).
182. S. I. Drakin, L. V. Lantukhova, and M. K. Karapet'yants, *Russian J. Phys. Chem.* **41**, 1436 (1967).

183. P. Dryjanski and Z. Kecki, *Roczniki Chem.* **43**, 1053 (1969).
184. P. Dryjanski and Z. Kecki, *Roczniki Chem.* **44**, 1141 (1970).
185. W. C. Duer and G. L. Bertrand, *J. Chem. Phys.* **53**, 3020 (1970).
186. L. A. Dunn, R. H. Stokes, and L. G. Hepler, *J. Phys. Chem.* **69**, 2808 (1965).
187. R. Dunsmuir and J. G. Powles, *Phil. Mag.* **37**, 61 (1946).
188. P. A. Egelstaff, "An Introduction to the Liquid State," Academic Press, New York (1967).
189. M. Eigen and E. Wicke, *Z. Elektrochem.* **55**, 354 (1951).
190. D. Eisenberg and W. Kauzmann, "The Structure and Properties of Water," Oxford Univ. Press, New York (1969).
191. M. Eisenstadt and H. L. Friedman, *J. Chem. Phys.* **48**, 4445 (1968).
192. D. D. Eley and M. G. Evans, *Trans. Faraday Soc.* **34**, 1093 (1938).
193. M. I. Emel'yanov, *Z. Strukt. Khim.* **6**, 295 (1965) [English transl. *J. Struct. Chem.* **5**, 270 (1965)].
194. J. W. Emsley, J. Feeney, and L. H. Sutcliffe, "High Resolution Magnetic Resonance Spectroscopy," Pergamon, Oxford (1965).
195. L. Endom, H. G. Hertz, B. Thül, and M. D. Zeidler, *Ber. Bunsenges. Physik. Chem.* **71**, 1008 (1967).
196. G. Engel and H. G. Hertz, *Ber. Bunsenges. Physik. Chem.* **72**, 808 (1968).
197. T. S. England and E. E. Schneider, *Physica* **17**, 221 (1957).
198. D. F. Evans and T. L. Broadwater, *J. Phys. Chem.* **72**, 1037 (1968).
199. D. F. Evans, G. P. Cunningham, and R. L. Kay, *J. Phys. Chem.* **70**, 2974 (1966).
200. D. F. Evans and P. Gardam, *J. Phys. Chem.* **73**, 158 (1969).
201. D. F. Evans and M. A. Matesich, *in* "Techniques of Electrochemistry" (E. Yeager, ed.), Vol. II, Wiley, New York (1971).
202. D. F. Evans, C. Zawoyski, and R. L. Kay, *J. Phys. Chem.* **69**, 3878 (1965).
203. G. Fabbri and S. Roffia, *Ann. Chim. (Rome)* **50**, 3 (1960).
204. G. Fabbri and S. Roffia, *Ann. Chim. (Rome)* **50**, 199 (1960).
205. B. P. Fabricand and S. S. Goldberg, *Mol. Phys.* **13**, 323 (1967).
206. B. P. Fabricand, S. S. Goldberg, R. Leifer, and S. G. Ungar, *Mol. Phys.* **7**, 425 (1964).
207. B. P. Fabricand, S. S. Goldberg, R. Leifer, and S. G. Ungar, *Mol. Phys.* **9**, 399 (1965).
208. K. Fajans, *Verh. dtsh. physik. Ges.* **21**, 709 (1919).
209. M. Falk and T. A. Ford, *Can. J. Chem.* **44**, 1699 (1966).
210. M. Falk and P. A. Giguère, *Can. J. Chem.* **35**, 1195 (1957).
211. E. Fatuzzo and P. R. Mason, *Proc. Phys. Soc.* **90**, 729, 741 (1967).
212. F. S. Feates and D. J. G. Ives, *J. Chem. Soc.* **1956**, 2798.
213. H. M. Feder, Ph.D. Thesis, Univ. of Chicago, 1954.
214. H. M. Feder and H. Taube, *J. Chem. Phys.* **20**, 1335 (1952).
215. R. Fernández-Prini and G. Atkinson, *J. Phys. Chem.* **75**, 238 (1971).
216. D. Fiat and R. E. Connick, *J. Am. Chem. Soc.* **90**, 608 (1968).
217. A. Finch, P. N. Gates, K. Radcliffe, F. N. Dickson, and F. F. Bentley, "Chemical Applications of Far Infrared Spectroscopy," pp. 58–60, Academic, New York (1970).
218. F. Fister and H. G. Hertz, *Ber. Bunsenges. Physik. Chem.* **71**, 1032 (1967).
219. P. J. Flory, *J. Chem. Phys.* **10**, 51 (1942).
220. D. W. Fong and E. Grunwald, *J. Phys. Chem.* **73**, 3909 (1969).

221. R. H. Fowler and E. A. Guggenheim, "Statistical Thermodynamics," Cambridge (1939).
222. E. U. Franck and K. Roth, *Disc. Faraday Soc.* **43**, 108 (1967).
223. H. S. Frank, *J. Am. Chem. Soc.* **63**, 1789 (1941).
224. H. S. Frank, *J. Chem. Phys.* **13**, 493 (1945).
225. H. S. Frank, *J. Chem. Phys.* **23**, 2023 (1955).
226. H. S. Frank, *Z. Phys. Chem.* (*Leipzig*) **228**, 364 (1965).
227. H. S. Frank, *Fed. Proc.* **24**, No. 2, Part III, Suppl. 15, 1 (1965).
228. H. S. Frank, *in* "Chemical Physics of Ionic Solutions," (B. E. Conway and R. G. Barradas, eds.), Wiley, New York (1966).
229. H. S. Frank, *Disc. Faraday Soc.* **43**, 137 (1967).
230. H. S. Frank and M. W. Evans, *J. Chem. Phys.* **13**, 507 (1945).
231. H. S. Frank and A. L. Robinson, *J. Chem. Phys.* **8**, 933 (1940).
232. H. S. Frank and W. Y. Wen, *Disc. Faraday Soc.* **24**, 133 (1957).
233. F. Franks, *in* "Hydrogen-Bonded Solvent Systems," (A. K. Covington and P. Jones, eds.), pp. 31–47, Taylor and Francis, London (1968).
234. F. Franks and D. J. G. Ives, *Quart. Rev.* (*London*) **20**, 1 (1966).
235. F. Franks and D. S. Reid, *J. Phys. Chem.* **73**, 3152 (1969).
236. F. Franks and H. T. Smith, *Trans. Faraday Soc.* **63**, 2586 (1967).
237. F. Franks and B. Watson, *Trans. Faraday Soc.* **65**, 2339 (1969).
238. A. Fratiello, Ph.D. Thesis, Brown Univ., 1962.
239. A. Fratiello, R. E. Lee, V. M. Nishida, and R. E. Schuster, *J. Chem. Phys.* **48**, 3705 (1969).
240. H. L. Friedman, *J. Chem. Phys.* **32**, 1134 (1960).
241. H. L. Friedman, *J. Chem. Phys.* **32**, 1351 (1960).
242. H. L. Friedman, "Ionic Solution Theory," Interscience, New York (1962).
243. H. L. Friedman and G. R. Haugen, *J. Am. Chem. Soc.* **76**, 2060 (1954).
244. H. L. Friedman and P. S. Ramanathan, *J. Phys. Chem.* **74**, 3756 (1970).
245. L. Friedman and V. J. Shiner, Jr., *J. Chem. Phys.* **44**, 4639 (1966).
246. H. Fröhlich, "Theory of Dielectrics," p. 8, Oxford Univ. Press, London (1958).
247. S. Fujiwara and H. Hayashi, *J. Chem. Phys.* **43**, 23 (1965).
248. R. M. Fuoss, *Proc. Natl. Acad. Sci. U.S.* **45**, 807 (1959).
249. R. M. Fuoss and F. Accascina, "Electrolytic Conductance," p. 196, Interscience, New York (1959).
250. R. M. Fuoss and K. L. Hsia, *Proc. Natl. Acad. Sci. U.S.* **57**, 1550 (1967).
251. R. M. Fuoss and K. L. Hsia, *Proc. Natl. Acad. Sci. U.S.* **58**, 1818 (1967).
252. A. B. Gancy and S. B. Brummer, *J. Electrochem. Soc.* **115**, 804 (1968).
253. A. B. Gancy and S. B. Brummer, *J. Phys. Chem.* **73**, 2429 (1969).
254. E. Ganz, *Z. Physik. Chem. Leipzig*, **B33**, 163 (1936).
255. E. Ganz, *Ann. Physik* **28**, 445 (1937).
256. E. Ganz, *Z. Physik. Chem. Leipzig* **B35**, 1 (1937).
257. R. Garnsey, R. J. Boe, R. Mahoney, and T. A. Litovitz, *J. Chem. Phys.* **50**, 5222 (1969).
258. B. B. Garret and L. O. Morgan, *J. Chem. Phys.* **44**, 890 (1966).
259. F. Garrick, *Phil. Mag.* **9**, 131 (1930).
260. F. Garrick, *Phil. Mag.* **10**, 76 (1931).
261. E. L. Gasner, Dissertation, Univ. of Chicago (1965).

262. L. M. Gedansky, E. M. Woolley, and L. G. Hepler, *J. Chem. Thermodynamics* **2**, 561 (1970).
263. E. E. Genser, Thesis, Univ. of California (1962).
264. P. George, G. I. H. Hanania, and D. H. Irvine, *J. Chem. Phys.* **22**, 1616 (1954).
265. P. George, G. I. H. Hanania, and D. H. Irvine, *Reucil.* **74**, 759 (1955).
266. P. George and D. S. McClure, *in* "Progress in Inorganic Chemistry (F. A. Cotton, ed.), Vol. 1, Interscience, New York (1959).
267. R. George and E. M. Woolley, *J. Solution Chem.* **1**, 279 (1972).
268. E. Gerdes, W. D. Kraeft, and M. Zecha, *Z. Phys. Chem.* **241**, 25 (1969).
269. P. Gerding, I. Leden, and S. Sunner, *Acta Chem. Scand.* **17**, 2190 (1963).
270. K. Giese, *Ber. Bunsenges. Physik. Chem.* **76**, 495 (1972).
271. K. Giese, U. Kaatze, and R. Pottel, *J. Phys. Chem.* **74**, 3718 (1970).
272. L. J. Gillespie, R. H. Lambert, and J. A. Gibson, *J. Am. Chem. Soc.* **52**, 3806 (1930).
273. R. M. Glaeser and C. A. Coulson, *Trans. Faraday Soc.* **61**, 389 (1965).
274. D. N. Glew, H. D. Mak, and N. S. Rath, *in* "Hydrogen-Bonded Solvent Systems" (A. K. Covington and P. Jones, eds.), p. 195, Taylor and Francis, London (1968).
275. R. E. Glick and K. C. Tewari, *J. Chem. Phys.* **44**, 546 (1966).
276. E. Glueckauf, *Trans. Faraday Soc.* **51**, 1235 (1955).
277. E. Glueckauf, *Trans. Faraday Soc.* **60**, 572 (1964).
278. E. Glueckauf, *Trans. Faraday Soc.* **61**, 914 (1965).
279. E. Glueckauf, *in* "Chemical Physics of Ionic Solutions," (B. E. Conway and R. G. Barradas, eds.), Wiley, New York (1966).
280. V. Gold, *Proc. Chem. Soc. (London)* **1963**, 141.
281. V. Gold and B. M. Lowe, *Proc. Chem. Soc. (London)* **1963**, 140.
282. V. Gold and B. M. Lowe, *J. Chem. Soc. B* **1967**, 936.
283. E. v. Goldammer and M. D. Zeidler, *Ber. Bunsenges. Physik. Chem.* **73**, 4 (1969).
284. R. N. Goldberg and L. G. Hepler, *J. Phys. Chem.* **72**, 4654 (1968).
285. M. Goldblatt and W. M. Jones, *J. Chem. Phys.* **51**, 1881 (1969).
286. S. Golden and C. Guttman, *J. Chem. Phys.* **43**, 1894 (1965).
287. J. D. S. Goulden and D. J. Manning, *Spectrochim. Acta* **23A**, 2249 (1967).
288. B. S. Gourary and F. J. Adrian, *Solid State Phys.* **10**, 127 (1960).
289. D. C. Grahame, *J. Chem. Phys.* **18**, 903 (1950).
290. D. C. Grahame, *J. Chem. Phys.* **21**, 1054 (1953).
291. I. Grenthe, H. Ots, and O. Ginstrup, *Acta Chem. Scand.* **24**, 1067 (1970).
292. J. Greyson, *J. Phys. Chem.* **71**, 2210 (1967).
293. J. Greyson and H. Snell, *J. Phys. Chem.* **73**, 3208 (1969).
294. J. Greyson and H. Snell, *J. Phys. Chem.* **73**, 4423 (1969).
295. J. Greyson and H. Snell, *J. Chem. and Eng. Data* **16**, 73 (1971).
296. E. Grunwald, *J. Phys. Chem.* **71**, 1846 (1967).
297. E. Grunwald, G. Baughman, and G. Kohnstam, *J. Am. Chem. Soc.* **82**, 5801 (1960).
298. E. Grunwald and E. K. Ralph, III, *J. Am. Chem. Soc.* **89**, 4405 (1967).
299. F. T. Gucker, C. L. Chernick, and P. R. Chowdhury, *Proc. Natl. Acad. Sci. U.S.* **55**, 12 (1965).
300. E. A. Guggenheim, "Thermodynamics," 5th rev. ed., North-Holland, Amsterdam (1967).
301. E. A. Guggenheim and J. C. Turgeon, *Trans. Faraday Soc.* **51**, 757 (1955).
302. R. W. Gurney, "Ionic Processes in Solution," McGraw-Hill, New York (1953).

303. B. Gutbezahl and E. Grunwald, *J. Am. Chem. Soc.* **75**, 565 (1953).
304. G. H. Haggis, J. B. Hasted, and T. J. Buchanan, *J. Chem. Phys.* **20**, 1452 (1952).
305. E. L. Hahn, *Physics Today* **6**, 4 (1953).
306. J. D. Hale, R. M. Izatt, and J. J. Christensen, *J. Phys. Chem.* **67**, 2605 (1963).
307. H. F. Halliwell and S. C. Nyburg, *Trans. Faraday Soc.* **59**, 1126 (1963).
308. S. D. Hamann, *J. Phys. Chem.* **67**, 2233 (1963).
309. M. A. Haney and J. L. Franklin, *J. Chem. Phys.* **50**, 2028 (1969).
310. D. Hankins, J. W. Moskowitz, and F. H. Stillinger, *Chem. Phys. Letters* **4**, 527 (1970).
311. D. Hankins, J. W. Moskowitz, and F. H. Stillinger, *J. Chem. Phys.* **53**, 4544 (1970).
312. L. D. Hansen, J. J. Christensen, and R. M. Izatt, *Chem. Comm.* **3**, 36 (1965).
313. L. D. Hansen and E. A. Lewis, *J. Chem. Thermodynamics* **3**, 35 (1971).
314. H. S. Harned, *J. Phys. Chem.* **63**, 1299 (1957).
315. H. S. Harned, *J. Phys. Chem.* **64**, 112 (1958).
316. H. S. Harned and L. D. Fallon, *J. Am. Chem. Soc.* **61**, 2374 (1939).
317. H. S. Harned and B. B. Owen, "The Physical Chemistry of Electrolytic Solutions," 3rd ed., Reinhold, New York (1958).
318. H. S. Harned and R. A. Robinson, "The International Encyclopedia of Chemistry and Chemical Physics," Vol. 2, Topic 15 (R. A. Robinson, ed.), Pergamon, New York (1968).
319. F. A. Harris and C. T. O'Konski, *J. Phys. Chem.* **61**, 310 (1957).
320. K. A. Hartman, Jr., *J. Phys. Chem.* **70**, 270 (1966).
321. J. B. Hasted and S. H. M. El Sabeh, *Trans. Faraday Soc.* **49**, 1003 (1953).
322. J. B. Hasted, D. M. Ritson, and C. H. Collie, *J. Chem. Phys.* **16**, 1 (1948).
323. J. B. Hasted and G. W. Roderick, *J. Chem. Phys.* **29**, 17 (1958).
324. G. R. Haugen and H. L. Friedman, *J. Phys. Chem.* **72**, 4549 (1968).
325. R. Hausser and G. Laukien, *Z. Physik.* **153**, 394 (1959).
326. R. Hausser and F. Noack, *Z. Physik* **182**, 93 (1964).
327. J. L. Hawes and R. L. Kay, *J. Phys. Chem.* **69**, 2420 (1965).
328. O. Haworth and R. E. Richards, *in* "Progress in Nuclear Magnetic Spectroscopy" (J. W. Emsley, J. Feeney, and L. H. Sutcliffe, eds.) Vol. 1, p. 1, Pergamon, Oxford (1966).
329. R. G. Hayes and R. J. Meyers, *J. Chem. Phys.* **40**, 877 (1964).
330. H. G. Hecht, "Magnetic Resonance Spectroscopy," Wiley, New York (1967).
331. K. Heinzinger and R. E. Weston, Jr., *J. Phys. Chem.* **68**, 744 (1964).
332. K. Heinzinger and R. E. Weston, Jr., *J. Phys. Chem.* **68**, 2179 (1964).
333. K. Heinzinger and R. E. Weston, Jr., *J. Chem. Phys.* **42**, 272 (1965).
334. G. Held and F. Noack, Proc. XVIth Colloque Ampère (1971), p. 620.
335. L. G. Hepler, *J. Am. Chem. Soc.* **85**, 3089 (1963).
336. L. G. Hepler, *J. Phys. Chem.* **68**, 2645 (1964).
337. L. G. Hepler, *J. Phys. Chem.* **69**, 965 (1965).
338. L. G. Hepler, *Can. J. Chem.* **47**, 4613 (1969).
339. L. G. Hepler, *Can. J. Chem.*, **49**, 2803 (1971).
340. L. G. Hepler and W. F. O'Hara, *J. Phys. Chem.* **65**, 811 (1961).
341. L. G. Hepler, J. M. Stokes, and R. H. Stokes, *Trans. Faraday Soc.* **61**, 20 (1965).
342. E. F. G. Herington, *J. Am. Chem. Soc.* **73**, 5883 (1951).
343. H. G. Hertz, *Z. Elektrochem.* **65**, 20 (1961).
344. H. G. Hertz, *Ber. Bunsenges. Physik. Chem.* **67**, 311 (1963).

345. H. G. Hertz, *Ber. Bunsenges. Physik. Chem.* **71**, 979 (1967).

346. H. G. Hertz, *Ber. Bunsenges. Physik. Chem.* **71**, 999 (1967).

347. H. G. Hertz, *in* "Progress in Nuclear Magnetic Resonance Spectroscopy" (J. W. Emsley, J. Freeney, and L. H. Sutcliffe, eds.), Vol. 3, p. 159, Pergamon, Oxford (1967).

348. H. G. Hertz, *Mol. Phys.* **14**, 291 (1968).

349. H. G. Hertz, *Ber. Bunsenges. Physik. Chem.* **74**, 666 (1970).

350. H. G. Hertz, *Ber. Bunsenges. Physik. Chem.* **75**, 183 (1971).

351. H. G. Hertz, *Ber. Bunsenges. Physik. Chem.* **75**, 572 (1971).

352. H. G. Hertz, *in* "Theorie der Elektrolyte" (H. Falkenhagen, ed.), Hirzel, Leipzig (1971).

353. H. G. Hertz, M. Holz, G. Keller, Ch. Y. Yoon, and H. Versmold, to be published.

354. H. G. Hertz, G. Keller, and H. Versmold, *Ber. Bunsenges. Physik. Chem.* **73**, 549 (1969).

355. H. G. Hertz and R. Klute, *Z. Physik. Chem. Frankfurt* **69**, 101 (1970).

356. H. G. Hertz, M. Holz, R. Klute, G. Stalidis, and H. Versmold, to be published.

357. H. G. Hertz, B. Lindman, and V. Siepe, *Ber. Bunsenges. Physik. Chem.* **73**, 542 (1969).

358. H. G. Hertz, G. Stalidis, and H. Versmold, *J. Chim. Physique* (Numero Special) **1969** (Octobre), 177.

359. H. G. Hertz, R. Tusch, and H. Versmold, *Ber. Bunsenges. Physik. Chem.* **75**, 1177 (1971).

360. H. G. Hertz, to be published.

361. H. G. Hertz and M. D. Zeidler, *Ber. Bunsenges. Physik. Chem.* **67**, 311, 774 (1963).

362. R. E. Hester, *Ann. Rep. Chem. Soc.* **66A**, 79 (1969).

363. R. E. Hester and W. E. L. Grossman, *Inorg. Chem.* **5**, 1308 (1966).

364. R. E. Hester and R. A. Plane, *Inorg. Chem.* **3**, 769 (1964).

365. R. E. Hester and R. A. Plane, *J. Chem. Phys.* **40**, 411 (1964).

366. R. E. Hester, R. A. Plane, and G. E. Walrafen, *J. Chem. Phys.* **38**, 249 (1963).

367. R. E. Hester, R. A. Plane, and G. E. Walrafen, *J. Chem. Phys.* **46**, 3405 (1967).

368. J. H. Hildebrand and R. L. Scott, "The Solubility of Nonelectrolytes," 3rd ed. Dover, New York (1964).

369. N. E. Hill, *J. Phys. C., Proc. Phys. Soc.* **3**, 238 (1970).

370. N. E. Hill, W. E. Vaughan, A. H. Price, and M. Davies, "Dielectric Properties and Molecular Behavior," p. 57, Van Nostrand, London (1969).

371. T. L. Hill, "Statistical Mechanics," McGraw-Hill, New York (1956).

372. T. L. Hill, "An Introduction to Statistical Thermodynamics," Addison-Wesley, Reading, Massachusetts (1960).

373. R. J. Hinchey and J. W. Cobble, *Inorg. Chem.* **9**, 917 (1970).

374. C. C. Hinckley and L. O. Morgan, *J. Chem. Phys.* **44**, 898 (1966).

375. J. C. Hindman, *J. Chem. Phys.* **36**, 1000 (1962).

376. J. F. Hinton and E. S. Amis, *Chem. Rev.* **67**, 367 (1967).

377. J. O. Hirschfelder, C. F. Curtiss, and R. B. Bird, "Molecular Theory of Gases and Liquids," Chapter 4, Wiley, New York (1954).

378. J. H. W. Hittorf, *Poggendorff's Annalen* **89**, 177 (1853).

379. J. R. Holmes, D. Kivelson, and W. C. Drinkard, *J. Chem. Phys.* **37**, 150 (1962).

380. A. Holtzer and M. F. Emerson, *J. Phys. Chem.* **73**, 26 (1969).

381. H. P. Hopkins, Jr., C. H. Wu, and L. G. Hepler, *J. Phys. Chem.* **69**, 2244 (1965).

382. D. F. Hornig, *J. Chem. Phys.* **40**, 3119 (1964).
383. P. S. Hubbard, *Proc. Roy. Soc. A* **291**, 537 (1966).
384. P. S. Hubbard, *J. Chem. Phys.* **52**, 563 (1970).
385. M. L. Huggins, *Ann. N. Y. Acad. Sci.* **41**, 1 (1942).
386. M. L. Huggins, *J. Am. Chem. Soc.* **64**, 1712 (1942).
387. M. L. Huggins, *J. Phys. Chem.* **46**, 151 (1942).
388. J. R. Hulston, *J. Chem. Phys.* **50**, 1483 (1969).
389. J. P. Hunt and H. Taube, *J. Chem. Phys.* **18**, 757 (1950).
390. J. P. Hunt and H. Taube, *J. Chem. Phys.* **19**, 602 (1951).
391. N. S. Hush, *Austral. J. Sci. Res.* **1**, 482 (1948).
392. D. E. Irish and H. Chen, *J. Phys. Chem.* **74**, 3796 (1970).
393. D. E. Irish and A. R. Davis, *Can. J. Chem.* **46**, 943 (1968).
394. D. E. Irish, A. R. Davis, and R. A. Plane, *J. Chem. Phys.* **50**, 2262 (1969).
395. D. E. Irish, B. McCarroll, and T. F. Young, *J. Chem. Phys.* **39**, 3436 (1963).
396. D. E. Irish and G. E. Walrafen, *J. Chem. Phys.* **46**, 378 (1967).
397. A. Isihara and R. V. Hanks, *J. Chem. Phys.* **36**, 433 (1962).
398. E. N. Ivanov, *Soviet Phys.—JETP* **18**, 1041 (1964).
399. D. J. G. Ives and P. D. Marsden, *J. Chem. Soc.* **1965**, 649.
400. D. J. G. Ives and P. G. N. Moseley, *J. Chem. Soc. B* **1970**, 1655.
401. D. J. G. Ives and D. Prasad, *J. Chem. Soc. B* **1970**, 1652.
402. R. M. Izatt and J. J. Christensen, *in* "Handbook of Biochemistry" (H. A. Sober, ed.), Chemical Rubber Co., Cleveland, Ohio (1968).
403. J. A. Jackson, J. F. Lemons, and H. Taube, *J. Chem. Phys.* **32**, 553 (1960).
404. B. Jacobson, *Nature* **172**, 666 (1953).
405. B. Jacobson, *Acta Chem. Scand.* **9**, 191 (1955).
406. B. Jacobson, *J. Am. Chem. Soc.* **77**, 2919 (1955).
407. B. Jacobson, W. A. Anderson, and J. T. Arnold, *Nature* **173**, 772 (1954).
408. G. A. Jeffrey and R. K. McMullan, "Progress in Inorganic Chemistry" (F. A. Cotton, ed.), Vol. 8, Interscience, New York (1967).
409. W. P. Jencks and J. Regenstein, *in* "Handbook of Biochemistry" (H. A. Sober, ed.), Chemical Rubber Co., Cleveland, Ohio (1968).
410. G. P. Johari and P. H. Tewari, *J. Phys. Chem.* **69**, 2862 (1965).
411. C. Jolicoeur and H. L. Friedman, *J. Phys. Chem.* **75**, 165 (1971).
412. C. Jolicoeur, P. R. Phillip, G. Perron, P. A. Leduc, and J. E. Desnoyers, *Can. J. Chem.* **50**, 3167 (1972).
413. C. Jolicoeur, P. Picker, and J. E. Desnoyers, *J. Chem. Thermodynamics* **1**, 485 (1969).
414. C. Jolicoeur, N. D. The, and A. Cabana, *Can. J. Chem.* **49**, 2008 (1971).
415. G. Jones and B. C. Bradshaw, *J. Am. Chem. Soc.* **55**, 1780 (1933).
416. G. Jones and M. Dole, *J. Am. Chem. Soc.* **51**, 2950 (1929).
417. G. T. Jones and J. G. Powles, *Mol. Phys.* **8**, 607 (1964).
418. J. C. Justice, *J. Chim. Phys. Physicochem. Biol.* **65**, 353 (1968).
419. M. Kaminsky, *Disc. Faraday Soc.* **24**, 171 (1957).
420. W. Kauzmann, *Advan. Protein Chem.* **14**, 1 (1959).
421. J. L. Kavanau, "Water and Solute–Water Interactions," Holden-Day, San Francisco, Calif. (1964).
422. R. L. Kay, *J. Am. Chem. Soc.* **82**, 2099 (1960).
423. R. L. Kay, Advan. in Chem. Series, No. 73, Am. Chem. Soc., Washington, D. C. (1968).

424. R. L. Kay and T. L. Broadwater, *Electrochim. Acta* **16**, 667 (1971).
425. R. L. Kay, G. P. Cunningham, and D. F. Evans, *in* "Hydrogen-Bonded Solvent Systems" (A. K. Covington and P. Jones, eds.), p. 249, Taylor and Francis, London (1968).
426. R. L. Kay and J. L. Dye, *Proc. Natl. Acad. Sci. U.S.* **49**, 5 (1963).
427. R. L. Kay and D. F. Evans, *J. Phys. Chem.* **69**, 4216 (1965).
428. R. L. Kay and D. F. Evans, *J. Phys. Chem.* **70**, 2325 (1966).
429. R. L. Kay, D. F. Evans, and G. P. Cunningham, *J. Phys. Chem.* **73**, 3322 (1969).
430. R. L. Kay, B. J. Hales, and G. P. Cunningham, *J. Phys. Chem.* **71**, 3925 (1967).
431. R. L. Kay and G. A. Vidulich, *J. Phys. Chem.* **74**, 2718 (1970).
432. R. L. Kay, G. A. Vidulich, and K. S. Pribadi, *J. Phys. Chem.* **73**, 445 (1969).
433. R. L. Kay, T. Vituccio, C. Zawayski, and D. F. Evans, *J. Phys. Chem.* **70**, 2325 (1966).
434. R. L. Kay, T. Vituccio, C. Zawoyski, and D. F. Evans, *J. Phys. Chem.* **70**, 2336 (1966).
435. R. L. Kay, C. Zawoyski, and D. F. Evans, *J. Phys. Chem.* **69**, 4208 (1965).
436. P. Kebarle, *in* "Mass Spectrometry in Inorganic Chemistry" (Advan. in Chem. Series, No. 72, R. F. Gould, ed.), Am. Chem. Soc., Washington, D.C. (1968).
437. P. Kebarle, *J. Chem. Phys.* **53**, 2129 (1970).
438. P. Kebarle, M. Arshadi, and J. Scarborough, *J. Chem. Phys.* **49**, 817 (1968).
439. P. Kebarle and A. M. Hogg, *J. Chem. Phys.* **42**, 798 (1965).
440. P. Kebarle, S. K. Searles, A. Zolla, J. Scarborough, and M. Arshadi, *J. Am. Chem. Soc.* **89**, 6393 (1967).
441. Z. Kecki, P. Dryjanski, and E. Kozlowska, *Roczniki Chem.* **42**, 1749 (1968).
442. Z. Kecki, J. J. Witanowski, K. Akst-Lipszyc, and S. Minc, *Roczniki Chem.* **40**, 919 (1966).
443. G. Keller, *Ber. Bunsenges. Physik. Chem.* **76**, 24 (1972).
444. B. E. Kerwin, Thesis, Univ. of Pittsburgh, Pittsburgh, Pennsylvania (1964).
445. E. J. King, "Acid–Base Equilibria," Pergamon, New York–London (1965).
446. E. J. King, *J. Phys. Chem.* **73**, 1220 (1969).
447. E. J. King, *J. Phys. Chem.* **74**, 4590 (1970).
448. J. G. Kirkwood, *J. Chem. Phys.* **2**, 351 (1934).
449. J. G. Kirkwood, *J. Chem. Phys.* **7**, 911 (1939).
450. J. G. Kirkwood and F. P. Buff, *J. Chem. Phys.* **19**, 774 (1951).
451. I. Kirschenbaum, "Physical Properties and Analysis of Heavy Water," p. 54, McGraw-Hill, New York (1951).
452. D. Kivelson, *J. Chem. Phys.* **45**, 1324 (1966).
453. O. Klein and E. Lange, *Z. Elektrochem.* **43**, 570 (1937).
454. I. M. Klotz, "Chemical Thermodynamics," Benjamin, New York (1964).
455. F. W. Kohlrausch, *Ann. Physik.* **26**, 108 (1855).
456. F. Kohlrausch, L. Halborn, and H. Dreselhorst, *Wied. Ann.* **64**, 425 (1898).
457. P. A. Kollman and L. C. Allen, *J. Chem. Phys.* **51**, 3286 (1969).
458. P. A. Kollman and L. C. Allen, *J. Am. Chem. Soc.* **92**, 6101 (1970).
459. J. J. Kozak, W. S. Knight, and W. Kauzmann, *J. Chem. Phys.* **48**, 675 (1968).
460. W. D. Kraeft, *Z. Physik. Chem.* **230**, 368 (1965).
461. W. D. Kraeft and E. Gerdes, *Z. Physik. Chem.* **228**, 331 (1965).
462. L. M. Krausz, *J. Am. Chem. Soc.* **92**, 3168 (1970).
463. M. M. Kreevoy and C. A. Mead, *J. Am. Chem. Soc.* **84**, 4596 (1962).

464. M. M. Kreevoy and C. A. Mead, *Disc. Faraday Soc.* **39**, 166 (1965).
465. R. W. Kreis and R. H. Wood, *J. Phys. Chem.* **75**, 2319 (1971).
466. A. J. Kresge and A. L. Allred, *J. Am. Chem. Soc.* **85**, 1541 (1963).
467. A. J. Kresge and Y. Chiang, *J. Chem. Phys.* **49**, 1439 (1968).
468. G. C. Kresheck, H. Schneider, and H. A. Scheraga, *J. Phys. Chem.* **69**, 3132 (1965).
469. G. A. Krestov, *J. Struct. Chem. (USSR)* **4**, 236 (1963).
470. R. Krishnaji and A. Mansingh, *J. Chem. Phys.* **41**, 827 (1964).
471. C. V. Krishnan and H. L. Friedman, *J. Phys. Chem.* **73**, 1572 (1969).
472. C. V. Krishnan and H. L. Friedman, *J. Phys. Chem.* **73**, 3934 (1969).
473. C. V. Krishnan and H. L. Friedman, *J. Phys. Chem.* **75**, 3606 (1971).
474. C. V. Krishnan and H. L. Friedman, *J. Phys. Chem.* **74**, 2356 (1970).
475. C. V. Krishnan and H. L. Friedman, *J. Phys. Chem.* **74**, 3900 (1970).
476. C. V. Krishnan and H. L. Friedman, *J. Phys. Chem.* **75**, 388 (1971).
477. C. V. Krishnan and H. L. Friedman, *J. Phys. Chem.* **75**, 3598 (1971).
478. A. Kruis, *Z. Physik. Chem.* **34B**, 1 (1936).
479. K. Krynicki, *Physica* **32**, 167 (1966).
480. R. W. Kunze, R. M. Fuoss, and B. B. Owen, *J. Phys. Chem.* **67**, 1719 (1963).
481. K. J. Laidler, *Can. J. Chem.* **34**, 1107 (1956).
482. K. J. Laidler, *J. Chem. Phys.* **27**, 1423 (1957).
483. K. J. Laidler and J. S. Muirhead-Gould, *Trans. Faraday Soc.* **63**, 953 (1967).
484. K. J. Laidler and C. Pegis, *Proc. Roy. Soc. A* **241**, 80 (1957).
485. L. H. Laliberte and B. E. Conway, *J. Phys. Chem.* **74**, 4116 (1970).
486. H. Lamb, "Hydrodynamics" (reprint), p. 602, Dover, New York (1949).
487. V. K. La Mer and E. Noonan, *J. Am. Chem. Soc.* **61**, 1487 (1939).
488. L. D. Landau and E. M. Lifschitz, "Quantum Mechanics," Pergamon, London (1958).
489. L. D. Landau and E. M. Lifschitz, "Statistical Physics," Addison-Wesley, Reading, Mass. (1958).
490. E. Lange and W. Martin, *Z. Physik. Chem. A* **180**, 233 (1937).
491. J. W. Larson, *J. Phys. Chem.* **74**, 685 (1970).
492. J. W. Larson and L. G. Hepler, *J. Org. Chem.* **33**, 3961 (1968).
493. J. W. Larson and L. G. Hepler, *in* "Solute–Solvent Interactions" (J. F. Coetzee and C. D. Ritchie, eds.), Dekker, New York (1969).
494. K. E. Larsson and V. Dahlborg, *React. Sci. Tech. (J. Nucl. Eng.)* **16**, 81 (1962).
495. W. M. Latimer, *J. Am. Chem. Soc.* **48**, 1236 (1926).
496. W. M. Latimer, *in* "Oxidation Potentials," 2nd ed., Prentice-Hall, Englewood Cliffs, N.J. (1952).
497. W. M. Latimer, *J. Chem. Phys.* **23**, 90 (1955).
498. W. M. Latimer, K. S. Pitzer, and C. M. Slansky, *J. Chem. Phys.* **7**, 108 (1939).
499. P. M. Laughton and R. E. Robertson, *in* "Solute–Solvent Interactions," (J. F. Coetzee and C. D. Ritchie, eds.), Dekker, New York (1969).
500. H. A. Lauwers and G. P. van der Kelen, *Inorg. Chim. Acta* **2**, 281 (1968).
501. H. A. Lauwers, G. P. van der Kelen, and Z. Eeckhaut, *Inorg. Chim. Acta* **3**, 612 (1969).
502. H. Lee and J. K. Wilmshurst, *Aust. J. Chem.* **17**, 943 (1964).
503. J. E. Leffler and E. Grunwald, "Rates and Equilibria of Organic Reactions," Wiley, New York (1963).
504. T. W. Leland, J. S. Rowlinson, and G. A. Sather, *Trans. Faraday Soc.* **64**, 1447 (1968).

505. J. Lennard-Jones and J. A. Pople, *Proc. Roy. Soc. (London) A* **205**, 155 (1951).
506. C. S. Leung and E. Grunwald, *J. Phys. Chem.* **74**, 687 (1970).
507. A. S. Levine, N. Bhatt, M. Ghamkbar, and R. H. Wood, *J. Chem. Eng. Data* **15**, 34 (1970).
508. S. Levine and D. K. Rozenthal, *in* "Chemical Physics of Ionic Solutions" (B. E. Conway and R. G. Barradas, eds.), Wiley, New York (1966).
509. G. N. Lewis and M. Randall, "Thermodynamics," McGraw-Hill, New York (1923).
510. G. N. Lewis and M. Randall, "Thermodynamics," 2nd ed. (rev. by K. S. Pitzer and L. Brewer), McGraw-Hill, New York (1961).
511. W. B. Lewis, M. Alei, Jr., and L. O. Morgan, *J. Chem. Phys.* **44**, 2409 (1966).
512. W. B. Lewis, M. Alei, Jr., and L. O. Morgan, *J. Chem. Phys.* **45**, 4003 (1966).
513. A. Liberti and T. S. Light, *J. Chem. Ed.* **39**, 236 (1962).
514. T. S. Light and K. S. Fletcher, III, *Anal. Chem.* **39**, 70 (1962).
515. S. F. Lincoln, *Coord. Chem. Rev.* **6**, 309 (1971).
516. J. E. Lind, Jr., J. J. Zwolenik, and R. M. Fuoss, *J. Am. Chem. Soc.* **81**, 1557 (1959).
517. S. Lindenbaum, *J. Phys. Chem.* **70**, 814 (1966).
518. S. Lindenbaum and G. E. Boyd, *J. Phys. Chem.* **71**, 581 (1967).
519. H. Linder, Ph.D. Dissertation, Univ. of Karlsruhe (1970).
520. B. Lindman, S. Forsen, and E. Forslind, *J. Phys. Chem.* **72**, 2805 (1968).
521. B. Lindman, H. Wennerström, and S. Forsen, *J. Phys. Chem.* **74**, 754 (1970).
522. U. Lindner, *Ann. Physik, 7 Flg.* **16**, 319 (1965).
523. C. L. Liotta, K. H. Leavell, and D. F. Smith, *J. Phys. Chem.* **71**, 3091 (1967).
524. L. Llopis, *in* "Modern Aspects of Electrochemistry," No. 6 (J. O'M. Bockris and B. E. Conway, eds.), Plenum, New York (1971).
525. W. Lohmann, *Z. Naturforsch.* **19a**, 814 (1964).
526. J. Long and B. Munson, *J. Chem. Phys.* **53**, 1356 (1970).
527. D. A. Lown, H. R. Thirsk, and W. F. K. Wynne-Jones, *Trans. Faraday Soc.* **64**, 2073 (1968).
528. W. A. P. Luck, *Ber. Bunsenges. Physik. Chem.* **67**, 186 (1963).
529. W. A. P. Luck, *Fortschr. Chem. Forsch.* **4**, 653 (1964).
530. W. A. P. Luck, *Ber. Bunsenges. Physik. Chem.* **69**, 626 (1965).
531. W. A. P. Luck, *Ber. Bunsenges. Physik. Chem.* **70**, 1113 (1966).
532. W. A. P. Luck, *Disc. Faraday Soc.* **41**, 141 (1967).
533. W. A. P. Luck, *J. Phys. Chem.* **71**, 459 (1967).
534. W. A. P. Luck, *in* "Physico-Chemical Processes in Mixed Aqueous Solvents" (F. Franks, ed.), Chapter 2, Heinemann, London (1967).
535. W. A. P. Luck and W. Ditter, *J. Mol. Struct.* **1**, 339 (1967-68).
536. R. Lumry and S. Rajender, *Biopolymers* **9**, 1125 (1970).
537. Z. Luz and R. G. Shulman, *J. Chem. Phys.* **43**, 3750 (1965).
538. W. C. McCabe and H. F. Fisher, *Nature* **207**, 1274 (1965).
539. W. C. McCabe and H. F. Fisher, *J. Phys. Chem.* **74**, 2990 (1970).
540. W. C. McCabe, S. Subramanian, and H. F. Fisher, *J. Phys. Chem.* **74**, 4360 (1970).
541. D. W. McCall and D. C. Douglass, *J. Phys. Chem.* **69**, 200 (1965).
542. D. W. McCall, D. C. Douglass, and E. W. Anderson, *Ber. Bunsenges. Physik. Chem.* **67**, 336 (1963).
543. H. M. McConnell, *J. Chem. Phys.* **28**, 430 (1958).
544. W. H. McCoy and W. E. Wallace, *J. Am. Chem. Soc.* **78**, 1830 (1956).

545. B. R. McGarvey, *J. Phys. Chem.* **61**, 1232 (1957).
546. J. D. E. McIntyre and M. Salomon, *J. Phys. Chem.* **72**, 2431 (1968).
547. J. L. Mackey, J. E. Powell, and F. H. Spedding, *J. Am. Chem. Soc.* **84**, 2047 (1962).
548. A. D. McLachlan, *Disc. Faraday Soc.* **40**, 239 (1965).
549. W. G. McMillan and J. E. Mayer, *J. Chem. Phys.* **13**, 276 (1945).
550. J. Malsch, *Physik. Z.* **29**, 770 (1928).
551. J. Malsch, *Physik. Z.* **30**, 837 (1929).
552. G. S. Manning, *J. Chem. Phys.* **43**, 4260, 4268 (1965).
553. R. A. Marcus, *J. Chem. Phys.* **24**, 966, 979 (1956).
554. R. A. Marcus, *J. Chem. Phys.* **38**, 1335, 1858 (1963).
555. R. A. Marcus, *J. Chem. Phys.* **39**, 460 (1963).
556. R. A. Marcus, *J. Chem. Phys.* **43**, 58 (1965).
557. K. N. Marsh, M. L. McGlashan, and C. Warr, *Trans. Faraday Soc.* **66**, 2453 (1970).
558. W. L. Marshall, *J. Phys. Chem.* **74**, 346 (1970).
559. M. Mastroianni and C. M. Criss, *J. Chem. Eng. Data* **17**, 222 (1972).
560. M. Mastroianni and C. M. Criss, *J. Chem. Thermodynamics* **4**, 321 (1972).
561. J. P. Mathieu and M. Lounsburg, *Disc. Faraday Soc.* **9**, 196 (1950).
562. J. E. Mayer, *J. Chem. Phys.* **18**, 1426 (1950).
563. J. E. Mayer and M. G. Mayer, "Statistical Mechanics," Wiley, New York (1940).
564. S. Meiboom, *J. Chem. Phys.* **34**, 375 (1961).
565. S. Meiboom and D. Gill, *Rev. Sci. Instr.* **29**, 688 (1958).
566. S. Meiboom, Z. Luz, and D. Gill, *J. Chem. Phys.* **27**, 1411 (1957).
567. R. E. Mesmer, C. F. Baes, Jr., and F. H. Sweeton, *J. Phys. Chem.* **74**, 1937 (1970).
568. W. J. Middleton and R. V. Lindsey, *J. Am. Chem. Soc.* **86**, 4948 (1964).
569. W. A. Millen and D. W. Watts, *J. Am. Chem. Soc.* **89**, 6051 (1967).
570. J. T. Miller and D. E. Irish, *Can. J. Chem.* **45**, 147 (1967).
571. F. J. Millero, *J. Phys. Chem.* **72**, 4589 (1968).
572. F. J. Millero, *J. Chem. Eng. Data* **15**, 562 (1970).
573. F. J. Millero, *J. Chem. Eng. Data* **16**, 229 (1971).
574. F. J. Millero, *J. Phys. Chem.* **75**, 280 (1971).
575. F. J. Millero, *Chem. Rev.* **71**, 147 (1971).
576. F. J. Millero, *in* "Water and Aqueous Solutions" (R. A. Horne, ed.), Wiley, New York (1971).
577. F. J. Millero and W. Drost-Hansen, *J. Chem. Eng. Data* **13**, 330 (1968).
578. F. J. Millero and W. Drost-Hansen, *J. Phys. Chem.* **72**, 1758 (1968).
579. F. J. Millero, W. Drost-Hansen, and L. Korson, *J. Phys. Chem.* **72**, 2251 (1968).
580. R. Mills, *Ber. Bunsenges. Physik. Chem.* **75**, 195 (1971).
581. K. P. Miscenko, *Zh. Fiz. Khim.* **26**, 1736 (1952).
582. T. Moeller, D. F. Martin, L. C. Thompson, R. Ferrús, G. R. Feistel, and W. J. Randall, *Chem. Rev.* **65**, 1 (1965).
583. S. C. Mohr, W. D. Wilk, and G. M. Barrow, *J. Am. Chem. Soc.* **87**, 3048 (1965).
584. L. O. Morgan and A. W. Nolle, *J. Chem. Phys.* **31**, 365 (1959).
585. L. R. Morss, *J. Phys. Chem.* **75**, 392 (1971).
586. J. S. Muirhead-Gould and K. J. Laidler, *Trans. Faraday Soc.* **63**, 944 (1967).
587. I. Nakagawa and T. Shimanouchi, *Spectrochim. Acta* **20**, 429 (1964).
588. G. H. Nancollas, *Coordination Chem. Rev.* **5**, 379 (1970).
589. T.-W. Nee and R. Zwanzig, *J. Chem. Phys.* **52**, 6353 (1970).
590. P. Nehmiz and M. Stockhausen, *Z. Naturforsch.* **24**, 573 (1969).

591. D. L. Nelson and D. E. Irish, *J. Chem. Phys.* **54**, 4479 (1971).

592. G. Nemethy and H. A. Scheraga, *J. Chem. Phys.* **36**, 3382 (1962).

593. G. Nemethy and H. A. Scheraga, *J. Chem. Phys.* **41**, 680 (1964).

594. D. Neumann and J. W. Moskowitz, *J. Chem. Phys.* **49**, 2056 (1968).

595. M. Newton and S. Ehrenson, *J. Am. Chem. Soc.* **93**, 4971 (1971).

596. E. R. Nightingale, *in* "Chemical Physics of Ionic Solutions" (B. E. Conway and R. G. Barrada, eds.), p. 87, Wiley, New York (1966).

597. E. R. Nightingale, *J. Phys. Chem.* **63**, 1381 (1959).

598. M. Noack and G. Gordon, *J. Chem. Phys.* **48**, 2689 (1968).

599. A. W. Nolle and L. O. Morgan, *J. Chem. Phys.* **36**, 378 (1962).

600. R. M. Noyes, *J. Am. Chem. Soc.* **84**, 513 (1962).

601. R. M. Noyes, *J. Am. Chem. Soc.* **86**, 971 (1963).

602. R. M. Noyes, *J. Am. Chem. Soc.* **86**, 971 (1964).

603. R. M. Noyes, *J. Am. Chem. Soc.* **86**, 972 (1964).

604. R. P. Oertel and R. A. Plane, *Inorg. Chem.* **7**, 1192 (1968).

605. R. A. M. O'Ferrall, G. W. Koeppl, and A. J. Kresge, *J. Am. Chem. Soc.* **93**, 1 (1971).

606. R. A. M. O'Ferrall, G. W. Koeppl, and A. J. Kresge, *J. Am. Chem. Soc.* **93**, 9 (1971).

607. W. F. O'Hara, T. Hu, and L. G. Hepler, *J. Phys. Chem.* **67**, 1933 (1963).

608. G. Öjelund and I. Wadsö, *Acta Chem. Scand.* **21**, 1408 (1967).

609. B. G. Oliver and G. J. Janz, *J. Phys. Chem.* **75**, 2948 (1971).

610. M. V. Olson, Y. Kanazawa, and H. Taube, *J. Chem. Phys.* **51**, 289 (1969).

611. L. Onsager, *Phys. Z.* **27**, 388 (1926).

612. L. Onsager, *Phys. Z.* **28**, 277 (1927).

613. L. Onsager, *J. Am. Chem. Soc.* **58**, 1486 (1936).

614. I. Oppenheim, *J. Phys. Chem.* **68**, 2959 (1964).

615. I. G. Orlov, V. S. Markin, Y. V. Moiseev, and U. I. Khurgin, *J. Struct. Chem.* **7**, 796 (1966).

616. I. Oshida and O. Horiguchi, *Bull. Kobayasi Inst. Phys. Res.* **5**, 61 (1955).

617. B. B. Owen and S. R. Brinkley, *Chem. Rev.* **29**, 461 (1941).

618. B. B. Owen and P. L. Kronick, *J. Phys. Chem.* **65**, 84 (1961).

619. M. Paabo and R. G. Bates, *J. Res. Natl. Bur. Std.* **74A**, 667 (1970).

620. M. Paabo, R. A. Robinson, and R. G. Bates, *Anal. Chem.* **38**, 1573 (1966).

621. J. Padova, *J. Phys. Chem.* **74**, 4587 (1970).

622. M. H. Panckhurst, *Rev. Pure Appl. Chem.* **19**, 45 (1969).

623. H. M. Papee, W. J. Canady, and K. J. Laidler, *Can. J. Chem.* **34**, 1677 (1956).

624. A. J. Parker, *Chem. Rev.* **69**, 1 (1969).

625. V. B. Parker, "Thermal Properties of Aqueous Uni-Univalent Electrolytes," U.S. Nat. Bur. Std., NSRDS-NBS 2, Washington, D.C. (1965).

626. R. Parsons, *in* "Modern Aspects of Electrochemistry" (J. O'M. Bockris, ed.), Butterworths, London (1954).

627. L. Pauling, "The Nature of the Chemical Bond," 3rd. ed., p. 452, Cornell Univ. Press, Ithaca, New York (1960).

628. L. Pentz and E. R. Thornton, *J. Am. Chem. Soc.* **89**, 6931 (1967).

629. F. Perrin, *J. Phys. Radium* **7**, 1 (1936).

630. E. Petreanu, S. Pinchas, and D. Samuel, *J. Inorg. Nucl. Chem.* **27**, 2519 (1965).

631. H. Pfeifer, *Ann. Physik.*, 7 Flg., **8**, 1 (1961).

632. H. Pfeifer, *Z. Naturforsch.* **17a**, 279 (1964).

633. H. Pfeifer, D. Michel, D. Sames, and H. Sprinz, *Mol. Phys.* **11**, 591 (1966).
634. G. C. Pimentel and A. L. McClellan, "The Hydrogen Bond," Freeman, San Francisco, Calif. (1960).
635. S. Pinchas and D. Sadeh, *J. Inorg. Nucl. Chem.* **30**, 1785 (1968).
636. K. S. Pitzer, *J. Am. Chem. Soc.* **59**, 2365 (1937).
637. R. F. Platford, *J. Chem. Thermodynamics* **3**, 319 (1971).
638. E. K. Plyler and E. S. Barr, *J. Chem. Phys.* **6**, 316 (1938).
639. J. A. Pople, *Proc. Roy. Soc. (London)* **A205**, 163 (1951).
640. J. A. Pople, W. G. Schneider, and H. J. Bernstein, "High-Resolution Nuclear Magnetic Resonance," McGraw-Hill, New York (1959).
641. O. Popovych, *Critical Reviews in Anal. Chem.* **1**, 73 (1970).
642. R. Pottel, *Ber. Bunsenges. Physik. Chem.* **69**, 363 (1965).
643. R. Pottel, *in* "Chemical Physics of Ionic Solutions" (B. E. Conway and R. G. Barradas, eds.), pp. 581–598, Wiley, New York (1966).
644. R. Pottel and U. Kaatze, *Ber. Bunsenges. Physik. Chem.* **73**, 437 (1969).
645. R. Pottel and O. Lossen, *Ber. Bunsenges. Physik. Chem.* **71**, 135 (1967).
646. A. D. Potts and D. W. Davidson, *J. Phys. Chem.* **69**, 996 (1965).
647. R. E. Powell, *J. Phys. Chem.* **58**, 528 (1954).
648. R. E. Powell and W. M. Latimer, *J. Chem. Phys.* **19**, 1139 (1951).
649. K. S. Pribadi, Ph.D. Thesis, Carnegie-Mellon Univ., 1971; *J. Solution Chem.* **1**, 455 (1972).
650. I. Prigogine, "The Molecular Theory of Solutions," North-Holland, Amsterdam (1957).
651. R. H. Provost and C. A. Wulff, *J. Chem. Thermodynamics* **2**, 793 (1970).
652. L. J. Puckett and M. W. Teague, *J. Chem. Phys.* **54**, 2564 (1971).
653. A. D. E. Pullin, *Spectrochim. Acta* **13**, 125 (1958).
654. A. D. E. Pullin, *Spectrochim. Acta* **16**, 12 (1960).
655. E. L. Purlee, *J. Amer. Chem. Soc.* **81**, 263 (1959).
656. J. W. Pyper, R. S. Newbury, and G. W. Barton, Jr., *J. Chem. Phys.* **46**, 2253 (1967).
657. A. S. Quist, *J. Phys. Chem.* **74**, 3396 (1970).
658. C. Rädle, Thesis, Karlsruhe.
659. A. Rahman and F. H. Stillinger, *J. Chem. Phys.* **55**, 3336 (1971).
660. P. S. Ramanathan and H. L. Friedman, *J. Chem. Phys.* **54**, 1086 (1971).
661. P. S. Ramanathan, C. V. Krishnan, and H. L. Friedman, *J. Solution Chem.* **1**, 237 (1972).
662. R. W. Ramette and R. F. Broman, *J. Phys. Chem.* **67**, 942 (1963).
663. R. W. Ramette and E. A. Dratz, *J. Phys. Chem.* **67**, 940 (1963).
664. J. E. B. Randles, *Trans. Faraday Soc.* **52**, 1573 (1956).
665. I. R. Rao, *Proc. Roy. Soc. A* **144**, 159 (1934).
666. N. R. Rao, *Indian J. Phys.* **15**, 185 (1941).
667. P. S. K. M. Rao and D. Premaswarup, *Ind. J. Pure Appl. Phys.* **5**, 581 (1967).
668. P. S. K. M. Rao and D. Premaswarup, *Ind. J. Pure Appl. Phys.* **7**, 68 (1969).
669. P. S. K. M. Rao and D. Premaswarup, *Trans. Faraday Soc.* **66**, 1974 (1970).
670. J. C. Rasaiah, *J. Chem. Phys.* **52**, 704 (1970).
671. J. Rasper and W. Kauzmann, *J. Am. Chem. Soc.* **84**, 1771 (1962).
672. D. G. Rea, *J. Opt. Soc. Am.* **49**, 90 (1959).
673. O. Redlich and J. Bigeleisen, *Chem. Rev.* **30**, 171 (1942).
674. O. Redlich and J. Bigeleisen, *J. Am. Chem. Soc.* **64**, 758 (1942).

675. O. Redlich and J. Bigeleisen, *J. Am. Chem. Soc.* **65**, 1883 (1943).
676. O. Redlich, E. K. Holt, and J. Bigeleisen, *J. Am. Chem. Soc.* **66**, 13 (1944).
677. O. Redlich and D. M. Meyer, *Chem. Rev.* **64**, 221 (1964).
678. F. Reif, "Fundamentals of Statistical Thermal Physics," McGraw-Hill, New York (1965).
679. P. J. Reilly and R. H. Wood, *J. Phys. Chem.* **73**, 4292 (1969).
680. J. Reuben and D. Fiat, *Inorg. Chem.* **6**, 579 (1967).
681. J. Reuben and D. Fiat, *J. Chem. Phys.* **51**, 4918 (1969).
682. O. K. Rice, "Statistical Mechanics, Thermodynamics and Kinetics," Freeman, San Francisco, Calif. (1967).
683. T. W. Richards and L. P. Hall, *J. Am. Chem. Soc.* **51**, 731 (1929).
684. T. W. Richards and A. W. Rowe, *J. Am. Chem. Soc.* **44**, 684 (1922).
685. R. E. Robertson, S. E. Sugamori, R. Tse, and C. Y. Wu, *Can. J. Chem.* **44**, 487 (1966).
686. R. A. Robinson, *J. Res. Natl. Bur. Std.* **74A**, 495 (1970).
687. R. A. Robinson and R. G. Bates, *J. Res. Natl. Bur. Std.* **70A**, 553 (1966).
688. R. A. Robinson and V. E. Bower, *J. Res. Natl. Bur. Std.* **70A**, 305 (1966).
689. R. A. Robinson, A. K. Covington, and C. P. Bezboruah, *J. Chem. Thermodynamics* **2**, 431 (1970).
690. R. A. Robinson and C. K. Lim, *Trans. Faraday Soc.* **49**, 1144 (1953).
691. R. A. Robinson and R. H. Stokes, *J. Am. Chem. Soc.* **70**, 1870 (1948).
692. R. A. Robinson and R. H. Stokes, "Electrolyte Solutions," Butterworths, London (1955).
693. R. A. Robinson and R. H. Stokes, "Electrolyte Solutions," 2nd ed., p. 125, Butterworths, London (1959).
694. R. A. Robinson and R. H. Stokes, "Electrolyte Solutions," 2nd ed., rev., Butterworths, London (1965).
695. R. A. Robinson, R. H. Wood, and P. J. Reilly, *J. Chem. Thermodynamics* **3**, 461 (1971).
696. R. E. Ronnick and E. D. Stover, *J. Phys. Chem.* **65**, 2074 (1961).
697. D. R. Rosseinsky, *Chem. Rev.* **65**, 467 (1965).
698. F. J. C. Rossotti, *in* "Modern Coordination Chemistry" (J. Lewis and R. G. Wilkins, eds.), Interscience, New York (1960).
699. J. S. Rowlinson, *Trans. Faraday Soc.* **47**, 120 (1951).
700. R. M. Rush, ORNL-4022, Oak Ridge National Laboratory.
701. R. M. Rush and J. M. Johnson, *J. Phys. Chem.* **72**, 767 (1968).
702. R. M. Rush and R. A. Robinson, *J. Tenn. Acad. Sci.* **43**, 22 (1968).
703. H. Rüterjans, F. Schreiner, U. Sage, and Th. Ackermann, *J. Phys. Chem.* **73**, 986 (1969).
704. L. Sacconi, P. Paoletti, and M. Ciampolini, *Ric. Sci.* **29**, 2412 (1959).
705. G. J. Safford, P. S. Leung, A. W. Naumann, and P. C. Shaffer, *J. Chem. Phys.* **50**, 4444 (1969).
706. G. Safford, P. C. Schaffer, P. S. Leung, G. F. Doebbler, G. W. Brady, and E. F. X. Lyden, *J. Chem. Phys.* **50**, 2140 (1969).
707. P. Salomaa and V. Aalto, *Acta Chem. Scand.* **20**, 2035 (1966).
708. M. Salomon, *J. Phys. Chem.* **74**, 2519 (1970).
709. D. Sames, *Ann. Physik*, 7 Flg., **15**, 363 (1965).
710. D. Sames and D. Michel, *Ann. Physik*, 7 Flg., **18**, 353 (1966).

711. O. Y. Samoilov, *Disc. Faraday Soc.* **24**, 141 (1957).
712. O. Y. Samoilov, *J. Struct. Chem.* **1**, 32 (1960).
713. O. Y. Samoilov, "Structure of Aqueous Electrolyte Solutions and Hydration of Ions," Consultants Bureau, New York (1965).
714. O. Y. Samoilov and V. G. Tsvetkov, *J. Struct. Chem.* (*USSR*) **9**, 142 (1968).
715. T. S. Sarma and J. C. Ahluwalia, *J. Phys. Chem.* **74**, 3547 (1970).
716. G. Sartori and C. Furlani, *Z. Phys. Chem.*, *Neue Folge* (*Frankfurter Ausgabe*) **15**, 336 (1958).
717. G. Sartori, C. Furlani, and A. Damiani, *J. Inorg. Nucl. Chem.* **8**, 119 (1958).
717a. G. Scatchard, *J. Am. Chem. Soc.* **83**, 2636 (1961).
718. G. Scatchard, *J. Am. Chem. Soc.* **90**, 3124 (1968).
719. G. Scatchard and R. G. Breckenridge, *J. Phys. Chem.* **58**, 596 (1954).
720. G. Scatchard and R. G. Breckenridge, *J. Phys. Chem.* **59**, 1234 (1955).
721. G. Scatchard and S. S. Prentiss, *J. Am. Chem. Soc.* **56**, 2320 (1934).
722. J. A. Schellman, *J. Chem. Phys.* **24**, 912 (1956).
723. J. A. Schellman, *J. Chem. Phys.* **26**, 1225 (1957).
724. J. Schiffer, Ph.D. Thesis, Princeton Univ., Princeton, N. J. (1963).
725. J. Schiffer, *Dissertation Abstr.* **25**, 2786 (1964).
726. P. Schindler, R. A. Robinson, and R. G. Bates, *J. Res. Natl. Bur. Std.* **72A**, 141 (1968).
727. J. W. Schultz and D. F. Hornig, *J. Phys. Chem.* **65**, 2131 (1961).
728. R. E. Schuster and A. Fratiello, *J. Chem. Phys.* **47**, 1554 (1969).
729. L. M. Schwartz and L. O. Howard, *J. Phys. Chem.* **74**, 4374 (1970).
730. P. C. Scott and Z. Z. Hugus, Jr., *J. Chem. Phys.* **27**, 1421 (1957).
731. W. A. Senior and R. E. Verrall, *J. Phys. Chem.* **73**, 4242 (1969).
732. R. W. Shearman and A. W. C. Menzies, *J. Am. Chem. Soc.* **59**, 185 (1937).
733. L. G. Sillen and A. E. Martell, "Stability Constants of Metal-Ion Complexes," The Chemical Society, London (1964).
734. L. Simeral and R. L. Amey, *J. Phys. Chem.* **74**, 1443 (1970).
735. L. Simons, *Soc. Sci. Fenniea Commentationes*, *Phys. Math.* **7**, No. 9 (1934).
736. O. Sinanoglu, *Advan. Chem. Phys.* **12**, 283 (1967).
737. C. P. Slichter, "Principles of Magnetic Resonance," Harper and Row, New York (1963).
738. W. M. Slie, A. R. Donfor, and T. A. Litovitz, *J. Chem. Phys.* **44**, 3712 (1966).
739. H. Snell and J. Greyson, *J. Phys. Chem.* **71**, 2148 (1970).
740. H. Snell and J. Greyson, *J. Phys. Chem.* **74**, 2148 (1970).
741. F. H. Spedding and M. J. Pikal, *J. Phys. Chem.* **70**, 2430 (1966).
742. H. O. Spivey and T. Shedlovsky, *J. Phys. Chem.* **71**, 2165 (1967).
743. G. Sposito and K. L. Babcock, *J. Chem. Phys.* **47**, 153 (1967).
744. J. Stecki, *Advan. Chem. Phys.* **6**, 413 (1964).
745. B. J. Steel, R. A. Robinson, and R. G. Bates, *J. Res. Natl. Bur. Std.* **71A**, 9 (1967).
746. E. O. Stejskal and J. E. Tanner, *J. Chem. Phys.* **42**, 288 (1965).
747. H. Stephen and T. Stephen, "Solubilities of Inorganic and Organic Compounds," Vol. 1, Part 1, p. 199, Macmillan, New York (1963).
748. J. H. Stern, C. W. Anderson, and A. A. Passchier, *J. Phys. Chem.* **69**, 207 (1965).
749. J. H. Stern and A. A. Passchier, *J. Phys. Chem.* **66**, 752 (1962).
750. J. H. Stern and A. A. Passchier, *J. Phys. Chem.* **67**, 2420 (1963).
751. D. P. Stevenson, *J. Phys. Chem.* **69**, 2145 (1965).

752. D. P. Stevenson, *in* "Structural Chemistry and Molecular Biology" (A. Rich and N. R. Davidson, eds.), p. 490, Freeman, San Francisco, Calif. (1968).

753. F. H. Stillinger, Jr., *J. Phys. Chem.* **74**, 3677 (1970).

754. F. H. Stillinger and A. Ben-Naim, *J. Phys. Chem.* **73**, 900 (1969).

755. G. G. Stokes, *Cambridge Phil. Soc. Trans.* **9**, 5 (1856).

756. R. H. Stokes, *J. Am. Chem. Soc.* **86**, 979 (1964).

757. R. H. Stokes, *J. Am. Chem. Soc.* **86**, 982 (1964).

758. R. H. Stokes and R. Mills, "Viscosity of Electrolytes and Related Properties," Pergamon, Oxford (1965).

759. N. W. B. Stone and D. Williams, *Appl. Optics* **5**, 353 (1966).

760. H. Strehlow, *in* "The Chemistry of Non-Aqueous Solvents," (J. J. Lagowski, ed.) Vol. 1, Academic, New York (1966).

761. S. Subramanian and J. C. Ahluwalia, *J. Phys. Chem.* **72**, 2525 (1968).

762. R. Suhrmann and F. Breyer, *Z. Physik. Chem. Leipzig* **B20**, 17 (1933).

763. R. Suhrmann and F. Breyer, *Z. Physik. Chem. Leipzig* **B23**, 193 (1933).

764. G. D. Swain and R. F. W. Bader, *Tetrahedron* **10**, 182 (1960).

765. C. A. Swenson, *Spectrochim. Acta* **21**, 987 (1965).

766. J. F. Swindells, J. R. Cole, Jr., and T. B. Godfrey, *J. Res. Nat. Bur. Std.* **48**, 1 (1952).

767. T. J. Swift and R. E. Connick, *J. Chem. Phys.* **37**, 307 (1962); Err. *J. Chem. Phys.* **41**, 2553 (1964).

768. K. Szczepaniak and M. Falk, *Spectrochim. Acta* **26A**, 883 (1970).

769. G. L. Szepesy, J. Csaszar, and L. Lehotai, *Acta Phys. et Chim. Szeged* **2**, 149 (1956).

770. K. Tamm and M. Schneider, *Z. Angew. Phys.* **20**, 544 (1966).

771. H. Taube, *J. Phys. Chem.* **58**, 523 (1954).

772. R. W. Terhune, P. D. Maker, and C. M. Savage, *Phys. Rev. Letters* **14**, 681 (1965).

773. J. Thomas and D. F. Evans, *J. Phys. Chem.* **74**, 3812 (1970).

774. H. W. Thompson (ed.), Discussion on the Effect of Environment upon Molecular Energy Levels, *Proc. Roy. Soc. Ser. A*, **1960**, 1–81.

775. W. K. Thompson, *Trans. Faraday Soc.* **61**, 2635 (1965).

776. W. K. Thompson, *Trans. Faraday Soc.* **62**, 2667 (1966).

777. E. R. Thornton, *in* "Annual Review of Physical Chemistry" (H. Eyring, C. J. Christensen, and H. S. Johnston, eds.) Annual Reviews, Inc., Palo Alto, Calif. (1966).

778. B. A. Timini and D. H. Everett, *J. Chem. Soc. B* **1968**, 1380.

779. M. Tinkham, R. Weinstein, and A. F. Kip, *Phys. Rev.* **84**, 848 (1951).

780. P. G. Tishkov and G. Vishnevskaya, *Soviet Phys.—JETP* **9**, 949 (1959).

781. H. C. Torrey, *Phys. Rev.* **92**, 962 (1953).

782. H. C. Torrey, *Phys. Rev.* **104**, 563 (1956).

783. C. Treiner and R. M. Fuoss, *J. Phys. Chem.* **69**, 2576 (1965).

784. H. C. Urey, *J. Chem. Soc.* **1947**, 562.

785. K. A. Valiev and M. I. Emel'yanov, *Zh. Strukt. Khim.* **5**, 670 (1964) [English transl. *J. Struct. Chem.* **5**, 625 (1964)].

786. A. E. van Arkel and J. H. de Boer, Die Chemische Bindung als Elektrostatische Erscheinung, p. 218, Leipzig (1931).

787. C. E. Vanderzee and J. A. Swanson, *J. Phys. Chem.* **67**, 2608 (1963).

788. C. L. v. P. van Eck, Thesis, Leiden (1958).

789. M. van Thiel, E. O. Becker, and G. C. Pimentel, *J. Chem. Phys.* **27**, 486 (1957).
790. V. P. Vasil'ev and G. A. Lobanov, *Russ. J. Phys. Chem.* **41**, 434 (1967).
791. F. Vaslow, *J. Phys. Chem.* **67**, 2773 (1963).
792. F. Vaslow, *J. Phys. Chem.* **70**, 2286 (1966).
793. V. M. Vdovenko and V. A. Shcherbakov, *Z. Strukturchem.* (*Russ.*) **1**, 28, 122 (1960). Engl. transl. *J. Struct. Chem.* **1**, 25, 111 (1960).
794. L. Verlet and J. J. Weis, *Phys. Rev. A* **5**, 939 (1972).
795. R. E. Verrall and B. E. Conway, *J. Phys. Chem.* **70**, 3961 (1966).
796. H. Versmold, Thesis, Karlsruhe, 1970.
797. H. Versmold, *Z. Naturforsch.* **25a**, 367 (1970).
798. E. J. W. Verwey, *Chem. Weekblad.* **37**, 530 (1940).
799. E. J. W. Verwey, *Rec. Trav. Chim.* **60**, 887 (1941).
800. E. J. W. Verwey, *Rec. Trav. Chim.* **61**, 127 (1942).
801. R. Viclu and F. Irinci, *J. Chem. Thermodynamics* **1**, 409 (1969).
802. R. Viclu and F. Stanciu, *Rev. Roumaine Chim.* **13**, 7 (1968).
803. A. Voet, *Trans. Faraday Soc.* **32**, 1301 (1936).
804. P. M. Vollmar, *J. Chem. Phys.* **39**, 2236 (1963).
805. D. D. Wagman, W. J. Evans, V. B. Parker, I. Halow, S. M. Bailey, and R. H. Schumm, "Selected Values of Chemical Thermodynamic Properties," U.S. Nat. Bur. Std. Tech. Note 270-3, U.S. Govt. Printing Office, Washington, D.C. (1968).
806. P. Walden, *Z. Physik. Chem.* **55**, 207, 246 (1906).
807. R. D. Waldron, *J. Chem. Phys.* **26**, 809 (1957).
808. W. Walisch and J. Barthel, *Z. Physik. Chem. N.F.* **34**, 38 (1962).
809. T. T. Wall and D. F. Hornig, *J. Chem. Phys.* **43**, 2079 (1965).
810. T. T. Wall and D. F. Hornig, *J. Chem. Phys.* **47**, 784 (1967).
811. D. Wallach, *J. Chem. Phys.* **47**, 5258 (1967).
812. G. E. Walrafen, *J. Chem. Phys.* **36**, 1035 (1962).
813. G. E. Walrafen, *J. Chem. Phys.* **40**, 3249 (1964).
814. G. E. Walrafen, *J. Chem. Phys.* **44**, 1546 (1966).
815. G. E. Walrafen, *J. Chem. Phys.* **46**, 1870 (1967).
816. G. E. Walrafen, *J. Chem. Phys.* **47**, 114 (1967).
817. G. E. Walrafen, *J. Chem. Phys.* **48**, 244 (1968).
818. G. E. Walrafen, *in* "Hydrogen-Bonded Solvent Systems," (A. K. Covington and P. Jones, eds.), Taylor and Francis, London (1968).
819. G. E. Walrafen, *J. Chem. Phys.* **50**, 560 (1969).
820. G. E. Walrafen, *J. Chem. Phys.* **50**, 567 (1969).
821. G. E. Walrafen, *J. Chem. Phys.* **52**, 4176 (1970).
822. G. E. Walrafen, *J. Chem. Phys.* **55**, 768 (1971).
823. B. Watson and R. L. Kay, submitted to *J. Solution Chem.* (1973).
824. T. J. Webb, *J. Am. Chem. Soc.* **48**, 2589 (1926).
825. J. D. Weeks, D. Chandler, and H. C. Anderson, *Phys. Rev. Letters* **25**, 1491 (1970).
826. J. D. Weeks, D. Chandler, and H. C. Anderson, *J. Chem. Phys.* **54**, 26 (1971).
827. E. Weiss, E. Gerdes, and H. J. Hoffmann, *Z. Physik. Chem.* **228**, 51 (1965).
828. W. Y. Wen, K. Miyajima, and A. Otsuka, *J. Phys. Chem.* **75**, 2148 (1971).
829. W. Y. Wen and K. Nara, *J. Phys. Chem.* **71**, 3907 (1967).
830. W. Y. Wen, K. Nara, and R. H. Wood, *J. Phys. Chem.* **72**, 3048 (1968).
831. W. Y. Wen and S. Saito, *J. Phys. Chem.* **68**, 2639 (1964).
832. M. S. Wertheim, *J. Chem. Phys.* **55**, 4291 (1971).

833. R. E. Weston, *Spectrochim. Acta* **18**, 1257 (1962).

834. E. Wicke, *Angew. Chem. Int. Ed. Engl.* **5**, 106 (1966).

835. D. Williams, *Nature* **210**, 194 (1966).

836. D. Williams and T. N. Gautier, *Phys. Rev.* **56**, 616 (1939).

837. D. Williams and W. Millet, *Phys. Rev.* **66**, 6 (1944).

838. H. E. Wirth, *J. Am. Chem. Soc.* **59**, 2549 (1937).

839. H. E. Wirth, *J. Marine Res.* **3**, 230 (1940).

840. H. Wirth, *J. Chem. Eng. Data* **13**, 102, 226 (1968).

841. H. E. Wirth and F. N. Collier, *J. Phys. Chem.* **72**, 5292 (1960).

842. H. E. Wirth, R. Lindstrom, and J. Johnson, *J. Phys. Chem.* **67**, 2339 (1963).

843. D. E. Woessner, *J. Chem. Phys.* **36**, 1 (1962).

844. D. E. Woessner, *J. Chem. Phys.* **37**, 647 (1962).

845. D. E. Woessner, *J. Chem. Phys.* **40**, 2341 (1964).

846. D. E. Woessner, B. S. Snowden, and G. H. Meyer, *J. Chem. Phys.* **50**, 719 (1969).

847. D. E. Woessner, B. S. Snowden, Jr., and A. G. Ostroff, *J. Chem. Phys.* **49**, 371 (1968).

848. D. E. Woessner, B. S. Snowden, Jr., and A. G. Ostroff, *J. Chem. Phys.* **50**, 4714 (1969).

849. R. H. Wood and H. L. Anderson, *J. Phys. Chem.* **70**, 992 (1966).

850. R. H. Wood and H. L. Anderson, *J. Phys. Chem.* **71**, 1869 (1967).

851. R. H. Wood and H. L. Anderson, *J. Phys. Chem.* **71**, 1871 (1967).

852. R. H. Wood, H. L. Anderson, J. D. Beck, J. R. France, W. E. de Vry, and L. J. Soltzberg, *J. Phys. Chem.* **71**, 2149 (1967).

853. R. H. Wood, M. Ghamkhar, and J. D. Patton, *J. Phys. Chem.* **73**, 4298 (1969).

854. R. H. Wood and R. W. Smith, *J. Phys. Chem.* **69**, 2974 (1965).

855. R. H. Wood, R. K. Wicker, II, and R. W. Kreis, *J. Phys. Chem.* **75**, 2313 (1971).

856. M. Woodhead, M. Paabo, R. A. Robinson, and R. G. Bates, *Anal. Chem.* **37**, 1291 (1965).

857. E. M. Woolley and L. G. Hepler, *Anal. Chem.* **44**, 1520 (1972),

858. E. M. Woolley, L. G. Hepler, and R. S. Roche, *Can. J. Chem.* **49**, 3054 (1971).

859. E. M. Woolley, D. G. Hurkot, and L. G. Hepler, *J. Phys. Chem.* **74**, 3908 (1970).

860. E. M. Woolley, J. Tomkins, and L. G. Hepler, *J. Solution Chem.* **1**, 341 (1972).

861. E. M. Woolley, R. W. Wilton, and L. G. Hepler, *Can. J. Chem.* **48**, 3249 (1970).

862. J. D. Worley and I. M. Klotz, *J. Chem. Phys.* **45**, 2868 (1966).

863. A. Wörmann, *Ann. Physik.* [4] **18**, 775 (1905).

864. Y. C. Wu and H. L. Friedman, *J. Phys. Chem.* **70**, 501 (1966).

865. Y. C. Wu and H. L. Friedman, *J. Phys. Chem.* **70**, 2020 (1966).

866. Y. C. Wu, M. B. Smith, and T. F. Young, *J. Phys. Chem.* **69**, 1868 (1965).

867. C. A. Wulff, *J. Chem. Eng. Data* **12**, 82 (1967).

868. K. Wüthrich and R. E. Connick, *Inorg. Chem.* **6**, 583 (1967).

869. K. Wüthrich and R. E. Connick, *Inorg. Chem.* **7**, 1377 (1968).

870. P. A. H. Wyatt, *Trans. Faraday Soc.* **65**, 585 (1969).

871. H. R. Wyss and M. Falk, *Can. J. Chem.* **48**, 607 (1970).

872. G. Yagil, *J. Phys. Chem.* **73**, 1610 (1969).

873. H. Yamatera, B. Fitzpatrick, and G. Gordon, *J. Mol. Spectry.* **14**, 268 (1964).

874. T. F. Young, L. F. Maranville, and H. M. Smith, *in* "Structure of Electrolyte Solutions," (W. H. Homer, ed.), p. 35, Wiley (1959).

875. T. F. Young and M. B. Smith, *J. Phys. Chem.* **58**, 716 (1954).

876. T. F. Young, Y. C. Wu, and A. A. Krawetz, *Disc. Faraday Soc.* **24**, 27, 77, 80 (1957).

877. R. Zana and E. Yeager, *J. Phys. Chem.* **71**, 521 (1967).
878. N. Zengin, *Commun. Fac. Sci. Univ. Ankara, Ser. A***11**, **1961**, 1.
879. J. R. Zimmerman and W. E. Brittin, *J. Phys. Chem.* **61**, 1328 (1957).
880. R. Zwanzig, *J. Chem. Phys.* **22**, 1420 (1954).
881. R. Zwanzig, *J. Chem. Phys.* **38**, 1603 (1963).
882. R. Zwanzig, *J. Chem. Phys.* **52**, 3625 (1970).

References added in proof

883. F. J. Bartoli and T. A. Litovitz, *J. Chem. Phys.* **56**, 413 (1972).
884. J. F. Coetzee and W. R. Sharpe, *J. Phys. Chem.* **75**, 3141 (1968).
885. H. L. Friedman, *J. Solution Chem.* **1**, 387 (1972).
886. E. A. Guggenheim, *Disc. Faraday Soc.* **15**, 66 (1953).
887. A. van Hook, *J. Phys. Chem.* **72**, 1234 (1968).

Subject Index

Absorption intensity, integrated, 216, 217
Absorptivity, molar, 244
Acid-base
  equilibrium, isotope effects on, 165
  ionization, hydration effects on, 170–172
Activation energy
  for molecular reorientation, 396
    in hydration cosphere, 375
  for reorientation of water molecule, 375
Angular velocity
  correlation function, 316
  correlation time, 317
Anisotropy in rotational motion, 321
  of chemical shift, 385
Asymmetry in thermodynamic mixing function, 121
Autocorrelation function, 403

Barclay–Butler rule, 50, 51, 101
  and length of alkyl chain of solute molecule, 51
  and solvent isotope effects, 102
Bending mode, 233, 239–241, 293, 295
Bloch equations, 306
Bond stretching mode
  effect of polyatomic anions on, 243
  in ternary solutions, 299
  temperature variation of, 289
Born model, 29
  for ion-water interactions, 227, 228
  of ion hydration, 26–30, 32, 33, 35, 40, 42, 44, 57–60, 61, 107
    and partial molal volume, 69
  of ion solvation, 9, 10, 108
Brownian motion, rotational, 406

Calorimetry, 147
  its application to the study of ionization, 149, 166
Cavity field, 408
Cavity resonator, 412
Chemical model
  and ionic entropy, 62–64
  for ion hydration, 47–53, 62, 81, 110, 118
Chemical shift, anisotropy of, 385
Clathrate hydrate, 53, 204
Cluster integral, 123
Combination band, 250
Common ion, in mixing processes, 120
Compressibility
  adiabatic, 5
  and sound velocity, 73
  isothermal, 5
  partial molal, 70, 73–76
  partial molal of alkylammonium ions, 76
Concentration scales, 4, 5, 7
Conductance
  ionic, 173, 414
  measurement at high pressures, 176, 177
  measurement of, 174
Contact interaction, 310
Correlation
  field coupling effect in infrared spectroscopy, 262
  function for scalar interaction, 330
  function, rotational, 356–358
  time
    in diamagnetic electrolyte solution, 371–375
    molecular orientation, 404, 409–414
    rotational, 302, 303, 320, 328, 337, 350, 358

457

Compound Index

Deuterated analogs are not listed separately in the Compound Index, i.e., CH_3CH_2OH, CH_3CD_2OH, CD_3CH_2OH, CH_3CH_2OD, etc. are all listed under "ethanol." Water of hydration is indicated only where more than one hydrate of a compound is referred to in the text.

Formula Index

The entries in the Formula Index are restricted to complex substances, mainly hydrates.